T0299182

Vehicular Networking

With this essential guide to vehicular networking, you'll learn about everything from conceptual approaches and state of the art protocols to system designs and their evaluation.

Covering both in-car and inter-vehicle communication, this comprehensive work outlines the foundations of vehicular networking and demonstrates its commercial applications, from improved vehicle performance to entertainment and traffic information systems. All of this is supported by in-depth case studies and detailed information on proposed protocols and solutions for access technologies and information dissemination, as well as topics on rulemaking, regulations, and standardization. Importantly, for a field that is attracting increasing commercial interest, you'll learn about the future trends of this technology, its problems, and solutions to overcome them.

Whether you are a student, a communications professional, or a researcher, this is an invaluable resource.

Christoph Sommer is Assistant Professor in the Distributed Embedded Systems Group at the University of Paderborn. He created and gave tutorials and keynotes about Veins, one of the best-known vehicular networking simulation frameworks. He has about 50 papers in this field alone and has been active in this community as a general chair of IEEE/IFIP WONS and as co-founder of FG-IVC.

Falko Dressler is Professor and head of the Distributed Embedded Systems Group at the University of Paderborn. He is a Senior Member of the IEEE and ACM, as well as an IEEE Distinguished Lecturer in the fields of inter-vehicular communication, self-organization, and bio-inspired and nano-networking. He was TPC and general chair of a dozen international conferences on vehicular networking, created a Dagstuhl seminar series, and has given tutorial lectures at all major IEEE conferences on this topic.

"Sommer and Dressler have done a painstakingly thorough and solid review of the highly dispersed field of vehicular networking – all the way from in-vehicle networks to V2X, and to privacy and security issues. Destined to be an authoritative book in this area."

Onur Altintas, Toyota InfoTechnology Center

"This book provides an excellent coverage of all the important aspects of vehicular networking. It is very well written and of great value for a broad spectrum of readers including researchers, engineers in the automotive industry, and people who are not in this area but who would like to learn the basics of this exciting new field."

Ozan K. Tonguz, Carnegie Mellon University

"This is the best comprehensive guide to existing and emerging automotive networks. Whether interested in connecting components inside vehicles or networking vehicles with the outside world – it is worth reading for anyone trying to understand technology options and research results in this field."

Marco Gruteser, Rutgers University

Vehicular Networking

CHRISTOPH SOMMER
University of Paderborn, Germany

FALKO DRESSLER
University of Paderborn, Germany

Shaftesbury Road, Cambridge CB2 8EA, United Kingdom

One Liberty Plaza, 20th Floor, New York, NY 10006, USA

477 Williamstown Road, Port Melbourne, VIC 3207, Australia

314–321, 3rd Floor, Plot 3, Splendor Forum, Jasola District Centre, New Delhi – 110025, India

103 Penang Road, #05–06/07, Visioncrest Commercial, Singapore 238467

Cambridge University Press is part of Cambridge University Press & Assessment, a department of the University of Cambridge.

We share the University's mission to contribute to society through the pursuit of education, learning and research at the highest international levels of excellence.

www.cambridge.org
Information on this title: www.cambridge.org/9781107046719

First published 2015

A catalogue record for this publication is available from the British Library

Library of Congress Cataloging-in-Publication data
Sommer, Christoph.
Vehicular networking / Christoph Sommer, Falko Dressler.
pages cm
Includes bibliographical references and index.
ISBN 978-1-107-04671-9 (Hardback)
1. Vehicular ad hoc networks (Computer networks) I. Dressler, Falko. II. Title.
TE228.37.S66 2015
629.2′72–dc23 2014036378

ISBN 978-1-107-04671-9 Hardback

Additional resources for this publication at www.cambridge.org/9781107046719

Contents

Preface *page* ix
Abbreviations xi

1 **Introduction** 1

1.1 Terms and definitions 3
1.2 Who is who 5
1.2.1 Rulemaking, regulation, and standardization 5
1.2.2 Research 6
1.3 How to use this book 7
1.3.1 Target audience 7
1.3.2 Overview for non-experts 8
1.3.3 In-depth studies for the experienced reader 9

2 **Intra-vehicle communication** 12

2.1 In-vehicle networks 13
2.2 Automotive bus systems 15
2.2.1 CAN 15
2.2.2 LIN 21
2.2.3 MOST 24
2.2.4 FlexRay 27
2.3 In-vehicle Ethernet 32
2.3.1 Background 32
2.3.2 Adaptations for vehicular networks 34
2.3.3 Introduction into cars 36
2.4 Wireless in-vehicle networks 37

3 **Inter-vehicle communication** 38

3.1 Applications 39
3.1.1 Traffic information systems 39
3.1.2 Intersection collision warning systems 46
3.1.3 Platooning 48
3.1.4 Traffic-light information and control 50

3.1.5 Entertainment applications 53
3.2 Requirements and components 56
3.2.1 Application demands 56
3.2.2 Metrics to assess IVC solutions 62
3.2.3 Communicating entities 65
3.2.4 Communication principles 68
3.3 Concepts for inter-vehicle communication 71
3.3.1 FM radio and DAB 72
3.3.2 Cellular networks 76
3.3.3 Ad-hoc routing 80
3.3.4 Broadcasting 85
3.3.5 Geographic routing 96
3.4 Fundamental limits 100
3.4.1 Towards heterogeneous networks 100
3.4.2 The broadcast storm problem 102
3.4.3 Scalability of VANETs 104

4 Access technologies 106

4.1 Cellular networks 107
4.1.1 GSM 110
4.1.2 UMTS 112
4.1.3 LTE 113
4.1.4 Future developments 115
4.1.5 Use of cellular networks for IVC 116
4.2 Short-range radio technologies 118
4.2.1 Wireless LAN 119
4.2.2 IEEE 802.11p 122
4.2.3 Higher-layer protocols 125
4.3 White spaces and cog radio 129
4.3.1 Cognitive radio 130
4.3.2 TV white space 131
4.3.3 Use of white space for IVC 132

5 Information dissemination 136

5.1 Ad-hoc routing 138
5.1.1 Proactive routing protocols 139
5.1.2 Reactive routing protocols 140
5.1.3 Application in VANETs 145
5.2 Geographic routing 152
5.2.1 Geographic routing 153
5.2.2 Virtual-coordinate-based routing 157
5.3 Beaconing 167
5.3.1 Self-organized traffic information system 167

5.3.2 Cooperative awareness messages 172
5.4 Adaptive beaconing 174
5.4.1 Adaptive traffic beacon 175
5.4.2 Decentralized congestion control 185
5.4.3 Dynamic beaconing 191
5.5 Geocasting 196
5.5.1 ETSI GeoNetworking 197
5.5.2 Decentralized environmental notification messages 200
5.5.3 Topology-assisted geo-opportunistic routing 201
5.6 Infrastructure support 205
5.6.1 Roadside units 206
5.6.2 Parked vehicles 211
5.7 DTN and peer-to-peer networks 217
5.7.1 Distributed vehicular broadcast 219
5.7.2 MobTorrent 222
5.7.3 PeerTIS 225

6 Performance evaluation 229

6.1 Performance measurements 229
6.1.1 Concepts and strategies 230
6.1.2 Field operational tests 231
6.1.3 Simulation techniques 243
6.2 Simulation tools 255
6.2.1 Network simulation 256
6.2.2 Road traffic simulation 259
6.2.3 IVC simulation frameworks 262
6.3 Scenarios, models, and metrics 264
6.3.1 Scenarios 265
6.3.2 Channel models 274
6.3.3 Driver behavior 285
6.3.4 Metrics 290

7 Security and privacy 302

7.1 Security primitives 303
7.1.1 Security objectives and technical requirements 303
7.1.2 Security relationships 307
7.1.3 Certificates 308
7.1.4 Security vs. privacy 311
7.2 Securing vehicular networks 311
7.2.1 Using certificates for IVC 311
7.2.2 Performance issues 313
7.2.3 Certificate revocation 315
7.2.4 Position verification 316

7.3 Privacy 317
7.3.1 Location privacy 318
7.3.2 Tracking options 319
7.3.3 Temporary pseudonyms 321
7.3.4 Exchanging pseudonyms 323

References 325
Index 348

Preface

The intensive use of networked embedded systems is one of the key success factors in the automotive industry, also triggering a massive shortening of innovation cycles. Hundreds of so-called electronic control units (ECUs), connected by kilometers of electrical wiring, operate in today's modern car, enabling a huge variety of new functionalities ranging from safety to comfort applications. All this functionality can be realized only if the ECUs are able to communicate and to cooperate using a real-time enabled communication network in the car.

Today we are at the verge of another leap forward: This in-car network is being extended to not only connect local ECUs but also to connect the whole car to other cars and its environment using inter-vehicle communication (IVC). Relying on existing wireless Internet access using cellular networks of the third (3G) or fourth generation (4G), or novel networking technologies that are being designed specifically for use in the vehicular context such as IEEE WAVE, ETSI ITS-G5, and the IEEE 802.11p protocol, it becomes possible to use spontaneous connections between vehicles to exchange information, promising novel and sometimes futuristic applications.

Using such IVC, safety-relevant information can be exchanged that could not have been obtained using local sensors, enabling a driver to virtually see traffic through large trucks or buildings. This new idea of networked vehicles creates opportunities to not only increase road traffic safety but also improve our driving experience. Traffic jams can be prevented altogether (or at least we would be informed of jams well in advance) – and we might even be able to enable the driver to enjoying fully automated rides in a train-like convoy of cooperating vehicles on the road.

Vehicular networking, the fusion of vehicles' networks to exchange information, is the common basis on which all of these visions build.

Being fascinated with all the opportunities and challenges related to vehicular networking, we have been a part of this research community for close to ten years. In this time, many new and sometimes crazy ideas have been formulated regarding how to connect the cars of the future. Many of these ideas have been found not suitable after thorough investigation – yet, several survived and paved the road for what are now close to market-ready solutions.

From a research perspective, we are able to identify many open challenges, both in in-car and in inter-vehicle communication systems. To investigate these further, we co-organized two Dagstuhl seminars inviting leading experts from all over the world and bringing together practitioners from industry and scientists from research institutes and universities. In this context, we were able to formulate directions guiding the ongoing research activities at least in the medium term.

We also established a complementary seminar series for newcomers to the field, which is being organized in the context of the international FG-IVC series of seminars by the German computer science and electrical engineering societies GI and ITG.

This textbook is based on a tutorial series on the same topic presented at all the major IEEE conferences including IEEE CCNC, IEEE ICC, IEEE GLOBECOM, and IEEE VTC, as well as in the scope of Falko Dressler's IEEE Distinguished Lecturer Tours in Europe, the USA, South America, and the Asia–Pacific region. We also designed a new graduate-level university class, which is being held at different universities in Europe.

This has inspired us to gather our experiences in the form of a textbook, collecting in one place the common concepts of past and future vehicular networking topics for a broad range of readers – from students who want to enter this exciting new field to practitioners looking for a comprehensive overview.

This book would not have been possible without the many people who have inspired and supported us over the last decade in our research activites on vehicular networking – first and foremost the community centering around the IEEE Vehicular Networking Conference, the premier conference in the field. In particular we'd like to name Professors Ozan K. Tonguz (CMU) and Mario Gerla (UCLA) who collaborated with us on investigating some of the aforementioned crazy ideas, and finally identifying valuable and lasting solutions. The tutorial lectures mentioned above were prepared together with Dr. Onur Altintas (Toyota ITC) and Professor Claudio Casetti (Politecnico di Torino). We also wish to express our appreciation for the support we received from the most helpful staff at Cambridge during the preparation of this book. Finally, we would like to sincerely thank our families, friends, and colleagues for their enduring help and support.

We hope you will enjoy reading this textbook as much as we enjoyed preparing its contents for you. We gladly welcome any feedback and invite you to leave us a note or peruse the supplementary material we are offering on this book's companion website http://book.car2x.org/.

Christoph Sommer and Falko Dressler
Paderborn, Germany

Abbreviations

3GPP Third Generation Partnership Project, *the group that specified GSM, UMTS, and LTE*
3GPP2 Third Generation Partnership Project 2, *a competing group that specified CDMAone, CDMA2000, and UMB*
AAA American Automobile Association
ABS Anti-lock braking system
AC Access category
ACC Adaptive cruise control
ACK Acknowledgement
ACO Ant colony optimization
ADAC German Automobile Association (Allgemeiner Deutscher Automobilclub)
ADAS Advanced driver assistance system
ADQR Adaptive dispersity QoS routing
AHS Automated highway system
AIFS Arbitration interframe space
ALDL Assembly-line diagnostic link
AODV Ad-hoc on-demand distance vector
AP Access point
API Application programming interface
AQOR Ad-hoc QoS on-demand routing
ARIB Association of Radio Industries and Businesses
ASTM American Society for Testing and Materials
ATB Adaptive traffic beacon
ATIS Alliance for Telecommunications Industry Solutions
AVB Audio/video bridging
BGP Border gateway protocol
BMBF Federal Ministry of Education and Research (Bundesministerium für Bildung und Forschung)
BMVBS Federal Ministry of Transport, Building and Urban Development (Bundesministerium für Verkehr, Bau und Stadtentwicklung)
BMWi Federal Ministry of Economics and Technology (Bundesministerium für Wirtschaft und Technologie)
BPSK Binary phase-shift keying

BSC	Base station controller
BSM	Basic safety message
BSS	Basic service set
BTP	Basic transport protocol
BTS	Base transceiver station
BYOD	Bring your own device
CA	Certificate authority
CACC	Cooperative adaptive cruise control
CAM	Cooperative awareness message
CAN	Content addressable network, *when referring to the distributed hash table*
CAN	Controller area network, *when referring to the bus protocol*
CAS	Collision-avoidance symbol
CCA	Clear channel assessment
CCH	Control channel
CCK	Complementary code keying
CDF	Cumulative distribution function
CDMA	Code-division multiple access
CEPT	European Conference of Postal and Telecommunications Administrations (Conférence Européenne des Postes et Télécommunications)
CME	Certificate management entity
CoCar	Cooperative Cars
CoCarX	Cooperative Cars extended
COM	Component object model
Converge	Communication network vehicle road global extension
CPU	Central processing unit
CRC	Cyclic redundancy check
CRL	Certificate revocation list
CSD	Circuit switched data
CSMA	Carrier sense multiple access
CSMA/BA	Carrier sense multiple access with bitwise arbitration
CSMA/CA	Carrier sense multiple access with collision avoidance
CSMA/CD	Carrier sense multiple access with collision detection
CTS	Clear to send
CW	Contention window
CWS	Collision warning system
D-FPAV	Distributed fair power adjustment for vehicular networks
D2B	Domestic data bus *later domestic digital bus*
DAB	Digital audio broadcasting
DCC	Decentralized congestion control
DCF	Distributed coordination function
DCH	Dedicated channel
DENM	Decentralized environmental notification message

DES	Discrete event simulation
DHT	Distributed hash table
DLL	Dynamic-link library
DNS	Domain name system
DoIP	Diagnostic communication over Internet protocol
DSA	Dynamic spectrum access
DSC	DCC sensitivity control
DSDV	Destination sequenced distance vector
DSR	Dynamic source routing
DSRC	Dedicated short-range communication
DSRC/WAVE	Dedicated short-range communications/wireless access in vehicular environments
DSSS	Direct-sequence spread spectrum
DTN	Delay/disruption-tolerant network
DV-CAST	Distributed vehicular broadcast
DVD	Digital video disc *also digital versatile disc*
DYMO	Dynamic MANET on demand
DynB	Dynamic beaconing
ECC	Electronic Communications Committee
eCDF	Empirical cumulative density function
ECU	Electronic control unit
EDCA	Enhanced distributed channel access
EDGE	Enhanced data rates for GSM evolution
eMBMS	Evolved multimedia broadcast/multicast service
EMI	Electromagnetic interference
ESP	Electronic stability program
ETSI	European Telecommunications Standards Institute
FACH	Forward access channel
FairAD	Fair and adaptive data dissemination
FairDD	Fair data dissemination
FCC	US Federal Communications Commission
FCD	Floating car data
FDD	Frequency-division duplex
FDMA	Frequency-division multiple access
FHWA	Federal Highway Administration
FOT	Field operational test
FPGA	Field-programmable gate array
GGSN	Gateway GPRS support node
GHT	Geographic hash table
GIDR	Geographical inter-domain routing
GLOSA	Green-light optimal speed advisory
GPRS	General packet radio service
GPS	Global positioning system
GPSR	Greedy perimeter stateless routing

GRWLI	Geographic routing without location information
GSM	Global system for mobile communications
GUI	Graphical user interface
HLA	High-level architecture
HMI	Human–machine interface
HSCSD	High-speed circuit switched data
HSDPA	High-speed downlink packet access
HSPA	High-speed packet access
HSUPA	High-speed uplink packet access
IBSS	Independent basic service set
ICWS	Intersection collision warning system
IDE	Integrated development environment
IEEE	Institute of Electrical and Electronic Engineers
IETF	Internet Engineering Task Force
IFS	Interframe space
ILOC	Intersection location
IoT	Internet of things
IP	Internet protocol
ISM bands	Industrial–scientific–medical radio bands
ISO	International Organization for Standardization
ITS	Intelligent transportation system
ITS-G5	Intelligent transportation systems access layer for the 5-GHz band, *as specified by the ETSI*
ITSA	Intelligent Transportation Society of America
ITU	International Telecommunication Union
ITU-R	ITU Radiocommunication Sector
ITU-T	ITU Telecommunication Standardization Sector
IVC	Inter-vehicle communication
IVHS	Intelligent vehicle–highway system
LAN	Local area network
LDM	Local dynamic map
LER	Last encounter routing
LIDAR	Light detection and ranging
LIN	Local interconnect network
LLC	Logical link control
LOS	Line of sight
LTE	Long Term Evolution
MAC	Medium access control
MANET	Mobile ad-hoc network
MBMS	Multimedia broadcast/multicast service
MBSFN	Multicast-broadcast single-frequency network
MCD	Minimum cost distribution
METIS	Mobile and wireless communications enablers for the 2020 information society

MFD	Multi-function display
MIC	Ministry of Internal Affairs and Communications
MIMO	Multiple-input multiple-output
mmW	Millimeter-wave
MNO	Mobile network operator
MOST	Media-oriented systems transport
MSC	Mobile switching center
MTU	Maximum transmission unit
MVNO	Mobile virtual network operator
NHTSA	National Highway Traffic Safety Administration
NLOS	Non-line-of-sight
NRZ	Non-return-to-zero
OAD	Obstacle-aware distribution
OBU	On-board unit
OCB	Outside the context of a BSS
OD	Origin–destination
OEM	Original equipment manufacturer
OFDM	Orthogonal frequency-division multiplexing
OFDMA	Orthogonal frequency-division multiple access
OLSR	Optimized link state routing protocol
OPEN	One-pair Ethernet
P2P	Peer-to-peer
PAPR	Peak-to-average power ratio
PATH	Partners for advanced transit and highways
PCF	Point coordination function
PCI	Peripheral component interconnect
PDR	Packet delivery ratio
PeerTIS	Peer-to-peer traffic information system
PHY	Physical layer
PKI	Public key infrastructure
PoE	Power over Ethernet
POF	Plastic optic fiber
PRNG	Pseudo-random-number generator
PSID	Provider service identifier
PSSME	Provider service security management entity
QAM	Quadrature amplitude modulation
QoS	Quality of service
QPSK	Quadrature phase-shift keying
RACH	Random access channel
RAN	Radio access network
RB	Resource block
RDS	Radio data system
RERR	Route error
RF	Radio frequency

RFC	Request for comments
RIP	Routing information protocol
RMSE	Root mean square error
RNC	Radio network controller
ROI	Region of interest
RREP	Route reply
RREQ	Route request
RSS	Received signal strength
RSU	Roadside unit
RTPGE	Reduced twisted pair Gigabit Ethernet
RTS	Ready to send
SAE	Society of Automotive Engineers
SAM	Service announcement message
SARTRE	Safe road trains for the environment
SCFDMA	Single-carrier FDMA
SCH	Service channel
SDR	Software-defined radio
SFN	Single frequency network
SGSN	Serving GPRS support node
SIFS	Short interframe space
SNR	Signal-to-noise ratio
SODAD	Segment oriented data abstraction and dissemination
SOTIS	Self-organizing traffic information system
SPAT	Signal phase and timing
SRP	Stream reservation protocol
SSU	Stationary support unit
SUMO	Simulation of urban mobility
TAC	Transmit access control
TCP	Transmission control protocol
TD-CDMA	Time-division CDMA
TDC	Transmit data rate control
TDD	Time-division duplex
TDMA	Time-division multiple access
TIC	Traffic information center
TIS	Traffic information system
TMC	Traffic messaging channel
TO-GO	Topology-assisted geo-opportunistic routing
TOPO	Road topology
TPC	Transmit power control
TPEG	Transport Protocol Expert Group
TPM	Trusted platform module
TraCI	Traffic control interface
TRC	Transmit rate control
TSF	Time synchronization function

TSN	Time-sensitive networking
TTCAN	Time-triggered CAN
TTL	Time to live
TVWS	TV white space
TXOP	Transmission opportunity
U-NII bands	Unlicensed national information infrastructure bands
UART	Universal asynchronous receiver/transmitter
UDP	User datagram protocol
UDS	Unified diagnostic service
UE	User equipment
UMB	Ultra-mobile broadband
UMTS	Universal mobile telecommunications system
US DOT	US Department of Transportation
USRP	Universal software radio peripheral
UTC	Coordinated universal time
UV-CAST	Urban vehicular broadcast
V2I	Vehicle to infrastructure
V2V	Vehicle to vehicle
V2X	Vehicle to X
VANET	Vehicular ad-hoc network
VCP	Virtual coordinate protocol
Veins	Vehicles in network simulation
VLAN	Virtual LAN
VoIP	Voice over IP
VRR	Virtual ring routing
VSimRTI	V2X simulation runtime infrastructure
VTL	Virtual traffic light
W-CDMA	Wideband CDMA
WAVE	Wireless access in vehicular environments
WiMAX	Worldwide interoperability for microwave access
WLAN	Wireless LAN
WME	WAVE management entity
WRAN	Wireless regional area network
WSA	WAVE service advertisement
WSM	WAVE short message
WSM-S	WAVE short message with safety supplement
WSMP	WAVE short message protocol
WSN	Wireless sensor network
WSU	Wireless safety unit
WUP	Wake-up pattern
XML	Extensible markup language

1 Introduction

Vehicular networking, the exchange of information in the car and between cars, has been on the mind of researchers since at least the often-cited 1939 New York World's Fair. Here, in its *Futurama* exhibit, General Motors revealed utopian visions of what highways and cities might look like twenty years later. In fact, many of the visions of intelligent transportation systems (ITSs) showcased there, as well as in the exhibit designer's 1940 book *Magic Motorways* (Bel Geddes 1940), such as that "car-to-car radio hook-up might be used to advise a driver nearing an intersection of the approach of another car or even to maintain control of speed and spacing of cars in the same traffic lane", are still being pursued today. Modern vehicles collect huge amounts of information from on-board sensors, and this information is made available to the in-car network and ready for sharing with other cars – not just for the described visions of intersection assistance systems and platooning, i.e., road-train applications, but also for a whole wealth of new applications. Today, with *in*-car networks merging into networks *of* cars, these early visions seem closer to reality than ever.

But why did we have to wait this long?

Hugely many research projects have been undertaken since *Magic Motorways* was written, all of which tried to make visions of ITS a reality (Jurgen 1991). Among the most notable of research initiatives were the Japanese CACS, US ERGS, and European ALI projects for urban route guidance in the late 1960s to late 1970s, the European Prometheus project for autonomous driving (1986–1995), and the US PATH project for cooperative driving (1986–1992). Evidently, the majority of these initiatives led to working prototypes and successful field operational tests; yet, commercial success failed to match the projects' promises.

A possible explanation is given by Chen & Ervin (1990): early approaches were simply too visionary for their time, commonly focusing on infrastructure-less solutions, which could not be supported by current technology. The 1980s then saw a shift of attention from the more long-term goals of complete highway automation to nearer-term goals such as driver-advisory functions. However, for the same reasons, attention shifted also from infrastructure-less to infrastructure-assisted solutions, resulting in what the authors called a *chicken-and-egg* type of standoff in the deployment of what were called intelligent vehicle–highway system (IVHS) solutions:

> The automotive and electronics industries are skeptical as to whether the public infrastructure for IVHS will materialize. (Without an infrastructure, of course, there will be no market for cooperative IVHS products on-board the vehicle or on the highway.)

> Highway agencies are skeptical as to whether IVHS technologies will deliver solutions to real highway problems. (Without a sound expectation of public benefit, of course, public investment is unjustified.)

In the years since this 1990 article, however, these premises have changed considerably, causing interest in inter-vehicle communication (IVC) research to re-ignite.

First, with the commercial deployment of latest-generation cellular communication technology, there is now an almost universal communication infrastructure available. In fact, commercially available versions of what could be described as early IVC systems are already on the market, e.g., *On Star* (1995), *BMW Assist* (1999), *FleetBoard* (2000), and *TomTom HD Traffic* (2007), after which the number of IVC systems all but exploded. The new-found optimism with regard to IVC communication research can also be seen expressed in countries' allocation of dedicated short-range communication (DSRC) bands for the sole use of vehicular short-range wireless communication (in 1999 by the US FCC, followed in 2008 by the European ECC). New technologies to work in this spectrum have been conceived and, together with cellular networks, will allow future vehicles to exchange information in order to *cooperate* on the road.

Second, computing power has increased many-fold, enabling modern vehicles to collect and process data under tight temporal constraints, allowing enhanced degrees of *automation*. Radar, laser scanners, infrared and three-dimensional (3D) surround-view systems, fully networked with other systems in the car, are all shipping in modern vehicles or are planned for upcoming generations. In mid 2011, the state of Nevada passed a law allowing autonomous (that is, driver-less) vehicles to drive on its roads, and the first car was licensed roughly one year later (though it is still requiring at least two people on board to take over manual control if needed).

These two trends, increasing automation and increasing cooperation of vehicles, are now promising to merge. Vehicles' in-car networks are extended beyond the boundaries of their bodies, creating truly *intelligent transportation systems*.

When designing future vehicular networks it is therefore important to keep the broad picture of both in-vehicle and inter-vehicle communications in mind, as well as to consider the whole range of available communication paradigms and technologies from wired bus systems to wireless in-car networks and from wireless cellular to short-range radio communication.

From this holistic point of view, the research on vehicular networking solutions has been re-initiated in the second decade of the twenty-first century. This research perspective establishes a multi-disciplinary effort. The vehicular networking community reconvened at Dagstuhl, an internationally renowned meeting and seminar place, in 2013 to reconsider the developments in our field from both an academic and an industrial point of view. The main outcomes include new research directions and a new focus on open challenges that need to be addressed in order to enable day-one IVC applications.

We are now entering an era that might change the game in road-traffic management. The vehicular networking field is shifting from early academic research ideas to application-oriented research and development to global deployment. Large-scale field operational tests (FOTs) are taking place both in Europe and in the USA, helping decision makers such as the National Highway Traffic Safety Administration (NHTSA) and the US Department of Transportation (US DOT) to prepare recommendations to make DSRC technology mandatory for new cars.

This textbook is the first of its kind to investigate the common concepts of all vehicular networking approaches, their limits, what unifies them, and how they differ, as well as to take a detailed look at how the challenges of vehicular networks are met by selected techniques and technologies – both with a look back on how they evolved and with one eye on their potential future.

We also address two issues that are at the heart of research into new vehicular networking approaches: First, how to conduct a rigorous scientific evaluation of their feasibility and performance – both in a way that produces insightful results close to real life and in a way that can help advance science; and second, how to design systems in a way that ensures both the security of the system and the privacy of its users.

Before we do so, however, it might be a good idea to take a brief look at the vocabulary of vehicular networking as well as at the key players in the field.

1.1 Terms and definitions

Much of the terminology that we will be using in this book has a very clear-cut definition, but in many cases the original meaning has been diluted in common use or in the media, mixed with other definitions or supplanted altogether. With this in mind, to avoid misunderstandings in the following, let us briefly go over some of the basic terminology that may lead to confusion, considering its use over the last few decades.

One of the most obvious families of terms that are being used in vehicular networking consists of those referring to the communicating network itself.

Vehicular networking is what we adopted as the most general classifier, referring to the field of computer communications and networking as applied to vehicles. Vehicular networking thus encompasses both in-car and inter-vehicle communication aspects as well as their fusion.

Inter-vehicle communication (IVC) restricts this to exclude wired communication as well as any network (wired or wireless) within vehicles. It thus refers to a system where vehicles are participants in a wireless network. Other participants such as roadside units (RSUs) can explicitly be part of this network.

Vehicular ad-hoc network (VANET) has its origins in the discipline of mobile ad-hoc networks (MANETs), casting VANETs as a novel application domain. Being the basis for what we call IVC today, the term is still somewhat synonymous with IVC, but focuses on spontaneously created ad-hoc networks, much less on pre-deployed infrastructure like using RSUs or cellular networks.

Intelligent transportation system (ITS) describes the overall goal of being able to make better use of transportation networks, for which road networks are one of many such networks and IVC is one means among many. Lately, other modes of transportation have faded into the background and ITS has become synonymous with intelligent *road* networks. The precursor goals to ITS were intelligent vehicle–highway systems (IVHSs), aiming more at making roads smarter, before smart vehicles were considered an option.

Vehicle to vehicle (V2V) as well as **vehicle to infrastructure (V2I)** and **vehicle to X (V2X)**, all refer to the end points of communication, indicating whether information is being exchanged with other vehicles, with infrastructure (also called vehicle-to-roadside), or with arbitrary nodes – independently of the technology being used. Colloquially, some use substitutes the somewhat shorter word *car* for *vehicle* (forming C2C, C2I, and C2X) to refer to the same concepts.

A second group of terms consists of those related to the communication technology employed:

IEEE WAVE and ETSI ITS are complete communication architectures comprising a protocol stack together with needed facilities and regulatory provisions for IVC.

IEEE 802.11 Wireless LAN (WLAN) is the base technology for the most commonly used short-range wireless communication protocols, both for consumer networking and for IVC. The term **WiFi** refers to that subset of IEEE 802.11 that is part of the certification the Wi-Fi Alliance provides for consumer networking devices.

IEEE 802.11p and dedicated short-range communication (DSRC) have become synonymous with the use of IEEE 802.11 for IVC. IEEE 802.11p is the amendment that introduced many of the missing parts of functionality in the IEEE 802.11 family needed for efficient IVC – most notably operation at around 5.9 GHz. This radio band is known in the USA as the dedicated short-range communications (DSRC) radio band.

Finally, we have to deal with generic terminology, which we include mainly for the sake of completeness.

A channel seems to be the most irritating term. In general, it refers to a means of transporting information over a medium, wired or wireless. Multiple channels can be provided on a single medium by subdividing it via multiple access schemes, most commonly into time slots or frequency bands. For convenience, channels provided by frequency-division multiple access (FDMA) are commonly referred to not by their frequency range or their center frequency, but by a channel number, assigned in the respective standards of communication technology. To give an example, IEEE 802.11 specifies for its orthogonal frequency-division multiplexing (OFDM) physical layer a channel number of 178 to refer to a channel centered at 5.890 GHz.

The penetration rate refers to which fraction of all vehicles are equipped with IVC technology. It indicates how vehicle density (on the road) relates to node density (in the network). Other names for this fraction are equipment rate or market penetration.

Self-organization describes a control paradigm supporting a self-governing approach of network participants. No global state is maintained in the network and no centralized coordination is required (Dressler 2007).

1.2 Who is who

Many businesses and organizations worldwide are stakeholders in vehicular networking. From rulemaking, via regulation and standardization, to manufacturing, there are many entities involved that we want to give a brief overview of. Furthermore, we outline the key research dissemination and networking platforms.

1.2.1 Rulemaking, regulation, and standardization

In the context of in-vehicle networking, little rulemaking is needed, since vehicle busses constitute closed systems and vehicle manufacturers are free to follow an industry standardization process. In the context of IVC, extensive rulemaking, regulation, and standardization are required, because one brand's vehicles naturally need to be able to communicate with other manufacturers' vehicles. Moreover, radio spectrum is heavily regulated by countries' governments. In fact, whole books, e.g., that by Williams (2008) to name just one, have been written on the processes of rulemaking and standardization in the context of ITS. In this book, we are focusing on organizations from those countries in which today's top automobile manufacturers are based: Japan (e.g., Toyota, Nissan, Honda), the USA (e.g., General Motors, Ford), and Europe (e.g., Volkswagen, Daimler, BMW).

In Japan, spectrum is managed by its Ministry of Internal Affairs and Communications (MIC). Spectrum allocation is the result of an involved public consultation process, after which spectrum is allocated to specific uses. For implementing studies and implementing standardization processes, the MIC (or, rather, one of its precursor ministries, the Ministry of Posts and Telecommunications), chartered the creation of the Association of Radio Industries and Businesses (ARIB). ARIB counts over 200 companies and organizations as its members: aside from all of the major telecommunication companies, and those involved with research and development of radio technologies, all major automobile manufacturers are members.

In the USA, spectrum allocation and rulemaking is the task of the US Federal Communications Commission (FCC), an independent agency of the government. The FCC frequently adopts standards by reference. Standards are created by independent organizations, such as ASTM International, which are comparable to other national standards bodies like ANSI or DIN. It frequently collaborates with other organizations, in our context most prominently with the IEEE Standards Association, an organization within the Institute of Electrical and Electronics Engineers (IEEE).

In the European Union (EU), spectrum allocation in its constituent countries is governed by the European Commission, which issues decisions that need to be implemented by EU member states and associated countries. For this, the European Commission issues mandates to the European Conference of Postal and Telecommunications Administrations (Conférence Européenne des Postes et Télécommunications, CEPT), one committee of which is the Electronic Communications Committee (ECC), which is tasked with developing rulemaking for communications. For developing standards, the CEPT also created the European Telecommunications Standards Institute (ETSI) as an independent, not-for-profit, standardization organization. Today, the ETSI numbers 750 member organizations from 63 countries around the world. Its output, harmonized standards, serves as input for ECC spectrum regulations, which are the input for European Commission decisions.

Higher-layer standards lend themselves better to international standardization. Among the most active formal international standardization bodies are the International Organization for Standardization (ISO) and the International Telecommunication Union (ITU). The ISO comprises national standards organizations from around the world, with standardization conducted by its technical committees (TCs). The ITU, in particular its ITU Telecommunication Standardization Sector (ITU-T), is a United Nations agency that investigates technical or operational questions with the goal of providing recommendations for standardization worldwide; that is, it does not issue standards itself, even though these ITU-T recommendations are used as *de facto* standards in many domains.

This is complemented by specialized national standardization bodies, e.g., the Society of Automotive Engineers (SAE), and industry special interest groups like the Wi-Fi Alliance (for a particular short-range wireless radio standard) and the AVnu Alliance (for a particular in-vehicle networking standard).

1.2.2 Research

Long-term research activities towards future vehicular networking concepts that are more visionary in nature are discussed in the context of an international research seminar series, the *Dagstuhl Seminar on Inter-Vehicular Communication*, which is hosted at the world-renowned center for computer science of the German computer science community. Its last meetings took place in 2010 and 2013 (Dressler *et al.* 2011a, 2014). Traditionally, this seminar series brings together practitioners from many branches of industry and researchers from computer science, electrical engineering, traffic sciences, business, and other fields.

Research results in the scope of vehicular networking are published in a variety of journals and conference proceedings. Cross-cutting developments are regularly published in one of the many journals and magazines covering general networking topics, such as the

- *IEEE Journal on Selected Areas in Communications*,
- *IEEE/ACM Transactions on Networking*,

- *IEEE Transactions on Mobile Computing,*
- *Elsevier Ad Hoc Networks,*
- *IEEE Communications Letters*, and
- *IEEE Communications Magazine.*

Of special note is that, twice a year, the *IEEE Communications Magazine* publishes an issue of its Automotive Series. Among the periodicals that are focused more on the vehicular domain are the

- *IEEE Transactions on Vehicular Technology,*
- *IEEE Vehicular Technology Magazine*, and
- *IEEE Transactions on Intelligent Transportation Systems.*

Active research is, as always, being presented and published in the context of international conferences. Again, more general approaches are most commonly published at conferences focusing on networking as a whole. These conferences regularly dedicate one or more of their tracks to vehicular networking or host specialized workshops on this topic. The leading conferences are the

- IEEE Conference on Computer Communications (INFOCOM),
- ACM International Conference on Mobile Computing and Networking (MobiCom),
- ACM International Symposium on Mobile Ad Hoc Networking and Computing (MobiHoc),
- IEEE International Conference on Communications (ICC), and
- IEEE Global Telecommunications Conference (GLOBECOM).

More focused on vehicular networking is the IEEE Vehicular Technology Conference (VTC), which hosts two complementary yearly editions, in spring and fall. Finally, in 2009 major vehicular networking workshops merged to form a conference, which is also seeing a major influx of attendees from previous flagship workshops such as ACM VANET. This conference is the IEEE Vehicular Networking Conference (VNC).

1.3 How to use this book

In the following, we briefly outline the structure of the book and comment on a recommended reading approach depending on the reader's profile.

1.3.1 Target audience

We decided to design the textbook for a broad range of readers, from graduate students to academics who need a comprehensive overview of the current vehicular networking approaches and future research issues, and practitioners from the communications and automotive industries.

The book is structured in a way that allows beginners and those moving into the field to first get a higher-level overview as well as allowing experts in the domain to use the book as a quick reference providing in-depth studies of selected approaches to vehicular networking. This textbook can readily be used for a university class dedicated to vehicular networking and as supplementary material in any other computer networking and communications class.

Our main focus is first on introducing both in-vehicle networking techniques and approaches for inter-vehicle communication. We summarize these chapters as an "overview for non-experts" in Section 1.3.2. Secondly, we investigate concepts for inter-vehicle communication in detail, also studying methods for evaluating the performance of these approaches as well as security and privacy related concerns. The related chapters are subsequently discussed as "in-depth studies for the experienced reader" in Section 1.3.3.

1.3.2 Overview for non-experts

We organize this overview into two chapters, one focusing on in-vehicle networking, the other on inter-vehicle communication. We highly recommend the reader to start with these chapters – both providing a comprehensive overview of the respective field but not investigating all the little technical details. For the expert reader, it is reasonable to skip these chapters and start with those described in Section 1.3.3.

Intra-vehicle communication (Chapter 2)

After motivating the need for actual networking (i.e., bus systems and message forwarding as opposed to dedicated wire connections), this chapter discusses the makeup and the roles of the individual networking components, with a special focus on electronic control units (ECUs) as parts of automotive bus systems. It continues by describing the evolution of these components as well as of on-board and off-board networking standards, from simple LIN to Gigabit Ethernet. The chapter highlights their fundamental commonalities and differences, as well as lessons learned from less successful standards. A look at recent trends and early implementation efforts concludes the chapter.

- Discussion of networking as such, components (ECUs, gateways)
- Bus systems (LIN, CAN, MOST, FlexRay, trends towards in-car Ethernet)
- Wireless in-vehicle networking

Inter-vehicle communication (Chapter 3)

Communication between vehicles and between vehicles and available infrastructure is the topic of this chapter. Essentially, a basic introduction to IVC is given, identifying the key application domains of safety, efficiency, and entertainment. This chapter derives all the relevant communication concepts and solutions for those applications and, most importantly, identifies the minimum requirements such as maximum delays, minimum dissemination range, or minimum data rates. Furthermore, an overview on the possible communication paradigms is presented, such as whether information is to be exchanged without the help of any available infrastructure or whether infrastructure elements – RSUs, parking vehicles, or even widely deployed cellular networks – can be used for the information exchange.

- Applications: safety, efficiency, entertainment. Sample use cases such as intersection warning/assistance systems, platooning management, traffic information systems, traffic-light information and control, multimedia streaming, multiplayer games.
- Concepts, solutions, requirements: Example applications and proposed solutions, requirements on the communication technology (delays, reliability, dissemination range, data rates).
- Infrastructure-based vs. infrastructure-less: Ad-hoc communication, store–carry–forward, roadside units, stationary support units, cellular networks.

1.3.3 In-depth studies for the experienced reader

Being already familiar with the concept of inter-vehicle communication, we now explore the protocols and concepts in more detail. This part is organized into two chapters on key networking aspects (access and information dissemination) and two additional chapters focusing on performance evaluation and on security and privacy issues. These chapters do not necessarily build upon each other. The material is composed to establish both a general reference and the basis for an in-depth study of vehicular networking solutions, e.g., in the scope of a graduate-level university class or an extended seminar on the topic.

Access technologies (Chapter 4)

Any application relying on inter-vehicle communication demands a particular set of network characteristics. In this chapter, we discuss access technologies that have been considered for IVC. This chapter outlines all relevant aspects of cellular communication technology and introduces dedicated short-range communications/wireless access in vehicular environments (DSRC/WAVE) as a key player in the IVC domain. In this field, specifically IEEE 802.11 is discussed in its WiFi versions as well as in the form of the IEEE 802.11p protocol, which is the basis for DSRC/WAVE. Furthermore, the use of white spaces for IVC as is currently being considered in Japan is discussed.

- Cellular networks: Introduction to GSM, UMTS, and LTE-A; approaches using cellular networks for IVC.
- Short-range radio: IEEE 802.11, EDCA, WAVE, and ITS-G5 concepts, frequencies, IEEE 802.11p.
- White spaces and cognitive radio: Principles, cognitive radio, using white spaces for IVC.

Information dissemination (Chapter 5)

The objective of this chapter is to discuss typical vehicular network layer protocols in detail. The key objective is information exchange between any two vehicles, from one vehicle to all neighboring ones, from one vehicle to infrastructure components, from infrastructure to one or all neighboring vehicles, and dissemination from a vehicle to all those which are interested in the content. This chapter details the basic principles underlying such IVC approaches, namely unicast, local broadcast, anycast, and multicast communication principles as used in vehicular networks. This also includes the current IEEE and ETSI standards such as cooperative awareness messages (CAMs) and decentralized environmental notification messages (DENMs) for local broadcast and geocast for multi-hop transmissions. The main focus is on the most recent beaconing solutions inluding extensions providing congestion control and fair resource sharing. Furthermore, more sophisticated re-broadcasting and delay/disruption-tolerant network (DTN) concepts are described.

- Routing: MANET routing, applicability to IVC, impact of RSUs.
- Beaconing: Static beaconing, adaptive beaconing periods, adaptive transmission power, ATB, ETSI ITS-G5 DCC, DynB.
- Geocast: Geographic routing, convex networks, coordinate transformation, virtual coordinate-based routing, geographically constrained broadcasting.
- DTN concepts: Store–carry–forward, roadside units, heterogeneous vehicular networks.

Performance evaluation (Chapter 6)

The chosen methodology greatly influences the quality of performance evaluation. Field operational tests are being conducted in many areas and have provided some very helpful first results. Yet, large-scale experimentation is conceptually infeasible. Thus, simulation is still the method of choice for most performance studies. This chapter concerns how such simulations should be performed in the field of IVC and which tools are freely available to aid researchers. The key concern is the correct and realistic modeling of the vehicles' mobility. Besides that, the 'correct' choice of the scenario has a strong influence on the expressiveness, the validity, and the comparability of simulation experiments. This chapter also studies the impact of radio signal propagation models, the influence of the human driver's behavior, and suitable metrics for finally assessing the performance.

- Field operational tests: Measurement strategies, examples of field operational tests.
- Network simulation: Basic models and tools for network simulation, the use of these tools for IVC, scalability.
- Road traffic simulation: Road traffic microsimulation, traffic flow simulation, scalability.
- Simulation frameworks: IVC simulation frameworks Veins, iTetris, VSimRTI.
- Scenarios: Intersections, highway scenarios, impact of number of lanes, suburban and urban scenarios.
- Channel models: Freespace, two-ray, Nakagami-*m*, shadowing by buildings and other vehicles.
- Metrics: Classical network simulation metrics (delay, throughput, etc.) vs. application-dependent metrics (travel time, CO_2 emission, safety).

Security and privacy (Chapter 7)

Especially with the engagement of the major industry players to bring IVC into the market, the need to secure the communication between vehicles and the infrastructure became an important issue. Aside from satisfying the demand for deploying closed-market systems that offer services only to paying customers, security is also necessary in order to prevent fraud and malicious attacks. It has been demonstrated that IVC-related systems are not as secure as necessary: attackers successfully took over control of car electronics via tire pressure measurement systems or they attacked electronic message boards along a highway. Even spoofed traffic information transmissions via the traffic messaging channel (TMC) have been demonstrated. This chapter studies possible security solutions and their impact on IVC. Furthermore, this chapter looks into the very critical balance between security and privacy: the more secure a system is made, the more the driver's privacy is impacted. Therefore, this chapter also investigates privacy systems such as temporary pseudonyms, time-varying pseudonym pools, and the exchange of pseudonyms.

- Security vs. privacy: Balance between security and privacy.
- Security: Securing IVC, group management strategies.
- Certificates: Basic certificate management, PKI, certificate management and revocation.
- Privacy: Temporary pseudonyms, time-varying pseudonym pools, exchange of pseudonyms.

2 Intra-vehicle communication

Electronics are playing an ever-increasing role in today's vehicles. They have gone from humble beginnings in the 1970s that saw features like electronic fuel injection and power door locks become commonplace or the mass-market introduction of anti-lock braking systems in the 1980s to today's 3 km of wiring and 50 kg of electrical systems operating in a typical modern car. With electrical systems contributing to at least one third of today's breakdowns, it is clear that modern cars just could not be driven without electronics.

The use cases that most readily come to mind might be engine control, or more recent advances in chassis electrification such as an anti-lock braking system (ABS) or an electronic stability program (ESP), maybe also modern infotainment systems like DVD players streaming video to the rear seats, navigation systems, or video cameras. Yet, modern cars contain a vast number of small electronic subsystems – from windshield wipers, door locks, power windows, and adaptive lights to seat and mirror adjustment, climate control, and dashboard displays. All these devices are controlled by electronic control units (ECUs).

Early on, this motivated the introduction of a way to diagnose faults in the electronics system, requiring digital data communication between embedded systems and diagnostic tools. Such systems were first designed for use exclusively on the vehicle assembly line and were very specific to not just the manufacturer but also the component being diagnosed. One of the earliest examples is the assembly line diagnostic link (ALDL) system used in the early 1980s for engine control module diagnosis.

The need for a digital link between ECUs and other components was further driven by the increasing complexity of connecting individual electric systems. In the early 1980s this was still easily possible using dedicated wires for each interconnection, or even for each task, but it became very clear that this would not much longer be the case; and rightly so: Today, switching on the left-turn signal involves no fewer than eight individual embedded systems. Such complexity can be handled only by using a data bus to interconnect many systems, that is, by true networking between the electric components of a vehicle.

Aside from decreasing the complexity, this move towards networked systems also brought down development, installation, and maintenance costs, and the connections' total weight and volume, as well as leading to a much higher flexibility in the distribution of functions across embedded systems.

In this chapter, we study all relevant networking technologies used to connect devices within the vehicle. Most of these are wired bus systems specialized for the real-time communication requirements in the vehicle. We also investigate the current trend towards more general network technologies such as Ethernet as well as wireless options.

This chapter is organized as follows.

- In-vehicle networks (Section 2.1) – We first discuss the general concepts of in-vehicle networks and their networked systems.
- Automotive bus systems (Section 2.2) – We continue by taking a more detailed look at four special-purpose bus systems used in automotive networking: CAN, LIN, MOST, and FlexRay.
- In-vehicle Ethernet (Section 2.3) – Having discussed special-purpose bus systems, we then turn towards a modern concept: adapting Ethernet for in-vehicle networks. We explore the background, potentials, and limitations, as well as steps towards the adoption of this technology.
- Wireless in-vehicle networks (Section 2.4) – We conclude this chapter with a discussion of a more visionary concept: wireless in-vehicle networking.

2.1 In-vehicle networks

Individual functions of a modern vehicle (like detecting seat occupancy) are served by individual embedded systems, called ECUs. Each is essentially a ruggedized miniature computer, typically based around a microcontroller, that can gather data from attached sensors or trigger attached actors. Depending on the task of the ECU, which can be anything from monitoring battery charge to triggering the airbag, hardware of different complexity and different features is being used. All but the smallest ECUs run a small embedded operating system, booting when the ignition is turned on, powering up when they are needed (e.g., rear camera), or staying powered on (e.g., for keyless entry). They typically boot from flash memory, which means that they need to be pre-programmed with firmware, which can be a time-intensive process.

Often, an ECU also includes a (separate or integrated) *bus guardian* that observes all bus operations and can disable communication if the node is deemed faulty (solving the *babbling idiot problem*, where one node keeps sending bad data or bad frames). Figure 2.1 depicts such an ECU.

Conceptually there is no big difference between networking for intra-vehicle communication, industrial networking, and networking home computers. In all three deployment scenarios the general system requirements are highly diverse.

Any bus system should (to varying degrees) be cheap to manufacture, allow for arbitrarily long cables, and at the same time offer very low latency. It should be able to give real-time guarantees, provide highly robust data transmissions, and deliver very high throughput for multimedia streaming.

Industrial networking adds additional requirements in terms of robustness: An industrial bus system should be physically robust against vibration, shock, and torn lines,

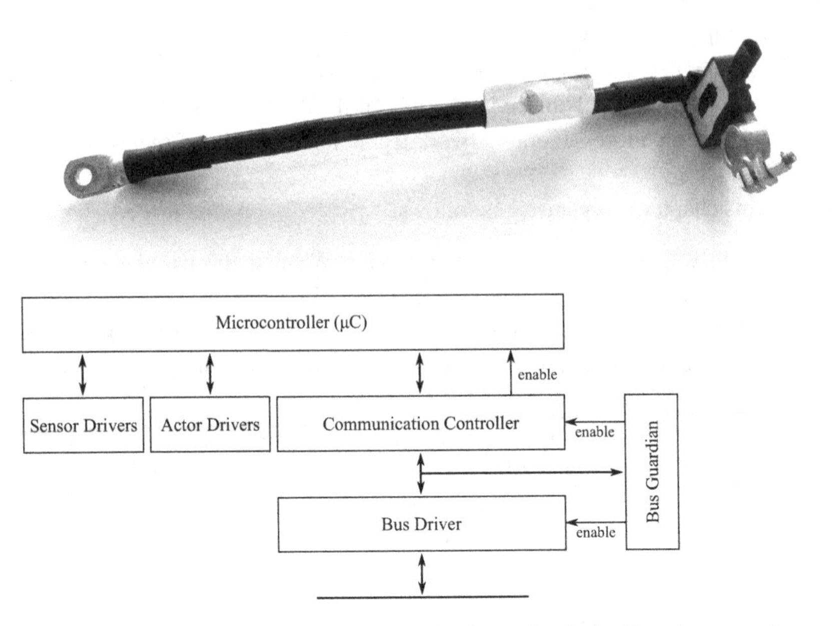

Figure 2.1 An ECU and a simplified schematic view: what looks like a bump on the negative battery lead in a modern car is a small networked system reporting charge information via a LIN bus.

as well as being robust against electromagnetic interference (and, in turn, it should minimize the interference it causes).

Automotive networking now further adds to all of these requirements, representing one of the most demanding deployment scenarios for a bus system. Ideally, an automotive bus system should weigh as little as possible, be operational within milliseconds after power up, and be able to run in a time-synchronized fashion (e.g., synchronizing frame transmissions to the sample rate of audio/video streams, minimizing – or, ideally, obviating – the need for buffering).

Further, most automotive bus systems need to be able to deal with both event-triggered and time-triggered data traffic. Event-triggered messages are those that are infrequently transmitted, e.g., when a turn signal indicator needs to be switched on, requiring the bus to be able to coordinate multiple concurrent transmission attempts. Time-triggered messages are those that need to be sent at regular intervals, such as engine control messages imposing tight deadline constraints. This particular mix requires special considerations in the design of a bus system.

It is clear that fulfilling any of these criteria will most likely involve a trade-off against other criteria, so modern vehicles typically implement many different bus systems, all interconnected and each serving a specific need.

Figure 2.2 shows a simplified example topology to illustrate this concept: Here, a seat-occupancy system is locally connected via LIN to the airbag control unit, which connects via a dedicated engine CAN bus to all systems sharing similar requirements and serving similar functions. A gateway connects the airbag control unit to systems on

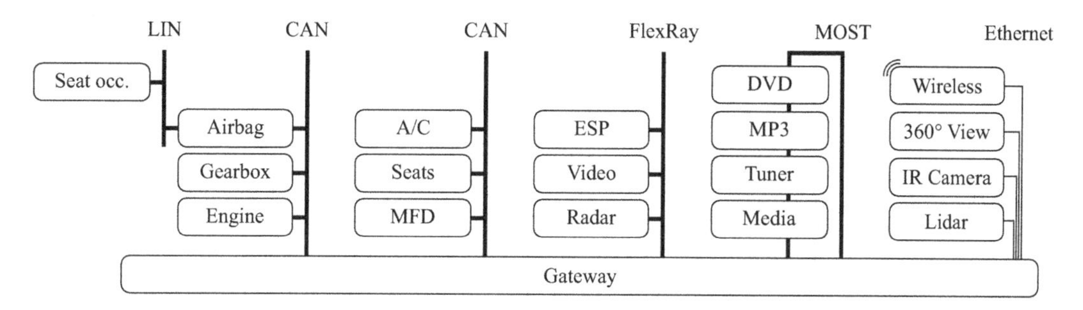

Figure 2.2 A simplified example illustrating how different bus systems are deployed in a vehicle: five example busses using four different technologies (CAN, FlexRay, MOST, and Ethernet), each serving a particular need, are interconnected via a gateway. A local LIN sub-bus interconnects a smaller set of systems.

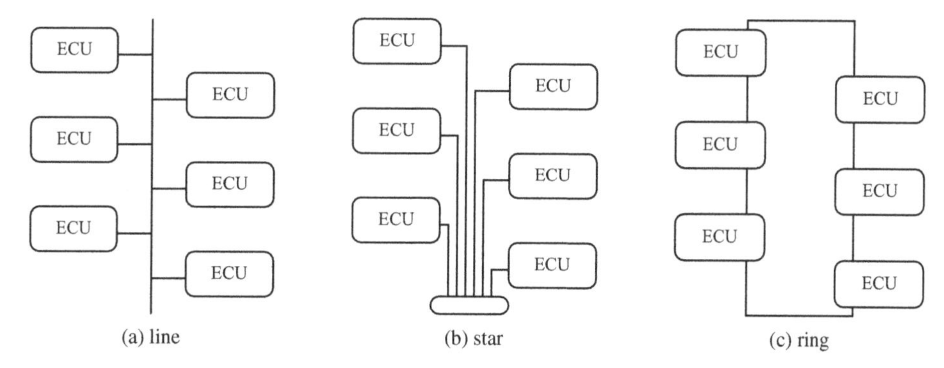

Figure 2.3 Bus topologies.

other buses, like the vehicle's multi-function display (MFD) indicating its status, which is reachable on a lower-speed CAN bus serving comfort systems.

Each bus system also relies on a bus topology (see Figure 2.3) that fulfills its specific need: Busses using a line topology are cheap and easy to deploy, but offer low robustness. A star topology is very robust against cable failure while still being easy to deploy, but comparatively costly. A ring topology, finally, is more complicated to deploy, but can offer a good trade-off between robustness and cost.

In the following, we will take a closer look at four bus systems that are tailored to in-vehicle networks: CAN, LIN, MOST, and FlexRay.

2.2 Automotive bus systems

2.2.1 CAN

The CAN bus is one of the oldest multipurpose bus specifications and one of the most ubiquitous bus systems still in operation. CAN (short for *controller area network*) was developed by Robert Bosch GmbH and released in its first version in 1986. Later

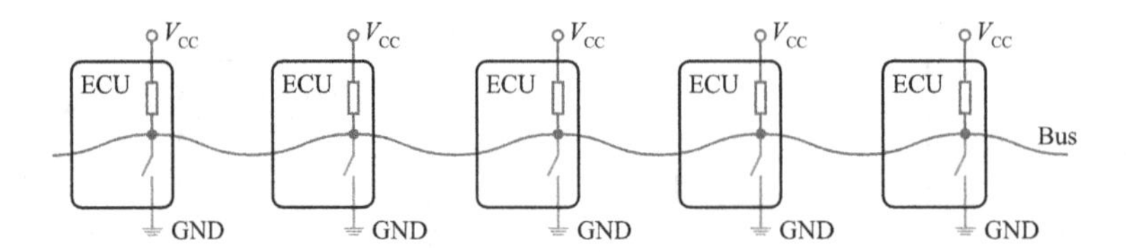

Figure 2.4 Electrical equivalent circuit of wired and. Pull-up resistors pull the bus to V_{CC} (logical 1). If any one of the switches closes, the bus gets shorted to ground (logical 0).

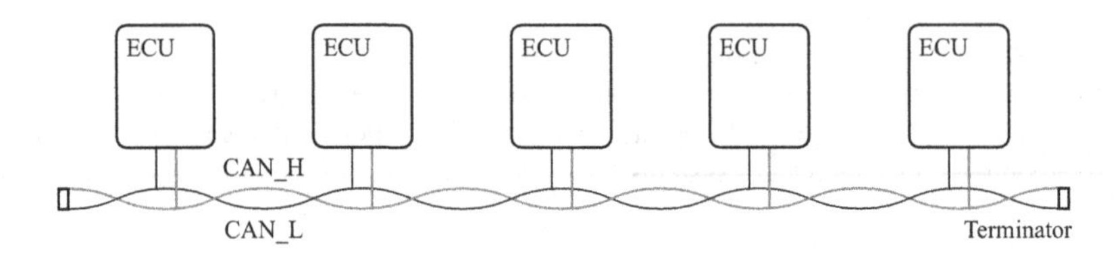

Figure 2.5 A schematic view of an automotive CAN bus according to ISO 11898.

updates to the standard have been released, with CAN 2.0 (1991) and CAN FD (2012) (Bosch 2012).

Beside vehicular networks, many different applications, from aviation to industrial automation, rely on a CAN bus to exchange data, leading to a wide variety in application- and network-layer protocols. Similarly, CAN can be run on a wide variety of different physical layers, depending on the needed bus speed, resilience against errors, and number of nodes. Typical configurations allow for up to 110 nodes and a total bus length of 500 m.

Common to all is that CAN operation relies on the physical layer being able to provide a single-channel serial bus with two different bus levels, of which one is *dominant* and one is *recessive*. What this means is that, if any device transmits a dominant bus level, the bus will read a dominant level, independently of what other devices are trying to transmit. For the sake of convenience we will refer to the dominant bus level as logical 0 (and to the recessive level as logical 1). This realizes a *wired and*, an equivalent circuit of which is depicted in Figure 2.4.

In the following, we will focus on the use of CAN in automotive environments where ISO 11898 specifies two different CAN variants: low-speed CAN with speeds of up to 125 kbit/s (ISO 11898-3:2006) and high-speed CAN with speeds of up to 1 Mbit/s (ISO 11898-2:2003). Both standards describe a physical layer using a twisted pair of wires designated *CAN_H* and *CAN_L* (CAN high and CAN low) to realize differential signaling, which maximizes resilience with respect to interference. Differential signaling means that bits are encoded as the difference of two complementary signals on this pair of wires, e.g., pulling CAN_L low and CAN_H high to apply a voltage difference between the two lines. The structure of this system is depicted in Figure 2.5.

In order to ease clock synchronization despite using non-return-to-zero (NRZ) encoding of bits, CAN uses *bit stuffing* for most protocol fields: After five consecutive identical bits have been sent, any transmitter must insert a complementary stuff bit

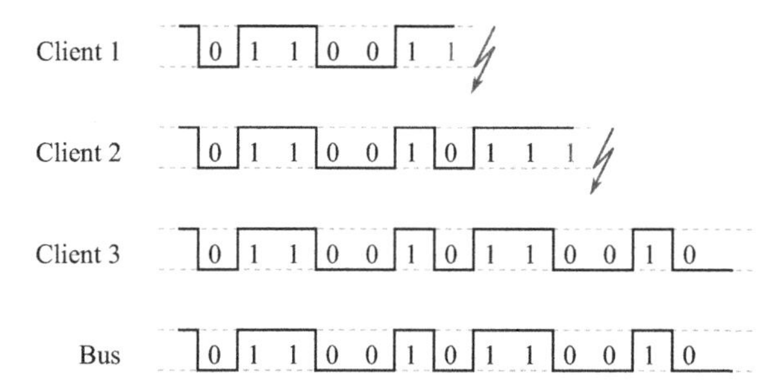

Figure 2.6 CSMA/BA CAN bus access (the dominant bus level is logical 0). Client 3 wins contention and continues transmitting uninterrupted. Clients 1 and 2 detect collision and back off.

(to be discarded by the receivers) so the signal level(s) on the bus will change at least once every five bits. Combined with the fact that there are dominant and recessive bus levels, this also gives nodes an easy opportunity to signal error conditions: transmitting six consecutive dominant bits – a signal that can never occur in regular data transmissions and is guaranteed to be decodable.

All devices on a CAN bus are peers, so no device fulfills a dedicated master role. To cope with this, CAN employs a unique carrier-sense multiple-access (CSMA) mechanism often referred to as carrier-sense multiple-access with bitwise arbitration (CSMA/BA). The mechanism is illustrated in Figure 2.6: When accessing the CAN bus, any ECU that wants to transmit waits for the bus to be idle, which can be deduced from the absence of stuff bits. It then starts its transmission by sending a message identifier reserved for this ECU. This message identifier can have both dominant (0) and recessive (1) bits. During the transmission of the message identifier, the ECU closely monitors the bus. For any transmitted recessive 1 bit, it might happen that another ECU is concurrently transmitting a message identifier with a dominant 0 bit, resulting in the bus carrying a 0 instead. If the ECU detects such a mismatch, it immediately stops transmitting, yields bus access to the other ECU, and retries later. The other ECU continues unimpaired – like any of the other ECUs on the bus, it is not even aware of any failed attempt to transmit a recessive 1 bit.

This bus access mechanism is both a huge benefit and a major drawback of CAN. On the plus side, it allows priority-controlled bus access without losing time or data in collisions, or even for the arbitration phase. On the other hand, bit arbitration and error signaling will work only if bit times are much longer than the time spent for signal propagation on the bus (to be precise, at least twice as long). This limits the bit rate and bus length: Longer buses need longer bit times (and, hence, can only employ lower bit rates).

Bus access is also perfectly integrated into the CAN frame structure: Each basic CAN frame consists of the following elements, illustrated in Figure 2.7. A single dominant bit indicates the start of a frame and is immediately followed by the arbitration and control fields. These fields can be in either of base format (11-bit message identifiers) and extended format (29-bit message identifiers, introduced in CAN 2.0 Part B) as well

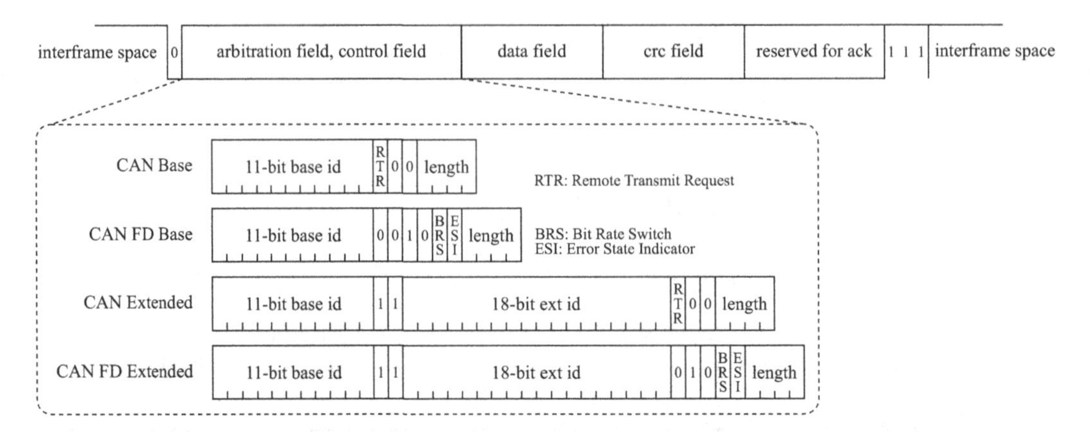

Figure 2.7 CAN bus protocol fields, shown for both CAN and CAN FD and for both base format and extended format.

as CAN format (single data rate) and CAN FD format (flexible data rate, introduced in CAN FD). Which format a transmitter has used can be reliably determined by a receiver by observing bit fields contained in the arbitration and control fields, as shown in Figure 2.7. Aside from the message identifiers, which are used for arbitration, the fields contain up to three flags: These flags can indicate whether this frame is merely used to poll remote data (RTR, remote transmit request flag), whether an alternative pre-configured higher data rate encoding has been used for the remainder of the frame (BRS, bit rate switch flag), or whether this ECU will actively signal any error state (ESI, error state indicator flag). The arbitration and control fields conclude with a data length field that encodes how long the data payload field will be (the four bits of this field correspond to predefined lengths in a look-up table). The remainder of a CAN frame is taken up by the higher-layer payload (of the indicated length), a frame checksum, and some reserved time for receivers to signal an *ack*. During this time, a receiver is allowed to put a dominant bit on the bus to indicate that it was able to decode the frame and the checksum matched.

TTCAN extension

One drawback of CAN is that deterministic delays (and, thus, real-time guarantees) can be given only for messages with an identifier of all 0 bits (or, in general, the one with the longest 0 prefix), since this message will always get priority access to the bus before any others. Any other message can be indefinitely delayed, should a higher-priority message be transmitted. Thus, to be able to give real-time guarantees, system designers normally need to closely control which ECUs will get assigned which message identifiers, and enforce restrictions on how often they can access the bus at the application level.

To alleviate this problem, ISO 11898-4 TTCAN (short for *time-triggered CAN*) was introduced as an optional extension that upgrades CAN with time-division multiple-access (TDMA) functionality. Here, each ECU synchronizes its local clock with that of a *time master* to keep track of which time slot is currently active. By consulting a pre-defined schedule table, called the system matrix, each ECU can decide whether the current slot is reserved for a particular ECU or whether the bus is available for regular CSMA/BA access.

Clock synchronization is performed via dedicated CAN messages which all nodes recognize as *reference messages*. These messages are sent by a *time master*, with up to seven fallback time masters competing for their transmission. This way, when the time master fails, the next fallback node will automatically take over.

Error detection and flow control

Error detection on a CAN bus happens on many levels. Each ECU checks the protocol conformance of each observed frame. Any receiver checks the correctness of the cyclic redundancy check (CRC) fields. The transmitter verifies that the bus is indeed carrying the transmitted signal (this is similar to what it does during CSMA/BA). Further, bus levels are monitored, even though CAN physical layers can operate under a wide range of conditions, extending (for some) even to partial shorts to ground or battery voltage. Finally, the transmitter verifies that it receives a positive acknowledgement within the time allotted in the frame. If any ECU detects an error, it will transmit an error frame: the already discussed six dominant 0 bits with no stuffing. This error frame will allow other nodes to immediately discard this frame and/or the transmitter to automatically repeat the message. Taken together, error detection and signaling measures can be expected to lower the occurrence of undetected corrupted messages to below one in 4.7×10^{11} messages (Bosch 2012). Special consideration is given in CAN to what happens if too many bus errors occur in a short time, with measures from temporarily stopping error signaling to stopping all bus access.

Rudimentary flow control in CAN is achieved by allowing receivers to transmit dedicated *overload frames* with six dominant 0 bits shortly before the bus becomes idle, to stop any sender from transmitting.

Finer-grained flow control as well as network- and transport-layer services such as fragmentation and reassembly or routing are not specified in CAN and are left to higher-layer protocols. To see how this can be achieved, we will briefly discuss two of these higher-layer protocols, ISO 15765-2 (also known as ISO-TP) and the proprietary Volkswagen TP 2.0, in the following.

ISO-TP

ISO-TP is specified in ISO 15765-2 and was conceived as a way of transmitting diagnostic messages over CAN. It is commonly used for the transmission of unified diagnostic service (UDS) messages. Access to the CAN bus via an external connector (an OBD-II connector, as depicted in Figure 2.8) is mandatory for external vehicle diagnosis in the European Union (as of 2004) and the USA (as of 2008). It also serves as the basis for many other application-layer protocols, such as those of the *Autosar* protocol stack.

ISO-TP is a very lightweight protocol, offering just a few, but essential, features: flow control, fragmentation, and higher-layer addressing. Figure 2.9 illustrates the four packet formats that ISO-TP introduces to achieve this. All packets can start with an optional additional address byte that can be utilized to distinguish higher-layer addresses from common CAN addresses. Packets that fit in one frame as well as fragmented

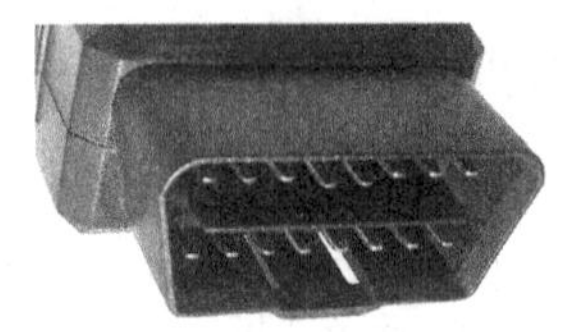

Figure 2.8 An OBD-II diagnosis plug. A socket is mandatory for all modern vehicles, and must be located in easy reach of the driver.

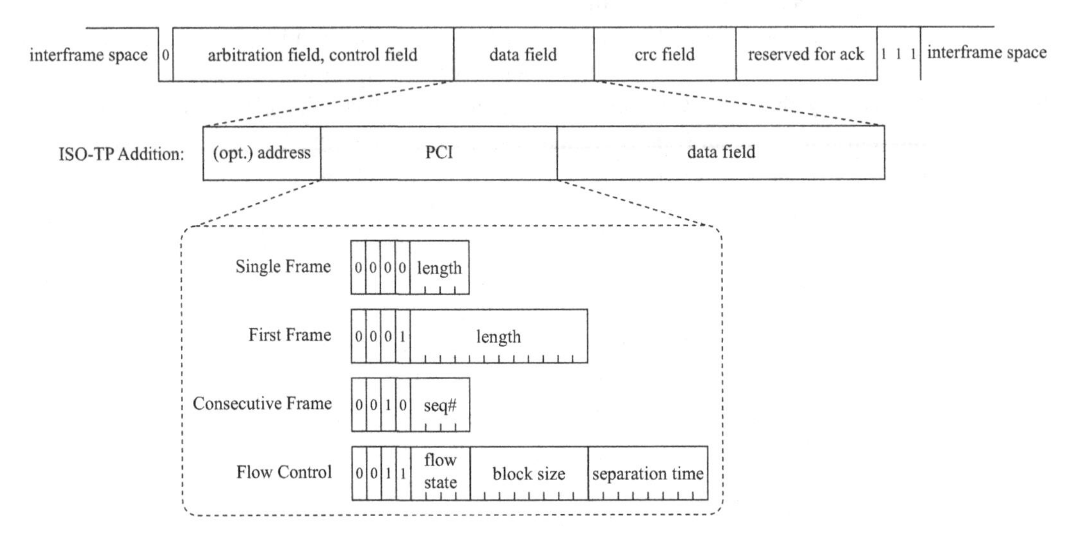

Figure 2.9 CAN with ISO-TP frame format.

packets are supported via the single frame, first frame, and consecutive frame types. Flow control between consecutive frames is achieved via explicit flow control frames. These are used by a receiver to indicate to the transmitter the flow state (clear to send or wait), the block size (how many consecutive frames may be sent before waiting for a flow-control frame again), and a separation time (the minimum time that must elapse between any two consecutive frames sent).

TP 2.0

TP 2.0, in contrast to ISO-TP, abstracts away from CAN primitives, introducing a connection-oriented approach. Communication in TP 2.0 is based on channels established between communicating ECUs via TP 2.0 and managed by means of dedicated TP 2.0 frame types. Individual ECUs or groups of ECUs are assigned logical addresses and CAN message identifiers are assigned dynamically, to established channels. Since these channels have dynamically assigned message identifiers they also have dynamically assigned CAN bus access priorities. Figure 2.10 illustrates a subset of defined TP 2.0 frame types, each serving multiple related purposes indicated via opcodes.

For example, a channel is established by sending a channel setup frame with an opcode of `0xc0` (channel request) and fields set as follows: No receiver CAN identifier

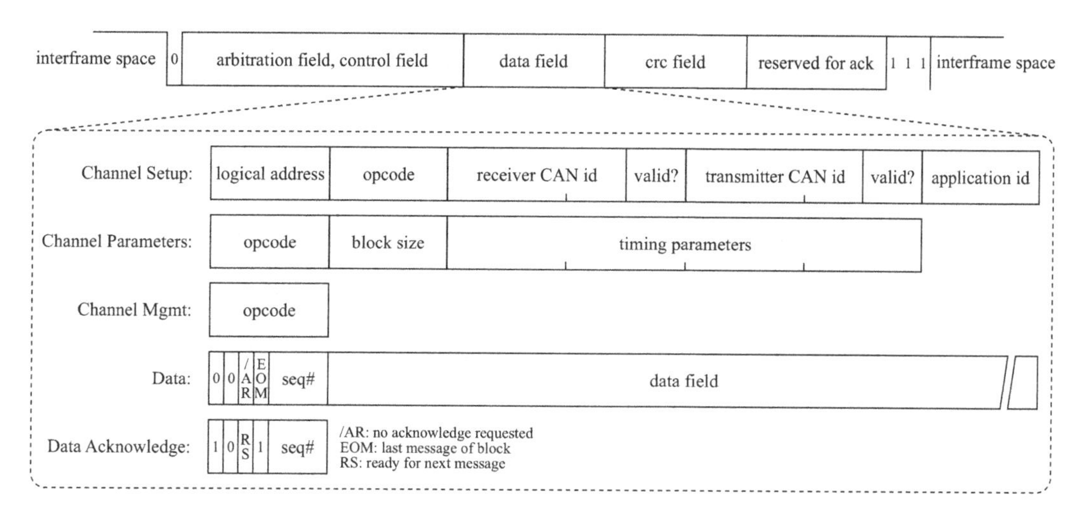

Figure 2.10 CAN with TP 2.0 frame format.

will be set (i.e., the *valid?* flag will read recessive 1), the transmitter CAN identifier will be set to a dynamically assigned, requested value, and the application type field will be set to indicate which of multiple application-layer services (similar to TCP ports) the channel should be established for. The receiver will then confirm the channel setup by sending back a channel setup frame with opcode `0xd0` (positive response), now confirming the requested CAN message identifier of the sender, and including a dynamically assigned CAN message identifier for itself.

After a successful exchange of channel setup frames, the ECUs can then set channel parameters for this connection. A dedicated frame type, the channel parameter frame, is used for this and will include a requested block size (the maximum number of frames until a sender will have to wait for an acknowledgement) and timing parameters, e.g., the minimal time between two CAN messages.

Channels can be torn down by the sender and the receiver sending channel management frames with an opcode of `0xa5`.

Packet transmissions can take place in very-lightweight data frames, using the dynamically assigned lower-layer CAN identifiers to identify the channel (and its associated parameters); thus, a data frame only includes a one-byte header indicating whether its reception needs to be acknowledged, whether this is the last frame in a block, and which sequence number the frame carries. Similarly, because all flow control information has already been set as channel parameters, an acknowledgement packet is only one byte long. It indicates whether the receiver is ready for more frames, as well as which frames have already been received.

2.2.2 LIN

Compared with the complexities of CAN, LIN (short for *local interconnect network*) is a simple bus system. It was designed as a low-cost sub-bus, that is, as a bus connecting

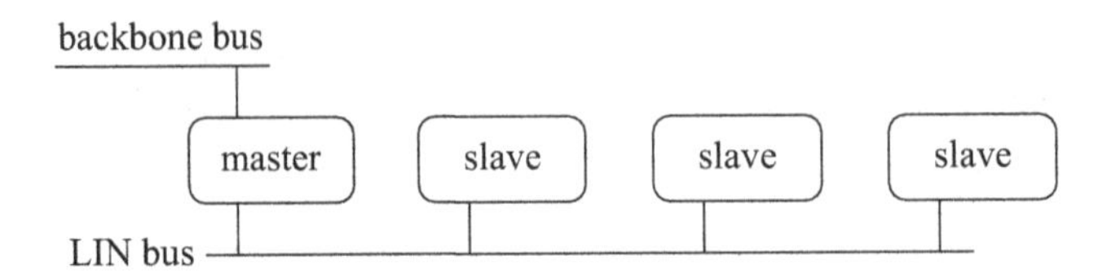

Figure 2.11 LIN bus topology.

multiple low-cost end devices to, e.g., a CAN device that is, in turn, connected to a backbone bus. Even though the initial goal of LIN being a highly cost-effective solution has not quite been reached, owing to a large part to CAN becoming ubiquitous and, therefore, prices dropping sharply, it has still become a widespread solution for the intended niche market, e.g., power door locks.

Unlike the peer-oriented CAN, LIN is a master–slave bus system: a single bus master (most commonly the device connected to the backbone, as illustrated in Figure 2.11) manages the whole LIN sub-bus with one or multiple connected slaves. A LIN specification was first published in 1999 and most recently updated as LIN 2.2A in 2010 (LIN Consortium 2010).

LIN follows a self-clocking concept requiring no expensive quartzes in slave devices. The LIN master alone decides when and which data is transfered on the bus. The LIN slaves supply this data.

LIN specifies a complete protocol stack, comprising a physical layer, a medium access layer, a transport layer for diagnostic and configuration services, and a lightweight application programming interface (API). To allow an extremely low manufacturing cost of LIN devices, not all functionality is required from all devices. Instead, the LIN specification classifies nodes into three diagnostic classes according to their capabilities. Class I devices, to give an example, do not support transport-layer services at all. Class II devices need not support the readout of other identifiers than the product ID. Class III devices support the full feature set of the transport layer.

LIN data exchange revolves around the concept of a signal, represented by a single scalar value (1-bit boolean, or 2–16-bit unsigned integer) or by a byte array of up to eight bytes. This signal represents the current state of a system. The LIN bus serves to get up-to-date values for these signals. Each frame carries either one or more signals, or diagnostic data.

On the physical layer, LIN uses a single-wire bus, transmitting universal asynchronous receiver/transmitter (UART) compatible symbols (1 start bit, 8 data bits, 1 stop bit) at up to 20 kbit/s. Error detection is done by monitoring the bus state and aborting transmissions on unexpected states. If a slave node detects a bus error, it will record this fact and make it available in a *response_error* signal, which can be queried by the master (querying this signal also resets it). No error correction is being done.

Each LIN frame encompases both a request part and a response part, as illustrated in Figure 2.12. A frame starts with a break field of at least 13 nominal bit times of dominant 0 followed by a recessive 1 (that is, the break field does not conform to the usual UART standard and can thus easily be detected). Next is a sync field (a byte

Table 2.1 A simplified example of a LIN schedule table.

Slot	Type	Signal
1	Unconditional	AC
2	Unconditional	Rain sensor
3	Unconditional	Tire pressure
4	Event triggered	Door state
5	Sporadic	(Unused) or fuel level or temperature

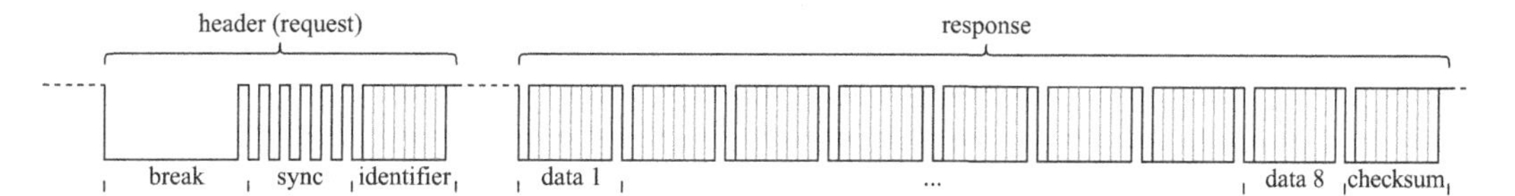

Figure 2.12 LIN frame format, consisting of request, response space, and response.

of value `0x55`, that is, alternating 0 and 1 bits) to synchronize the bit timing of the slave. An identifier and parity field concludes the request part of the frame. The six-bit identifier is used to specify which signal (in the range of `0x00` to `0x3b`) is requested. Aside from these identifiers for regular signals, two identifiers (`0x3c` and `0x3d`) are used to indicate diagnostic data transfers, and the two remaining identifiers (`0x3e` and `0x3f`) are reserved for future protocol versions. The response part follows after a short response space and encompasses the signal value (that is, one to eight bytes, as required by the requested signal) and a checksum byte (calculated, depending on the LIN version, over the signal only or over both identifier and signal).

While, technically, LIN specifies only one type of frame, its concept distinguishes between four different use cases of pre-configured time slots in what is called the *schedule table* of the bus master (see Table 2.1 for an example).

Unconditional frames are requested by the master periodically, according to its schedule table. Each triggers the transmission of one signal by the responsible slave node. Since the transmission of unconditional frames follows a rigid schedule, this allows fully deterministic timing (and, hence, real-time guarantees).

Sporadic frames can share a common position in the master's schedule table and are sent only seldom or only when needed. To give an example, a LIN master might poll for changes in the position of the front left power window only when it knows that it is currently opening or closing. Otherwise it might poll for the front right power window, or none at all. Although this also introduces some non-determinism, prioritization of sporadic frames by the master provides real-time guarantees with regard to the highest-priority frame.

Event-triggered frames serve to poll multiple slave nodes with uncommonly occuring events. To give an example, a LIN master might poll for changes in the state of door contacts for all doors simultaneously. Only a door that actually opened or closed will respond to this request. To distinguish between multiple possible responses, the first data byte of the response part of the frame is used for the identifier of the corresponding signal. Obviously, this use of event-triggered frames also opens up the possibility of collisions on the bus, introducing non-determinism that needs to be carefully planned for. In the case of a collision (that is, multiple slave nodes trying to respond to the event-triggered frame), the master will poll for each of the multiple possible responses once.

Diagnostic frames can carry ISO-TP transport-layer messages to be exchanged over the LIN bus, e.g., to allow the routing of diagnostic CAN messages via the LIN master to LIN slave nodes, as follows.

To keep the conversion overhead between ISO-TP over CAN and LIN low, the same packet format as shown in Figure 2.9 is used for single frames, first frames, and consecutive frames. Flow-control frames are unsupported. Relaying of ISO-TP packets is handled as follows. Relaying diagnostic messages *to* the LIN bus is straightforward: If a frame arrives at the LIN master, its transport-layer payload is passed on to the LIN bus with a LIN identifier of `0x3c`. Relaying diagnostic messages *from* the LIN bus is less straightforward. It requires the LIN master to poll the bus by sending frames with a LIN identifier of `0x3d`. This gives LIN slaves an opportunity to reply with an ISO-TP payload in the response part of the frame.

2.2.3 MOST

Another niche in the intra-vehicle bus systems is filled by the MOST (short for *media oriented systems transport*) bus (MOST Cooperation 2010). As the name implies, this bus was designed with different requirements in mind than those of CAN or LIN. Unlike those two bus systems, where low delay and high fault tolerance were leading goals, MOST was designed to be a multimedia and infotainment bus system. This means that two very different requirements, high data rate and low jitter as well as easy aftermarket extensibility (i.e., no reliance on pre-computed schedule tables), were of primary concern.

The history of MOST goes back to the domestic data bus (D2B) which was later, in the 1990s, specified as IEC 61030. This bus, however, saw little adoption by vehicle manufacturers, so soon thereafter a successor was being worked on by the then newly formed MOST Cooperation, initially consisting of Harmann/Becker, BMW, Daimler-Chrysler, and SMSC. In December 2009 MOST 3.0E1 was published and today the MOST Cooperation numbers 60 original equipment manufacturers (OEMs) and 15 vehicle manufacturers.

MOST was designed to run over plastic optic fiber (POF), although later variants standardized electrical physical layers as well. Bits are Manchester-coded, so bit transmission is self-clocking; electromagnetic interference (EMI) is no concern for POF. Data rates range from 25 Mbit/s (for MOST 25) to 150 Mbit/s (for the most recent MOST 150 physical layer).

MOST operates on a (logical) unidirectional ring topology of up to 64 ECUs with one dedicated bus master ECU (see Figure 2.13). Any data sent is (actively) passed on

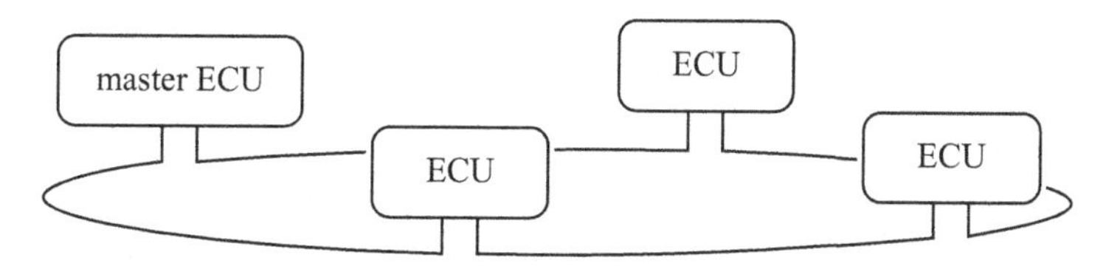

Figure 2.13 MOST bus topology.

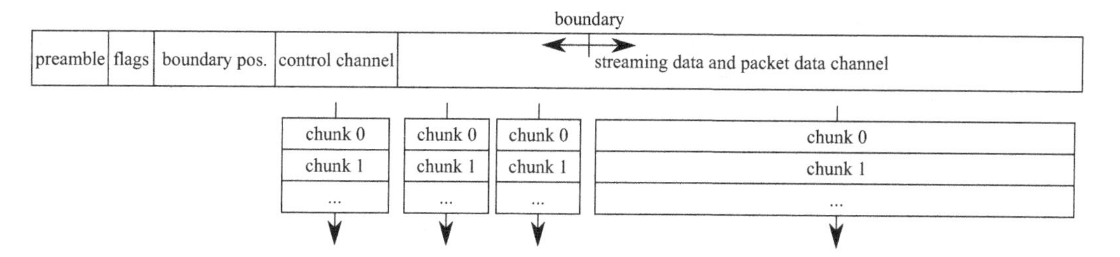

Figure 2.14 MOST frame format, consisting of a header and control channel, as well as packet data and streaming data channels, each transporting a small chunk out of a stream of data per frame.

from one node to the next until it has traversed the ring once and reaches the original sender again.

Data transmission takes place in frames of equal, fixed size (64 Byte for MOST 25 or 384 Byte for MOST 150). The frame rate of MOST is static, but fully configurable for each bus, with the standard recommending 48 kHz (i.e., the sampling rate of DVD audio).

Frames are used to transmit chunks of multiple independent continuous bit streams, called channels. For this, a TDMA concept as illustrated in Figure 2.14 is employed.

Part of each frame is allocated to *streaming data* channels. Any ECU on the bus can reserve a range of bytes of each frame for the transmission of (a small chunk of) a bit stream (e.g., a video or audio stream). This guarantees perfect throughput, no jitter, and perfect determinism for data transmissions, all key design goals of the MOST bus.

Part of each frame (as indicated by a boundary descriptor included in each frame) is allocated to the transmission of packet data. This channel carries a chunk of a bit stream of messages that are put on the ring in a random-access fashion. Each message is prefixed with a priority, which allows ECUs on the ring to decide whether to pass on the message unmodified or to replace it with a higher-priority message before passing it on. Aside from the payload, each message also contains fields for source and target address, message length, and a checksum. Any acknowledgement or the repeating of failed transmissions needs to be performed by upper layers, requiring protocols like TCP/IP to be run over a MOST packet data channel (the standard specifies either the use of *Ethernet over MOST* or the use of *MOST high*, a variant of TCP).

Finally, a small part of the frame (4 Byte for MOST 150) is allocated to a *control* channel that is in many aspects similar to the packet data channel, but it carries a bit stream of control messages only. Message transmissions are encoded using the application message service (AMS) protocol. It uses acknowledgements, so failed transmissions can be automatically retried. Each control message is 32 Byte long and can be used, among others, to reserve, release, and query ownership of streaming data channels – or to read from and write to registers and configuration data of ECUs.

Device addressing uses 16-bit addresses that are classified into four distinct groups: First, an address can refer to the physical position of an ECU in the ring (starting with `0x0400` for the master ECU and counting up). Secondly, addressing can use a logical address, assigned by the master (upwards of its own logical address, `0x0200`). Thirdly, multiple destinations can be addressed via groupcast (upwards of `0x0300`). Finally, MOST supports broadcasts to all attached devices, either with high priority and blocking the control channel during its transmission (`0x03c8`) or as a regular message on the control channel (`0x03ff`).

On a higher layer, addressing in MOST follows an object-oriented concept that facilitates aftermarket extensibility: Each MOST device contains one or more function blocks (FBlocks), of which one or more instances might be available. Each instance offers methods that can be called and properties that can be written to or read from via MOST.

As an example, changing to the 10th track of the first CD player of the CD changer (CDC) device (and getting confirmation that the change was successful by reading back the new value) can be accomplished by sending a command that can be visualized as follows: `CDC.CD.1.Track.SetGet(10)`. Function blocks have well-known identifiers (e.g., the MOST standard lists `0x31` for a CD player). Likewise, the IDs of available properties and methods are well known. Instance identifiers for a particular function block are ensured to be unique (the MOST bus master does a system scan and re-assigns duplicate instance identifiers during boot). If the device identifier is not known, MOST allows one to substitute a wildcard device ID (`0xffff`) for requests of the following form: `???.CD.1.Track.SetGet(10)`. In any case, the sending device is then informed about the changed property (the *track* property now has a value of 10). Other devices might also have subscribed to receive notifications when this property changes. They then also receive the new property value.

Error detection and correction in MOST is highly dependent on the physical layer used for the communication between two ECUs, but mostly straightforward since the individual connections are unidirectional and connect exactly two devices. MOST is, however, sensitive to a global error condition, a ring break. If the POF gets bent or otherwise damaged, or if an ECU fails, the whole MOST system stops working. Either messages are not transmitted to the recipient, or replies do not reach the sender. While the fact that a ring break occurred is easily detectable (from these symptoms), the underlying reason and the affected ECUs are often impossible to determine via MOST alone. Workarounds for this problem exist, but they are vendor-dependent and proprietary. Often, an additional single-wire bus is used for ring-break diagnosis.

Taken together, this means that MOST is indeed a prime candidate bus for high-speed and low-jitter communication and that it caters very well to the need for aftermarket extensibility of multimedia and infotainment components – albeit at the cost of low robustness of the bus system.

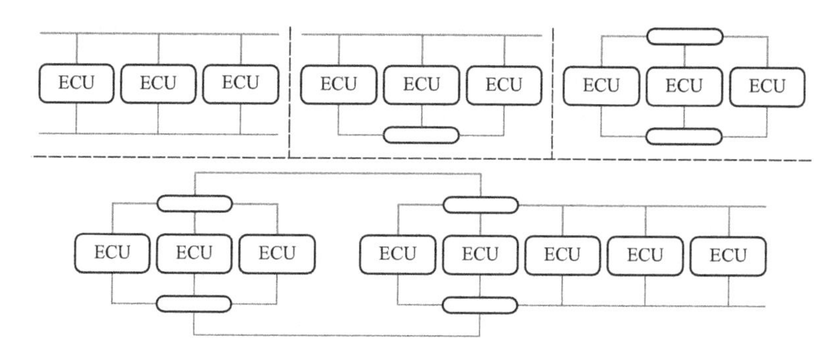

Figure 2.15 Sample FlexRay topologies: passive bus, hybrid, and active star topologies, each using two channels (top row); and cascaded two-level star/bus hybrid (bottom).

2.2.4 FlexRay

FlexRay was conceived to alleviate the drawbacks of earlier bus systems, particularly their low robustness, while still providing fully deterministic low-latency and high-speed transmissions. For this it specifies many possible bus topologies (see Figure 2.15). FlexRay also explicitly specifies that controllers need to supply a way for ECUs to directly tune the frame rate or frame offset, most prominently to support time synchronization across more than one FlexRay cluster, or with a different bus system.

The first specification of FlexRay, version 2.0, was published in 2004, by a consortium of automotive concerns (BMW, DaimlerChrysler, General Motors, Volkswagen), semiconductor manufacturers (Freescale, Philips), and an OEM (Robert Bosch GmbH). The first vehicles that relied on FlexRay entered the market in 2006. Use of FlexRay was limited to members of the FlexRay consortium (or companies that had a standing agreement with the consortium), which grew to include some hundred members and associated companies until the consortium adopted a more open approach to development and disbanded in 2009. Today, FlexRay is developed as an ISO standard series (ISO 17458).

The FlexRay data-link layer is defined in ISO 17458-2:2013. Similarly to that of MOST, it employs a TDMA scheme for channel access, but can use up to two physical connections (channels A and B). Channel B can be used either to send the same message (serving as a redundant communication path) or to send a different message (increasing the effective throughput).

Conceptually, time passes in increments of what FlexRay calls *macroticks*, which are a few (e.g., 50) bits long and synchronized across all ECUs of a FlexRay cluster. Time is divided into 64 communication cycles of pre-configured length (up to 16 000 macroticks), each of which is further subdivided into slots. The assignment of message IDs (that is, frame IDs) to channels, slots, and cycles is fixed and known to every node. Because there are 64 different cycles, messages can be sent at much slower rates than once per cycle (which can only be up to 16 ms long and is typically even shorter, e.g., 5 ms) and still be guaranteed a time slot. This also frees up slots for different messages to be sent in the remaining cycles. FlexRay refers to this sending of a message only

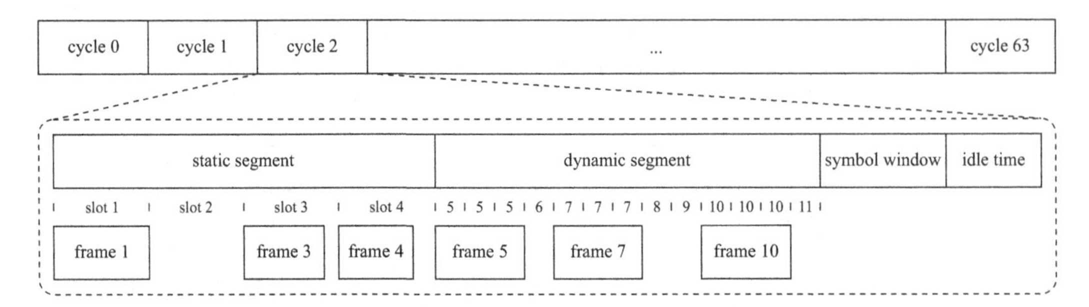

Figure 2.16 FlexRay communication cycles, each subdivided into a static segment further divided into equal-sized slots, an (optional) dynamic segment further divided into minislots, and an (optional) symbol window.

every other cycle (or, in general, every 2^n cycles) as *cycle multiplexing*. The schedule table needs to be carefully designed by engineers during the network-design phase, particularly to closely align message rates on FlexRay with their rate of generation by ECUs.

FlexRay further affords some non-determinism in exchange for better use of the channel capacity, by allowing communication cycles' schedules to contain a *dynamic segment* utilizing what FlexRay calls Flexible TDMA. This concept is illustrated in Figure 2.16.

While in the *static segment* each slot is of equal length and assigned to exactly one message, the dynamic segment utilizes a concept called *dynamic slots*. Here, the slot counter (i.e., the current slot number) for a given channel (channel A or channel B) is incremented only after either a message has finished transmitting on the channel or a *minislot* has passed without a message being transmitted. This means that the slot counters for channels A and B might advance at different rates. It also means that lower-priority messages (that are assigned higher slot numbers) may be postponed arbitrarily, or suppressed altogether when postponed until after the end of the dynamic segment of a communication cycle. Typically, such *minislots* are configured to be much shorter (FlexRay allows up to 63 macroticks per minislot, 5 being a more practical value) than slots in the static segment (which can be configured to be up to 2047 macroticks long).

Each communication cycle may also have allocated space for a *symbol window*, an opportunity to transmit a single symbol. At this time, only a *media access test* symbol for testing the bus guardian is specified.

On the physical layer, one of the core design concepts of FlexRay is its high wiring flexibility. An electrical physical layer is specified in ISO 17458-4:2013 (although alternative, proprietary physical layers exist). It specifies the use of either one or two pairs of unshielded twisted copper wires to connect ECUs, each carrying data at a default rate of 10 Mbit/s. Either channel can be wired in a passive bus configuration, a star configuration using an active star coupler, or as a mix of these configurations. Star couplers can be nested two levels deep. Figure 2.15 again gives an overview of these sample bus topologies.

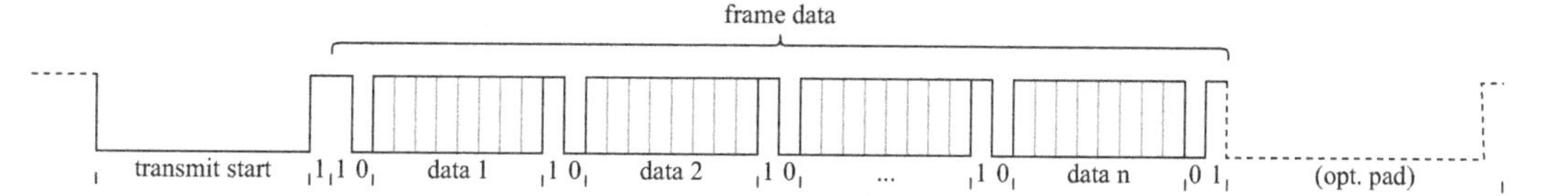

Figure 2.17 The FlexRay physical-layer concept. Shown is a frame consisting of a transmission start sequence, a frame start sequence, a flexible number of data bytes separated by byte start sequences, a frame end sequence, and an optional padding to the next full macrotick (dynamic trailing sequence) for frames sent in a dynamic segment.

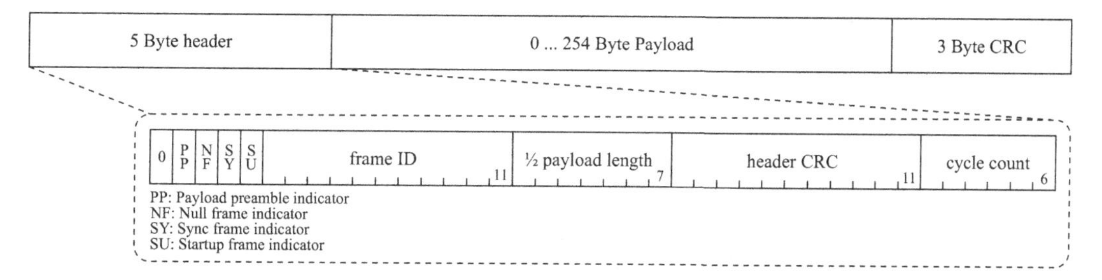

Figure 2.18 The FlexRay frame format, consisting of header, payload, and CRC.

Bits on each wire pair are encoded using differential NRZ encoding, as in common CAN physical-layer implementations. No bit stuffing is employed, so the message length remains deterministic. Instead, clock synchronization is helped by framing each transmission, each frame, and each single byte, as illustrated in Figure 2.17. Extended periods (11 bits) of the signal level on a channel remaining in the high state are interpreted as the channel being idle.

Figure 2.18 illustrates the contents of each FlexRay frame, which consists of a 5-Byte header, up to 254 Byte of payload, and a 3-Byte CRC computed over the complete header and payload. Each FlexRay header starts with five single bits, the first of which is a reserved bit, set to 0. It is followed by a payload preamble bit, which indicates whether the payload starts with additional control information that is opaque to FlexRay (for frames in the static segment this is assumed to be a fixed-length *network management information* block; for frames in the dynamic segment this is assumed to be a two-byte message ID that allows one to send any of a set of message types using the same frame ID). A null frame bit is used to signal that no actual payload is contained in the frame and the contents of the payload field should be discarded (this is of importance for frames sent in the static segment, which need to be of fixed length). A sync frame bit indicates that this frame is a viable frame to use for clock synchronization, i.e., that it was sent by an ECU that is deemed reliable. The startup frame bit is used during system startup and will be discussed later. After these five single bit fields, the header contains the frame ID, which identifies the slot in which a frame should have been transmitted. This field is followed by a half-length field, which (when multiplied by two) indicates how many bytes of payload the message contains. A header CRC field follows these fields and carries a checksum protecting the sync frame and startup frame bits, the frame

Table 2.2 The FlexRay cold-start procedure: Nodes A, B, and C are configured as cold-start nodes and try to wake the bus. Node A detects no wake-up pattern (WUP) on the channel and transmits one; node B detects a WUP and backs off; node C transmits a WUP, detects a collision, and backs off; node A transmits a collision-avoidance symbol (CAS) and begins bus operation, transmitting frames. After synchronizing their macroticks for four cycles, nodes B and C join the bus; after two more cycles all remaining nodes join.

Node	Cold-start node?	Symbol on the channel or communication cycle										
A	Yes	WUP	WUP	CAS	0	1	2	3	4	5	6	7 ...
B	Yes	↯							4	5	6	7 ...
C	Yes		WUP ↯						4	5	6	7 ...
D	No										6	7 ...
E	No										6	7 ...

ID, and the payload length field. The header concludes with a cycle-count field that carries the current value of the transmitting node's bus cycle counter.

The protocol fields of FlexRay allow fine-grained detection of errors and erroneous controllers, e.g., by checking the cycle-counter value, frame ID, header or frame CRC, or the overall frame timing. Since any frame collision must constitute a bus access by an erroneous controller, a FlexRay bus guard is easily able to determine faulty operation of the ECU and swiftly either switch it to a passive, listen-only state to allow re-synchronization or disable the ECU altogether. It should also be noted that no automatic repetition of failed messages is ever attempted by FlexRay, since this would undermine the determinism of the bus.

Because of its tight time-synchronization requirements and globally synchronized macrotick and slot counter, powering on a FlexRay cluster requires the system to undergo a well-defined cold-start procedure, which is illustrated in Table 2.2.

- At system-design time, at least three nodes are designated as *cold-start nodes.* These are the nodes that will compete for conducting the cold-start procedure.
- Before a cold start can be attempted, the bus (i.e., all attached ECUs) needs to be woken up. This is achieved by transmitting a special wake-up pattern (WUP) that can be detected by the attached communication controllers. Any number of nodes can simultaneously try to transmit a WUP; the pattern is designed so that collisions can be detected. After colliding nodes have backed off, only one node will transmit the WUP.
- After waking up the bus (if needed), these nodes will try to detect a special collision-avoidance symbol (CAS) on the bus and, if none is detected, send one. A node that manages to transmit the CAS is deemed the *leading cold-start node* (although, due to unresolved collisions, potentially more than one node might consider itself to be the leading cold-start node at this point).
- After successfully transmitting the CAS the leading cold-start node will start regular operation and start transmitting frames (with the startup frame bit set), beginning with the first communication cycle (cycle 0). For four cycles, the leading cold-start node will keep monitoring the bus for potential CASs (which would

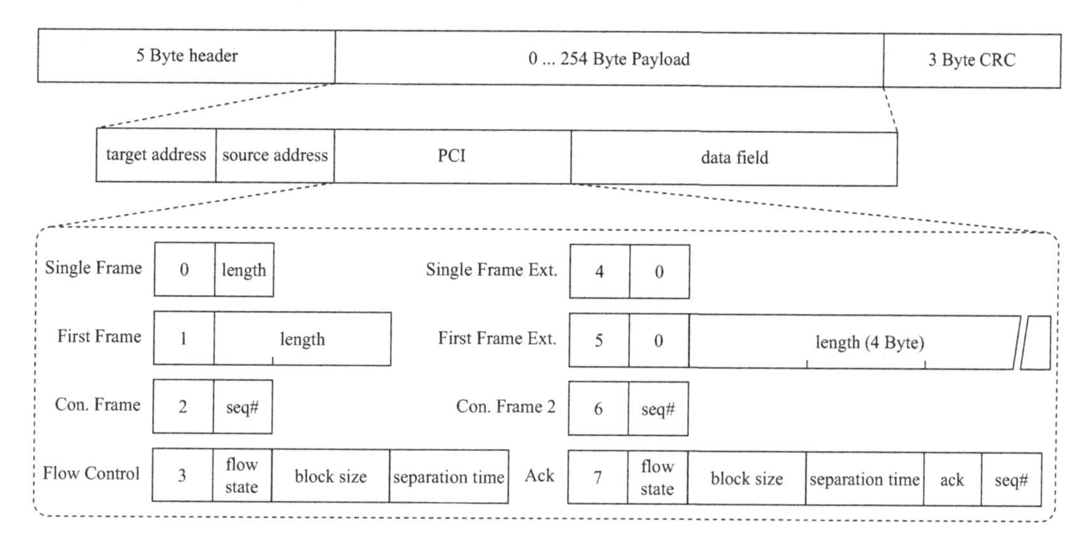

Figure 2.19 The extended ISO-TP packet format for FlexRay.

indicate that another node considers itself to be the leading cold-start node, triggering an abort of its own operation).

- At the same time, any other cold-start node will wait for at least the same four cycles, while adjusting its macrotick timekeeping to the received frames, before it starts regular operations, now also sending frames with the startup frame bit set.
- All other nodes wait for two frames by two different nodes that have the startup frame bit set before starting regular operations.
- The bus and all attached ECUs are then fully operational.

On the transport layer, an adapted version of ISO-TP exists in the Autosar protocol stack called FlexRay Transport Layer (FrTp) (Autosar 2011), which enables FlexRay to carry ISO-TP messages: While ISO-TP for CAN allowed one to transmit part of a higher-layer address in a CAN message ID, addresses are now encoded only (and completely) in the transport protocol header, which now carries an up to two-Byte-long target address and source address field. It also needs to work around the absence of a lower-layer *Ack* mechanism (CAN offered a way for the receiver to transmit a dominant bit in the sender's frame). Further, it introduces four new frame types to cope with larger amounts of data and the shorter required separation times required between blocks of data.

The new frames and fields are illustrated in Figure 2.19: Extended versions of the ISO-TP *single frame* and *first frame* had their payload length fields increased from 0.5–1 Byte and from 1.5–4 Byte, respectively, allowing a first frame to announce up to 4 GByte of data. The *flow control* frame was extended in three respects. First, the *flow state* takes new codes to allow a receiver to request that a sender aborts its transmission. Second, values for the *separation time* of `0xf1` through `0xf9` can now be used to specify separation times of 100–900 μs (instead of the field's usual meaning of multiples of ms). Finally, the flow-control frame was supplemented with additional fields to form an *acknowledgement frame*, which can signal whether a transmission was successful (ACK = 0) or whether a consecutive frame in the last block of frames (indicated by its sequence number) was not successfully received (ACK = 1). The *consecutive frame* was supplemented with an identical *consecutive frame 2*, which can be used for every other frame to simplify the detection of retried transmissions (which are allowed to be sent at the sender's earliest convenience).

2.3 In-vehicle Ethernet

Modern advanced driver assistance systems (ADASs) use data from different domains. Other than older applications, where these were strictly separated and communication was rare, such data sources will now often be in different domains and need to exchange data at high rates and with low latency while maintaining strict synchronization requirements to reduce or obviate the need for buffering.

Such systems include, for example, adaptive cruise control (ACC) systems, which need to integrate data from many sources in order to be able to carry out their function. They need regular odometry, high-resolution video, radar, and light detection and ranging (LIDAR) systems, that is, laser scanners. Taking one step further, cooperative adaptive cruise control (CACC), which we will discuss in Section 3.1.3 on page 48, will additionally need to be able to integrate data received wirelessly from other cars under tight real-time constraints. This concept is referred to as *sensor data fusion*.

Since different domains are typically using different, domain-specific bus systems, the strict synchronization and latency requirements can often be maintained across domain boundaries only if substantial resources are statically allocated to such systems. Thus, a migration of vehicular networks towards a unified bus system might be beneficial.

This is further compounded by two trends, mass customization of cars and aftermarket customization (Ziomek *et al.* 2013), especially when seen together with the recently emerging trend towards bring your own device (BYOD). All this makes it hard to stick with the classic splitting of vehicular networks into separate domains and a static configuration of these. A solution to these problems is seen in the introduction of a well-established general-purpose standard, as opposed to yet another domain-specific one, as the basis for vehicular networks: in-car Ethernet.

2.3.1 Background

The history of Ethernet goes back to research conducted in the early 1970s at Xerox PARC and published by Bob Metcalfe and David Boggs (Metcalfe & Boggs 1976) – this was not the first proposal for a local area network (LAN), but the most successful. Their original proposal already afforded a transmission speed of 3 Mbit/s and allowed 256 nodes to share a common transmission medium, a single 1-km-long coaxial cable called the *Ether* (or multiple cables connected via repeaters).

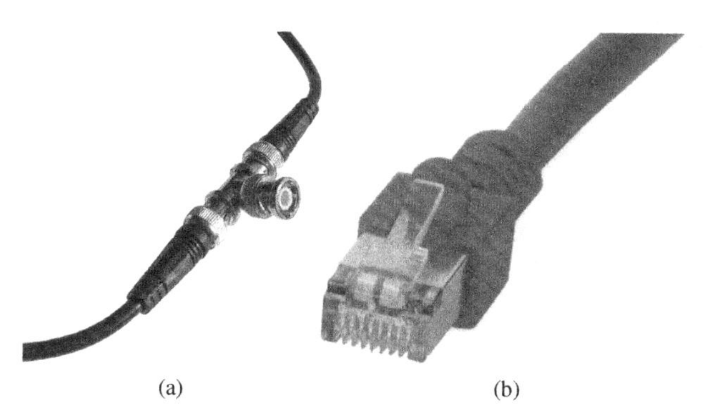

Figure 2.20 Cables and connectors for 10BASE2 and 1000BASE-T, two widely successful physical layers for Ethernet. (a) 50-Ω coaxial cable with BNC connectors and T-shaped connector (10BASE2 Ethernet). (b) Category 5e cable with 8P8C connector (1000BASE-T Ethernet).

Standardization was carried out by the IEEE and, to date, many different extensions and variants have been standardized for Ethernet in the IEEE 802.3 working group. The first standardized physical layer was specified in IEEE Std 802.3-1985, prescribing the sending of a Manchester-coded signal with a typical rise of approximately 2 V at 10 Mbit/s over a coaxial cable of length up to 500 m. Nodes could attach to this cable by *tapping* into it, that is, the Ethernet cable was opened and short wires connected to its shielding and core. This physical layer was named *10BASE5*, or *thicknet*. A later physical-layer standard and the first that was widely successful, *10BASE2*, was designed for thinner cables and moved from a single physical cable to multiple cable segments running between T-shaped cable connectors (see Figure 2.20(a)) that served as a way to tap into the Ethernet cable, albeit at the cost of reducing the maximum range to a little under 200 m.

Because in these physical layers the medium offers only one channel and is shared among all nodes, all nodes must adhere to a carrier sense multiple access with collision detection (CSMA/CD) scheme. The presence of a carrier on the channel indicates to all nodes on the same bus segment that the channel is currently busy. A frame may be transmitted only after an interframe space of 12 Byte since the last transmission on the channel has elapsed (excepting burst transmissions by the same station, where allowed). For at least the first two thirds of the interframe space the channel is continuously observed and, if the channel is sensed to be busy again, the waiting time for an idle channel restarts. If a collision is detected by a physical layer, it transmits a jam signal to make all nodes on the bus segment aware of this. Transmissions that suffered from a collision are then retried according to a binary exponential backoff procedure.

More recent versions of the IEEE 802.3 standards specify physical layers that offer more than one channel and network topologies with bus segments that are no longer shared by multiple nodes, allowing the segment to switch from half-duplex transmission with CSMA/CD to full-duplex transmission. At the time of writing, the most recent Ethernet standard, IEEE Std 802.3-2012, includes physical layers for the transmission of 100 Gbit/s over 40 km of optical fiber (*100GBASE-ER4*). The most well known of

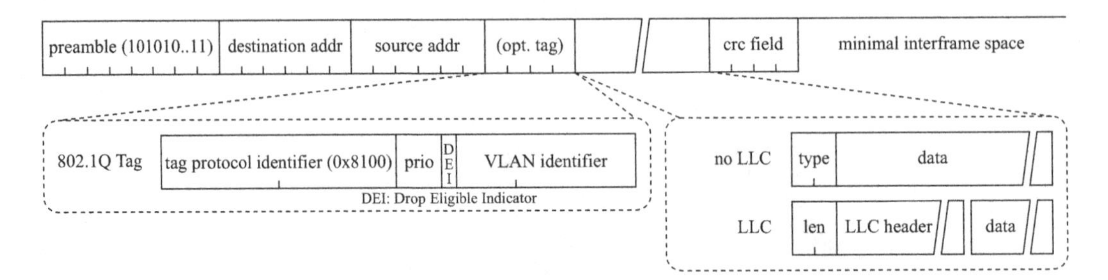

Figure 2.21 The Ethernet frame format; special values in the type/length field indicate the presence of an IEEE 802.1Q tag and/or IEEE 802.2 LLC information.

all physical layers, however, is the now very common *1000BASE-T* physical layer. It uses four pairs of twisted wires, typically run in *Category 5* cable or better (see Figure 2.20(b)), to transmit 1 Gbit/s over up to 100 m.

Figure 2.21 shows the makeup of an Ethernet frame, which is very lightweight: Each frame starts with an 8-Byte preamble and start frame delimiter field of alternating one and zero bits, with the last pair being two ones to enable easy frame synchronization, even if higher-layer decoding started late. The preamble is followed by two 6-Byte (48-bit) fields containing the destination and source medium access control (MAC) address. These can optionally be followed by a 4-Byte-long IEEE 802.1Q tag containing frame-priority information, whether the frame is eligible for dropping on congestion, and a virtual LAN (VLAN) tag. Just before the data, a 2-Byte type/length field contains either an indication of the type of payload that follows (e.g., a value of `0x86dd` indicates IPv6) or the length of an IEEE 802.2 logical link control (LLC) payload. In either case, this field is followed by the higher-layer payload. The standard specifies a minimum payload length (to allow CSMA/CD to work; 42 Byte for common configurations) and a maximum payload length of 1982 Byte (after encapsulation of higher-layer protocols, with 1500 Byte being the recommended maximum upper-layer payload size). The frame concludes with a 4-Byte (32 bit) frame-check sequence and must be followed by an idle period of at least 12 Byte.

2.3.2 Adaptations for vehicular networks

It is clear that, compared with any of the automotive bus systems discussed, Ethernet is an extremely lightweight protocol that offers little more than error-checked best-effort delivery of data. Thus, in order to fulfill the numerous demands listed in Section 2.1 on page 13 that we have discussed in this chapter so far, additional protocols need to be run on top of Ethernet to meet, e.g., delay and synchronization demands. Further, none of the classic physical-layer specifications is a good fit for the demands of robust yet lightweight cabling.

In order to meet physical-layer demands, a special interest group called the OPEN (short for *One-Pair Ether-Net*) Alliance Special Interest Group, formed in late 2011, has rallied behind a new physical-layer technology. The special interest group was initially formed by BMW, Broadcom, Freescale, Harman, Hyundai, and NXP, but quickly grew to include 150 members by early 2014. The physical layer is based on a single unshielded twisted pair of wires capable of transmitting power (conforming to the IEEE 802.3at PoE, short for *Power over Ethernet*, standard) and data at 100 Mbit/s. The technology is manufactured by Broadcom, marketed as *BroadR-Reach*, and licensed to members of the alliance. In parallel, a new IEEE 802.3bp task force, the *Reduced Twisted Pair Gigabit Ethernet* (RTPGE) task force, first met in early 2013 to standardize a new physical layer for transmission of 1 Gbit/s over no more than 15 m of a single twisted pair of wires for automotive use.

Novel upper layers complement these new physical layers for Ethernet and are needed in order to meet the demands of vehicular networks. Many proprietary solutions exist to this problem, but only a few have gained any traction so far (most notably SAE AS6802, time-triggered Ethernet).

A comprehensive set of upper layers has been specified by the IEEE 802.1 Time-sensitive networking (TSN) task group. The group has its root in an IEEE 802.3 study group and, until November 2012, had its first incarnation as the *Audio/Video Bridging* (AVB) task group. Standardization is promoted by the *AVnu Alliance* industry special interest group (not unlike how the *Wi-Fi Alliance* promotes IEEE 802.11).

TSN operation revolves around the concept of dynamic resource reservation on paths through the network called *streams*, together with tight global time synchronization. It defines two priority classes of streams, which differ in their presumed requirements: Class A and Class B specify latency constraints of 2 ms and 50 ms over seven hops. Additionally, *best-effort* legacy traffic is still permitted on the network.

Time synchronization, stream reservation, and the enforcement of guarantees is performed by a new layer-2 time-synchronizing service (IEEE 802.1AS) and extensions to the IEEE 802.1Q standard for Ethernet frame tagging. In the following, we will take a closer look at these higher-layer standards.

Time synchronization is performed by a new time-synchronizing service specified in IEEE 802.1AS, which is based on a subset of the IEEE 1588 precision time protocol (PTP) to synchronize the clocks (that is, clock value and frequency) of all participating nodes. Participating nodes are TSN-capable Ethernet bridges (i.e., switches).

With a pre-configured node priority, nodes can compete for being elected as the source of a grandmaster clock. A single node is selected to provide this grandmaster clock both to supply the absolute value of a global clock and to serve as a reference for clock frequency. After the election of a grandmaster, a spanning tree (rooted at the grandmaster) is constructed, along which timing information will be synchronized. Any node synchronized to the grandmaster clock can in turn serve as a timing master to nodes further away (in terms of hops) in the network. The achievable synchronization accuracy could be shown in simulations (Lim *et al.* 2011) to remain within 1 μs.

Stream reservation is performed by a new stream reservation protocol (SRP) specified in IEEE 802.1Qat. For this TSN introduces the concepts of talkers and listeners in a network. Talkers advertise new streams in a network, defining their quality-of-service (QoS) priority class along with parameters such as maximum packet size and packet interval, by disseminating an SRP stream advertisement through the network. These messages also include a field containing the cumulative worst-case latency. All intermediate nodes that receive this advertisement then check whether enough resources are available to satisfy the requirements of this stream, and if there are, temporarily reserve these resources, update the worst-case latency, and pass the SRP advertisement message on to the next hop. A listener that receives an SRP advertisement message can thus be sure that sufficient resources (e.g., available outbound bandwidth and buffer space) have been (temporarily) reserved in the network to guarantee the QoS and can check how much latency the stream would incur in the worst case. It can then accept the stream by sending a registration message back to the talker, which causes all intermediate nodes to keep reserving all needed resources for the duration of the stream. This methodology also serves to provide packet de-duplication if multiple listeners are interested in a stream: Each intermediate node is informed about currently admitted streams running through it and where to unicast (or multicast) incoming frames.

Traffic shaping is performed both at the sender (i.e., the talker) and on each intermediate node, according to extensions that can make use of tagged streams. The main tasks are to forward frames according to their priority, but at the same time avoid starvation of lower priority (or best-effort frames), and to avoid frame bursts that would make it hard to enforce policy constraints on downstream nodes and might overwhelm their buffers. A simple *token-bucket*-like shaper is described in IEEE 802.1Qav, and improvements (e.g., for lower latencies or easier detection of shaper failures) are being proposed on a regular basis.

Further improvements to the described concepts include media redundancy (IEEE 802.1Qcb), i.e., the transmission over multiple parallel paths to improve resilience with respect to link breaks, as well as frame preemption (IEEE 802.1Qbu). The latter enables TSN-capable bridges to selectively cancel ongoing transmissions of lower-priority frames when a higher-priority frame arrives for forwarding. This cuts down even further on transmission latencies, albeit at the cost of a slightly increased load for retries.

2.3.3 Introduction into cars

With all the benefits that new bus technologies can bring, an all-out immediate switch from CAN, LIN, and other bus technologies to, e.g., Ethernet is out of the question, since this would essentially mean scrapping all existing ECUs.

Instead, automobile manufacturers willing to switch to Ethernet have adopted a gradual approach. This started with the use of Ethernet for outside access to the vehicle's ECUs for diagnosis and firmware programming, standardized as ISO 13400 (DoIP, short for *Diagnostic Communication over Internet Protocol*). Later steps can include the attachment of Ethernet as another bus connected to the car's central gateway (connecting, e.g., video cameras or multimedia components), then the replacement of the central gateway infrastructure by an Ethernet backbone network.

With respect to replacing existing bus systems, Ethernet (with vehicle-specific physical and higher layers) could be demonstrated to share a lot of use cases of MOST and could be shown to interoperate very well with FlexRay, achieving tight time synchronization. CAN and LIN, on the other hand, fill very specific niches where Ethernet is currently unable to compete (particularly in terms of cost), and thus are commonly considered unlikely to be replaced anytime soon.

2.4 Wireless in-vehicle networks

Weighing up to 30 kg, both the cost and the weight of a modern automobile's cable harness are substantial, and, even though many modern bus systems are using redundant connections, link breaks are always a concern. Further, cable connections need space that has to be carved out of, e.g., doors, columns, or the firewall between the engine and the passenger compartment. In contrast, all that an ECU connected via a wireless link would need is access to the common 12-V rail. Wirelessly connected ECUs can also be envisioned to improve flexibility in outfitting (or retro-fitting) cars. Similar deliberations have been made in the past also in the fields of aeronautics and space flight (CCSDS 2010).

This has repeatedly spawned research efforts into possibilities to reduce parts of the extensive cabling by using wireless links. Efforts range from the investigation of general-purpose technologies like Bluetooth (Leen & Heffernan 2001) or IEEE 802.11g WiFi (Rahmani *et al.* 2009), to more specific technologies like a full ZigBee stack (Tsai *et al.* 2007) or plain IEEE 802.15.4 (Matischek *et al.* 2011), up to more recent investigations into the use of 60-GHz links (Nakamura & Kajiwara 2012). However, aside from tire-pressure sensors (which, by their very nature, need a wireless connection), no wireless technology has yet gained widespread acceptance for in-car networking.

One reason can be found in the fact that, while they fulfill most of the industry demands we discussed in Section 2.1 on page 13, such systems are commonly either based on niche technologies and are thus expensive or they cannot fulfill the demands of automotive networking in terms of QoS guarantees. Additionally, proprietary solutions are prone to security flaws, as was demonstrated when researchers hijacked the communication between a tire-pressure sensor and the car (Rouf *et al.* 2010). If such systems were connected to, e.g., the CAN bus, the security of the car could be further compromised. The impact of such an attack was demonstrated, e.g., by researchers who were able to disable a car's brakes via the CAN bus (Koscher *et al.* 2010), and further attacks paint an even bleaker picture (Wright 2011).

A way out might be offered by the time-sensitive networking (TSN) series of standards, which we discussed in Section 2.3.2 on page 35, when combined with IEEE 802.11 consumer WiFi, which we discuss in Section 4.2.1 on page 119. Since such a network would be based on well-established technologies, existing security measures could be expected to translate well to this domain. At the same time, TSN could supply the necessary QoS measures. Even though support for common IEEE 802.11 technology is limited at the moment, extensions made in the scope of IEEE 802.11aa can be expected to interoperate very well with the stream-reservation approach of TSN (Maraslis *et al.* 2012).

3 Inter-vehicle communication

Communication between vehicles (and between vehicles and available infrastructure) is the topic of this chapter. Essentially, we give a basic introduction to inter-vehicle communication (IVC) identifying the key concepts in the safety as well as in the non-safety application domains. For this, we derive all the relevant communication concepts and solutions for those applications and, most importantly, identify the requirements such as maximum delays, minimum dissemination range, or minimum data rates. Furthermore, we present an overview on the possible communication paradigms, such as whether information is to be exchanged without the help of any available infrastructure or whether infrastructure elements – roadside units (RSUs), parked vehicles, or even widely deployed cellular networks – can be used for the information exchange.

The scope of this chapter is to introduce IVC as an active research field. The main motivation is to become familiar with the field to a level that helps one to understanding the fundamental concepts and their limitations. All the communication principles outlined will be studied in greater detail in the following chapters.

This chapter is organized as follows.

- Applications (Section 3.1) – In this section, we introduce the field starting with typical applications for IVC. We will show that the scope and character of these applications vary widely, which complicates the development of common and generalized IVC protocols.
- Requirements and components (Section 3.2) – Starting from knowledge about IVC applications, we derive requirements on IVC solutions and study metrics to assess their effectiveness. In a second part, we introduce all the communication entities involved and possible mechanisms for information exchange.
- Concepts for inter-vehicle communication (Section 3.3) – This section can be regarded as the main part of this chapter. We broadly study all the communication principles and protocols that have been considered for IVC. This overview explains why the different protocols have been studied and what their main advantages and disadvantages are.
- Fundamental limits (Section 3.4) – We conclude this chapter by discussing fundamental limits for IVC. This is first related to some inherent restrictions of specific approaches to IVC. Second, there are more fundamental limitations of wireless communication in general and its application in the domain of vehicular networking in particular.

3.1 Applications

Before we start outlining the requirements on IVC, let us investigate selected example applications that fully rely on IVC. Please note that the main focus is not to study communication principles in detail at this stage. Instead, we aim to set the scene for a better-structured approach to identify the applications' requirements on IVC. We decided to select applications from all ranges starting with well-known traffic information systems, looking at safety-critical applications such as intersection assistance systems, but also at entertainment solutions such as multimedia streaming and multiplayer gaming. We also outline more details about the solutions as absolutely necessary to understand the application concepts. This allows us to demonstrate that quite different communication principles have been considered for supporting the selected applications.

3.1.1 Traffic information systems

Navigation systems such as TomTom and Navigon, among many others, have become, at least to many of us, quite helpful devices improving our driving experience. Typically, they operate based on global positioning system (GPS) position information and a locally installed map. The quality of the navigation systems differs not only in the quality of the map information and the routing algorithms used, but mainly in their ability to dynamically update the route depending on the current road traffic situation. This information has to be collected and transmitted to our devices. We are talking about so-called traffic information systems (TISs).

In the end, a TIS is most probably the best-known application that relies on communication. Our navigation systems make use of IVC in its broadest sense to retrieve dynamic updates about traffic jams, congestion, accidents, and many other events. The information can be collected from a central server, which is the case for our TomTom or Navigon systems, but also for those of us using their smart phones for navigating with current Google Maps information, just to name some examples. The communication link between the central server and our systems can thus be of very different nature, as long as it provides a downlink with sufficient capacity. We will discuss this version as a *centralized* TIS.

The alternative is to exchange road traffic information directly among our cars. The need for such a distributed approach becomes obvious if we ask the question of how and from whom the central servers actually get their data. Even though many events are still reported by roadside sensors or by administrative entities, including the highway operator and, of course, the police, it turned out that the best way to get relevant road traffic information, and to get the data in a timely way, is to rely on our cars themselves. Using local sensor data correlated with map information, each car knows whether it is just stopped in a traffic jam, currently involved in stop-and-go traffic, or driving at cruising speed with no problems at all. Assuming that each car would be able to communicate this information to the aforementioned central TIS, this can already improve the information quality. The reported information is called floating car data (FCD). Of course, the cars can also exchange this information directly among themselves, as

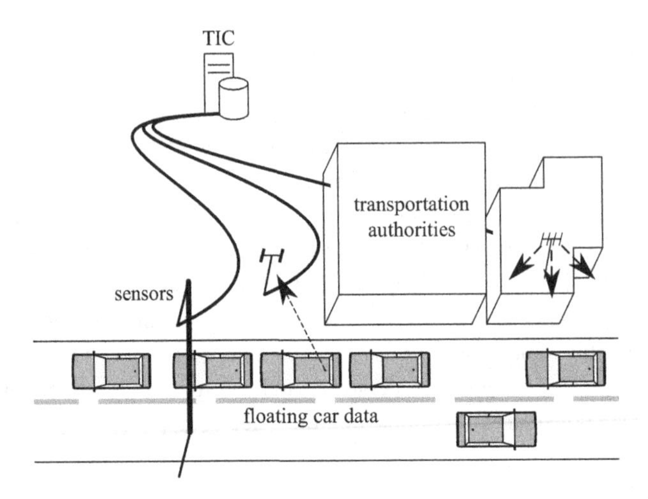

Figure 3.1 Basic building blocks for a centralized TIS.

long as they have a shared communication channel. Depending on the communication technology used, this might substantially improve the quality of the TIS. We will discuss this concept as a *distributed* or *self-organizing* TIS.

Centralized TIS

The concept of centralized TIS is depicted in Figure 3.1. As can be seen, the main component is the central traffic information center (TIC). At this point, all traffic data will be aggregated and maintained for later delivery to the participating cars. Our objective is to explore the communication channels, thus we will not dig into how information is processed in the TIC. There are quite a few aspects to be considered here, including data quality control, adequate aggregation and filtering, and also cross-validation if similar data is available from multiple sources. Looking at the networking aspects, we have to differentiate two quite different communication channels: uploading new information to the TIC and downloading or broadcasting available data to the vehicles.

Data collection – uplink channel

For the uplink, we have to look at the different data sources to understand which channels are used and what their limitations are. As depicted in Figure 3.1, we have to differentiate between various data sources. In the following, we summarize these into two main categories.

- Infrastructure – All our freeways are equipped with sensors, including roadside sensors, toll bridges, and video cameras, among others. In urban environments, induction sensors are used in addition to spot cars in front of traffic lights. All these sensors have in common that they help one to determine the density of vehicles on the respective streets. They are usually connected to a maintenance center. Also the police may operate patrols on the streets as well as helicopters to identify the current level of road traffic congestion. In addition, motor clubs such as the American Automobile Association (AAA) and the German Automobile Association (Allgemeiner Deutscher Automobilclub, ADAC) constitute very accurate and timely data sources. Whenever called, they learn about new accidents and resulting traffic jams. All these data sources have in common that they provide coarse-grained measurements that are frequently not real-time data.

- Floating car data (FCD) – The alternative is to use our cars as mobile sensors. Using local sensors ranging from ultrasonic to radar and possibly to cameras, the vehicle can even automatically assess the situation. The main advantage is that each car can perfectly determine the traffic conditions in its surroundings. Assuming we turn all the vehicles on our roads into sensors and collect the generated sensor data, we end up with very fine-grained information with a high time accuracy. This FCD not only complements infrastructure-generated data but can act, in theory, as a full replacement for it. Of course, questions about how to achieve privacy-preserving uploads and protection against malicious attackers need to be solved.

On studying the two categories, we end up with two very different uplink channels that are currently being used. All the infrastructure-generated data is collected and aggregated using complex and, in most cases, proprietary network protocols. This may cause unnecessary delays and even information loss due to aggregation. Assuming that these issues can be dealt with, the only remaining problem is the granularity of the data both in space and in time.

FCD relies on open protocols for the uplink. There needs to be some data channel from the car to the TIC. In most current systems, this is realized using 3G or 4G data networks. These provide, in most cases, sufficient capacity to upload information to the TIC. We will discuss alternatives such as mobile ad-hoc network (MANET) routing concepts in more detail later, in Section 3.3 on page 71.

Of course, the central TIC needs to be quite powerful to manage all the received data. This, however, is not part of our discussion, and this problem can most likely be solved using concepts from cloud computing. Furthermore, we are not talking about only a single TIC supporting all vehicles on the roads. Different companies and operators will maintain their own systems. So, the problem turns more into a synchronization issue, which, again, may cause unnecessary delays.

Data dissemination – downlink channel

Assuming we already have our TIC (or multiple ones), we can now look at the even more challenging question of how to download the relevant information to a car.

There are very different options and technologies available for data dissemination on the downlink channel. The most obvious idea is to simply send all information using a broadcast channel to all the vehicles at once. Such broadcasts can be realized using satellite downlinks, FM radio or digital audio broadcasting (DAB), and even based on 3G/4G networks.

Among these technologies, FM radio and DAB have become the *de facto* standards for all currently deployed navigation units. We discuss the access technology in more detail in Section 3.3.1 on page 72. The technology being used for these transmissions is known as the traffic messaging channel (TMC). TMC controls the (binary) encoding of the messages that are broadcast to the vehicles. Each message contains an event code, location information, an expected incident duration, and other details. The key problem faced by TMC is the very low data rate on the downlink channel. We are talking about a few hundred bits per second. Obviously, the number of events that can be transmitted on the downlink is very limited. Also, the security of TMC has been very controversial. In the end, Barisani & Daniele (2007) demonstrated how easy it is to send faked data messages.

The main drawback of TMC is the low data rate, yet, another, very important, aspect needs to be considered, too: Which data should be sent at which time? Given the nature of a broadcast channel, all the vehicles in transmission range (and that can be quite large for radio stations) will receive the same information. It is not possible to differentiate between information needed in a certain location and that needed elsewhere. Thus, the typical approach follows a carousel model. Available traffic information is inserted into an empty space in the carousel, which is broadcast all the time to the receivers. Each round, data is repeatedly sent, which solves two technical problems. First, the transmission itself is unreliable, thus, duplicated transmissions help to improve the reliability of the transmission. Second, cars may join and leave (i.e., by being turned on, being temporarily in a zone with weak reception quality such as a tunnel, etc.) the network at any time. This carousel model further reduces the overall capacity of the downlink channel. Overall, the granularity of transmitted (and received) events in time and space is very limited.

These problems can be overcome using cellular networks for the downlink channel. TMC has been extended by providing capabilities to use much faster download channels via 3G and also the possibility to download data relevant for the specific location only. We discuss this technology in detail in Section 4.1 on page 107. This concept has been further investigated in a multitude of research projects. Just to name one example, the Cooperative Cars (CoCar) project[1] (see Section 4.1.5 on page 116) investigated the use of XML encoded messages retrieved from a central TIC over a 3G UMTS network (Gläser *et al.* 2008). It turned out that the main bottleneck was indeed the central processing of all the received events. Regarding the communication channel, the system works very well for traffic information that is not safety-critical.

With the upcoming interest in direct vehicle-to-vehicle (V2V) communication using WiFi, or its derivative for vehicular networks IEEE 802.11p, the research community investigated its use also for supporting centralized TIS. We present and discuss the technical details in Section 4.2 on page 118. WiFi and IEEE 802.11p provide high data-rate wireless communication between multiple vehicles or vehicles and available infrastructure elements such as access points (APs) or RSUs.

After more than 20 years of experience with MANETs, it seems obvious to connect to a central TIC using ad-hoc networking protocols. This is also the origin of the term vehicular ad-hoc network (VANET). MANET routing is typically based on the

[1] In the context of the German BMBF project Aktiv.

establishment of a network topology for unicast routing among arbitrary nodes in the network. In mobile networks, this topology will likely be valid only for a short time, thus, constant updates are needed in order to establish and to maintain routes (see Section 3.3.3 on page 80). To make a long story short, it turned out that MANET algorithms work well only in a very specific configuration. This includes the parametrization of the protocol as well as the specific scenario, i.e., the number of vehicles in communication range, their relative speed, and so on.

A study revealing interesting effects on this was published by Sommer *et al.* (2008). Here, a centralized TIS was assumed to be connected to RSUs. The vehicles are able to download messages from the TIC by establishing an ad-hoc network to the closest RSU. In this way, it is possible to download location-specific data. Unfortunately, the scalability of this approach is very limited for increasing numbers of vehicles and larger distances in the network to bridge between the RSU and the vehicle. This limitation of scalability had already been predicted for MANETs by Gupta & Kumar (2000) and also characterized for VANETs by Scheuermann *et al.* (2009).

Of course, this is not the end of the story. There have been discoveries identifying new opportunities for ad-hoc routing at least in a smaller context, i.e., between a limited number of vehicles. We will discuss this in more detail in Chapter 5.

Finally, a technology that gained a lot of interest in the vehicular networking research community should be mentioned: the delay/disruption-tolerant network (DTN) (Fall 2003). Assuming a vehicular network to have a well-maintained topology is, as discussed, very difficult. Thus, the idea of understanding it as being more like nodes that are able to exchange information only from time to time, when in direct communication range, arose (Ott & Kutscher 2005). If managed successfully, such a delay- or disruption tolerant network would be able to overcome all the scalability problems. The general idea is to follow the *store–carry–forward* concept, i.e., not to forward immediately to the final destination (e.g., if such a path currently does not exist) but to carry the message until a proper forwarder has been identified. We investigate DTN approaches in more detail in Section 5.7 on page 217.

Distributed or self-organizing TIS

The key difference between the previously discussed centralized TIS and a so-called distributed or self-organizing TIS is that conceptually the TIC is relocated from a central server to a fully distributed data storage. The basic concepts are depicted in Figure 3.2. Each participating vehicle maintains a local knowledge base. The content is quite similar to what a centralized TIC is managing. Obviously, this local knowledge will be incomplete by definition. If it were possible to manage such a local knowledge base such that all vehicles have a complete and synchronized view, they would maintain what would essentially be a replicated decentralized database of traffic information.

As can be seen, we end up again with two orthogonal challenges: how to manage the distributed data and how to exchange information between vehicles. However, these two problems need to be addressed together in order to establish efficient solutions. In order to outline conceptual approaches to this problem, let's have a look at three selected

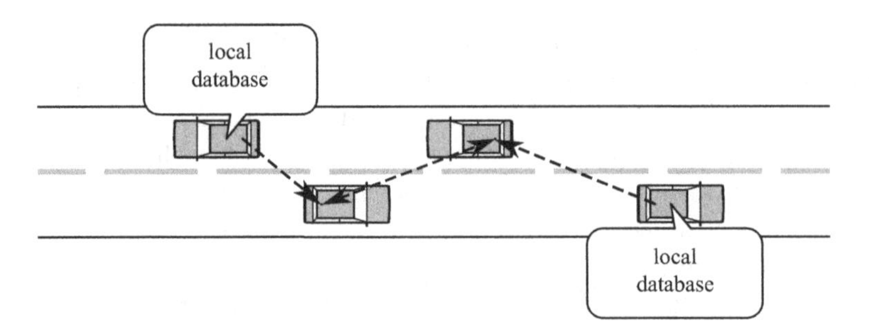

Figure 3.2 A distributed traffic information system (TIS).

solutions. All of them have their advantages and limitations. The interesting point is that very different approaches both to IVC and for the data management are used.

Self-organized traffic information system

A self-organizing traffic information system (SOTIS) (Wischhof *et al.* 2003, 2005) was one of the earliest developments towards a fully distributed TIS. It follows a very simple but powerful design: All the cars maintain the already mentioned local knowledge base in the form of an annotated map. Periodically (the original SOTIS papers talk about 10 Hz) this knowledge base is broadcast to all neighboring cars, which, in turn, integrate the received information with their already available knowledge-base entries. In this way, each car learns eventually about all relevant information. Obviously, this process is not time-bounded. Thus, there is no guarantee that a car will learn about some relevant traffic information in time or at all. SOTIS deals with the data management issue in a smart way. To understand the details, which are further elaborated in Section 5.3.1 on page 167, we need to investigate the IVC approach first. Periodic broadcasting means a full knowledge-base exchange with every single data transmission. Thus, all the information must fit into a single broadcast packet – the maximum transmission unit (MTU) is, depending on the medium access control (MAC) protocol used, limited to something on the order of 1500 B. However large the local knowledge base gets, this limit still holds. SOTIS relies on data aggregation to solve the problem. The smart approach is that data is aggregated in an intelligent way: The farther apart some stored events are from the local position, the more aggressive the aggregation. This leads to fine-grained information about events in the local neighborhood and to very coarse-grained information about events beyond some distance boundary. The remaining problem is the periodic broadcast, which, depending on the current road traffic density, will be too fast (given many vehicles in broadcast range) or too slow (assuming a very low density), i.e., overloading the wireless channel or not making use of the available capacity, respectively.

Adaptive traffic beacon

Adaptive traffic beacon (ATB) is conceptually extending SOTIS by addressing the broadcast interval (Sommer *et al.* 2010e, 2011c). In the literature, such broadcast-based protocols have frequently been named beaconing protocols; one beacon represents a single broadcast transmission (see Section 3.3.4 on page 85). Thus, also SOTIS is a beaconing protocol. Let's get back to the main weaknesses of SOTIS (and similar static period beaconing concepts as discussed in detail in Section 3.3.4 on page 85). The wireless channel being used for IVC has a limited capacity and static period beaconing either underutilizes the channel (which might not be that critical for TIS applications) or completely overloads the channel, leading to severe network congestion. ATB addresses exactly these problems by focusing on three main questions.

- How frequently can the protocol send beacons?
- How frequently should the protocol send beacons?
- What information should be included in a single beacon?

The ATB protocol follows a smart approach to answer the first two questions (we will return to all three of them in Section 5.4.1 on page 175). It uses a rather sophisticated model to analyze two parameters: the *channel quality* and the *message priority*. The channel quality describes the available (or remaining) channel capacity, i.e., answering the question of how frequently beacons can be sent. If the channel is overloaded, the beaconing period is extended, resulting in fewer beacons per time. In contrast, if the channel is still underutilized, the beaconing frequency is increased up to a certain upper limit – it does not make sense to re-broadcast too frequently by creating too many duplicated messages. Even without getting into the details at this time, it should be obvious that this concept, given that the parameters can be determined correctly, allows one to carefully maintain the resources on a wireless networking channel. The second parameter, the message priority, is used to describe the importance of traffic information to be sent. Again, without dealing too much with the internal structure, this priority can be estimated given information such as the distance to an event, the age of a message, or the type of the message (traffic accident warning vs. parking place information). This helps to answer the second question, i.e., how frequently beacons should be repeated: The answer is the more frequently the more important the messages are.

Besides making the beaconing approach more adaptive and carefully observing the available channel capacity in order to become congestion-aware, ATB also makes use of available infrastructure. With the term infrastructure, we refer to available roadside units (RSUs) along the streets and similar access points. Such RSUs may be connected to a wired network backbone but can also be standalone devices simply powered by a solar panel and contributing to the vehicular network. ATB allows one to run exactly the same protocol with some optimizations on the RSUs. This concept was around already for a while in the literature, but ATB integrated for the first time a beaconing protocol operating in a fully distributed manner with such infrastructure elements, helping also to overcome the problem of initial deployment before a certain minimum penetration rate of IVC ready devices can be reached (see Section 6.3.1 on page 271).

PeerTIS

The distributed TIS concepts introduced so far are based on direct V2V communication, partly supported by vehicle-to-infrastructure (V2I) communication. The only IVC primitive used was WiFi or IEEE 802.11p-based radio communication, and no specific organization of the distribution of data among the vehicles has been used. Even though this is extremely efficient in terms of use of the wireless communication channels, there is no guarantee that all the traffic information data is finally available for use during route (re-)planning. A conceptually completely different concept is to make use of more structured data management. In massively distributed systems, the use of so called distributed hash tables (DHTs) has been identified as a very efficient way to organize data and access to data. This concept has been investigated for a long time in the context of peer-to-peer systems. Its direct application has been studied, for example, in a system called MobTorrent (Chen & Chan 2009), which focuses on mobile Internet access. We return to this peer-to-peer-based system in Section 5.7.2 on page 222.

At this point, another concept needs to be introduced that, even though still limited in many aspects, opens up the road for many new concepts: the peer-to-peer traffic information system (PeerTIS) (Rybicki *et al.* 2007, 2009). PeerTIS exploits the capabilities of DHT systems to manage traffic information in an optimized manner. The key challenge in all DHTs is to come up with a suitable hash algorithm. PeerTIS approaches this issue very smartly by making use of the obviously available map information. Without going too much into the details (see Section 5.7.3 on page 225), PeerTIS simply uses a divide-and-conquer approach. The map is split into tiles, each being managed by a specific vehicle. This means that the respective car stores all traffic information related to its tile of the map. Whenever a car is interested in certain information, it queries the vehicle responsible for the specific part of the map. Communication is managed using cellular networks but can also make use of short-range radio such as WiFi or IEEE 802.11p. Without the help of a central repository for mapping vehicle ID (i.e., its network address) and the associated map information, this seems to be very inefficient. PeerTIS therefore integrated so-called fingers, i.e., pointers to certain map tiles. Essentially, each vehicle maintains pointers to all vehicles responsible for map tiles directly linked to its own tile. When following a route calculated by the navigation unit, the vehicle can then query car by car just following the route and each car can provide the address of the next responsible vehicle along the route. Both the scalability and the reliability of this approach are very high given the nature of the unicast communication in the 3G or 4G network. However, in very congested road areas, an overload situation might occur: first, in the area itself, since many cars will almost simultaneously query other parts of the map for possible re-routes; and secondly, at the car being responsible for that part, since a high number of updates and queries can be expected for this specific car.

3.1.2 Intersection collision warning systems

The second application domain for IVC we discuss is an intersection collision warning system (ICWS). This brings us into a new field of applications, namely safety-critical operations. Looking into the road traffic safety domain, we have to understand the reasons for accidents at intersections. It has been discovered that the driving performance is one of the most important factors associated with accidents (Evans 1991, Stutts *et al.* 1998). According to Evans (1991), the driving performance is determined by the driver's

Figure 3.3 An intersection collision warning system (ICWS).

perceptual and motor skills, including the ability to judge a vehicle's speed, to control the vehicle at that speed, and to react to hazards.

In order to support the driver's reaction and, thus, to reduce the likelihood for accidents, several driving assistance systems have been developed. Among these, we can distinguish different types of collision warning systems (CWSs). The key idea is to detect critical obstacles using a wide range of sensor technologies such as radar, ultrasonics, or video cameras. Using the sensor data, CWSs determine for example an appropriate safety distance for adaptive cruise control (ACC) or the distance to an overtaking car on another lane for lane-change assistance.

A very specific use case is an ICWS (Atev *et al.* 2004, Penny 1999). Here, sensor information is used to warn drivers about other vehicles approaching an intersection, with fully automated reactions. For example, Chang *et al.* (2009) have shown that ICWSs using audio-based warnings are able to reduce drivers' reaction time and hence reduce the accident rate.

IVC is considered to complement other sensor information in this very challenging environment. Figure 3.3 depicts the most critical scenario for an ICWS. As can be seen, buildings deny direct visual contact to the vehicles approaching from a side street. Thus, radar or a camera system cannot be used in this case. Yet, radio communication can be used to bridge this gap. In general, this can be realized using very different technologies, ranging from WiFi to 4G networks.

The impact of different warning systems has been studied by Chen *et al.* (2011a), and the results for each investigated type clearly indicate that there is a substantial safety advantage. These early results show the potential of ICWSs; typically it is claimed that intersection crashes could be reduced by 40%–50%. Looking at the network performance, there have been quite a few studies investigating ICWSs. For example, Le *et al.* (2009) had a look at the busy time fraction of IEEE 802.11p systems for intersection safety. Tang & Yip (2010) investigated timings for collision-avoidance systems assuming IEEE 802.11p transmission delays of 25 ms and 300 ms in normal and poorer channel conditions, respectively. They introduced the *time to avoid collision* metric, which represents the time from detecting a potential collision to the point of barely avoiding a collision and concentrated on the events (when to warn a driver early and latest, the reaction of the driver, and different deceleration rates) within this time interval.

Research that focuses on estimating, predicting, and/or reducing the likelihood of crashes at an intersection provides various approaches to model intersection-approaching vehicles. This research goes as far as to include threat assessment for avoiding arbitrary collisions with bicycles. For this, lateral as well as longitudinal movements and vehicle dynamics have been modeled (Brannstrom *et al.* 2010). From this, we can conclude that specific safety metrics are needed, answering more relevant questions like *How many crashes can be avoided?* and *Can the impact of crashes be significantly reduced?* (Dressler *et al.* 2011a).

Joerer *et al.* (2014) approached this question by investigating the likelihood of crashes and the potential of beaconing-based IVC systems. IVC can be used to announce the presence of a car, e.g., periodically using beacon messages. Given the current speed and heading of the approaching car, it is possible to calculate the resulting collision probability. Essentially, the likelihood of a crash can be determined by investigating all possible future trajectories a vehicle and its opponent can follow while approaching an intersection. It is simply a measure of what fraction of trajectories result in a crash. The actual calculation requires some simplifications to be tractable, but the results are very clear. We will discuss these issues in more detail in Section 6.3.4 on page 294.

The problem is that new collision probability values can be calculated only if new information about the other vehicle becomes available. This is exactly the point at which new messages are received via IVC. We can thus now continue the discussion from the previous section in which we introduced static period beaconing as well as an adaptive solution. The more frequently information is sent (in our ICWS scenario this translates to "the more quickly it can be transmitted"), the more precise is the calculated probability and the higher is the probability of being able to inform the driver early if a crash becomes likely to be unavoidable. However, the capacity of the wireless communication channel is still bounded, i.e., we cannot send as fast as would theoretically be possible. Joerer *et al.* (2014) investigated this application in detail and reported that static period beaconing with a frequency of 10 Hz or faster allows vehicles to reliably predict collisions.

3.1.3 Platooning

One of the applications being cited as most visionary as well as most demanded is platooning or cooperative adaptive cruise control (CACC). Platooning can enhance the travel experience in many ways. Most importantly from a commercial point of view, it provides huge potential to improve the road traffic flow and to reduce the fuel consumption of both lightweight vehicles and trucks (van Arem *et al.* 2006, Davila & Nombela 2012). This comes, of course, with a decrease in emissions. From the driver's perspective, improved safety might be the key argument for platooning. Statistics show that, for modern vehicles, a mechanical or electronic fault is less likely than a human

(a) Platooning experiment

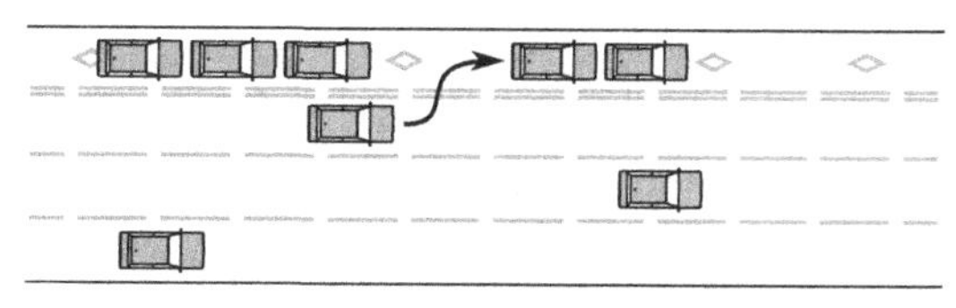

(b) Join operation

Figure 3.4 Platooning or CACC in action.

error, which is the main cause of accidents (Davila & Nombela 2012). As a side effect, the autonomously driving cars permit the driver to relax, as shown by the SARTRE project (Bergenhem *et al.* 2010).

Platooning has been investigated since the 1980s, for example within the framework of the PATH project (Shladover 2006). From a research point of view, platooning is extremely challenging, since it involves several research fields, including control theory, communications, vehicle dynamics, and traffic engineering. The platooning research community focused initially on the problems connected to the automated control of vehicles, because the design of such a system is a non-trivial task. The concept is illustrated in Figure 3.4, which shows a picture taken during a platooning experiment as well as the principle of a simple maneuver, the join operation.

Vehicular networking supports a controller system that needs frequent and timely information about vehicles in the platoon to avoid instabilities which might lead to vehicle collisions (Rajamani *et al.* 2000, Ploeg *et al.* 2011). In general, it is assumed that a platooning system requires an update frequency of about 10 Hz to enable shorter distances between the vehicles. In the system presented by Ploeg *et al.* (2011), the distance that can be maintained by the controller has to be speed-dependent as for ACC. If vehicular networking is used to disseminate speed information to the direct follower, the headway time can be reduced to 0.5–0.7 s compared with a radar-only ACC requiring 1.5 s to avoid compromising the stability of the system.

If, however, the controller is able to base its decisions on information from both the vehicle in front and the platoon leader, the system can be proven to be stable under a constant-spacing policy, i.e., the inter-vehicle distance does not need to be speed-dependent. This means that the inter-vehicle gap can be chosen (almost) arbitrarily small, e.g., on the order of 5–7 m, as demonstrated by field operational tests (FOTs) in the PATH and SARTRE projects (Rajamani *et al.* 2000, Bergenhem *et al.* 2010).

Segata *et al.* (2014) adopted these ideas, concentrating primarily on the communication protocol design. In particular, they assume that each vehicle is aware of its position in the platoon, and uses this to decide *how* and *when* to send a beacon. Since information is required both from the platoon leader and from the car immediately ahead, the transmit power can thus be reduced for the latter communication in order to increase spatial reuse. Leaders can instead use high transmit power in order to reach all vehicles within the platoon. The concept for dissemination scheduling is to have the leader sending first, then the others can follow in a cascading fashion.

3.1.4 Traffic-light information and control

Let's stay in the field of intersections and concentrate on one of the best known elements to substantially improve the driver's safety: traffic lights. It is an established fact that traffic lights help to prevent crashes, yet the optimized control of traffic lights is a research area of its own. Not only are we frequently annoyed when we have to stop in a series of red lights, but also this is a major source of unnecessary emissions. Interconnecting the traffic lights with our cars would be a first step to improve the situation. In the following, we discuss three approaches to make the road traffic flow more efficient by integrating traffic lights into the vehicular network.

Travolution

One of the first approaches based on large-scale wireless networking was the Travolution project back in 2006 (Braun *et al.* 2009), which was mainly planned and deployed by Audi in the city of Ingolstadt, Germany. The idea is to equip each traffic light with a radio transmitter to enable wireless communication with approaching cars. Essentially, each traffic light continuously broadcasts its status to all vehicles in communication range. This status report includes a description of the traffic-light system at the intersection, the colors of the individual lights for the respective directions, and information on when the next status change is to be expected. The on-board unit (OBU) in the car can process the received information and provide the driver with suggestions about the optimal speed with which to reach the intersection when the traffic light is showing a green light. In a second phase of the project, even automated configuration of the ACC was considered.

The operation principle is depicted in Figure 3.5. The traffic light uses radio communication (in the scope of the project, WiFi was used, but it was planned to replace this technology by IEEE 802.11p) to disseminate its status information. The car's OBU can then prepare an optimal speed advisory. Of course, the speed will be in a suitable range between some minimum and a maximum allowed limit on the street. It is the driver's obligation to follow the advice.

If the car is stopped at a red traffic light, the OBU informs the driver about how long the red phase is expected to last. If the car cannot manage it to a green light at the next intersection, the speed advisory will show the allowed speed limit to get as quickly and efficiently as possible to the red traffic light. In both situations, a start–stop automatic can use the information from the traffic light to turn off the engine for as

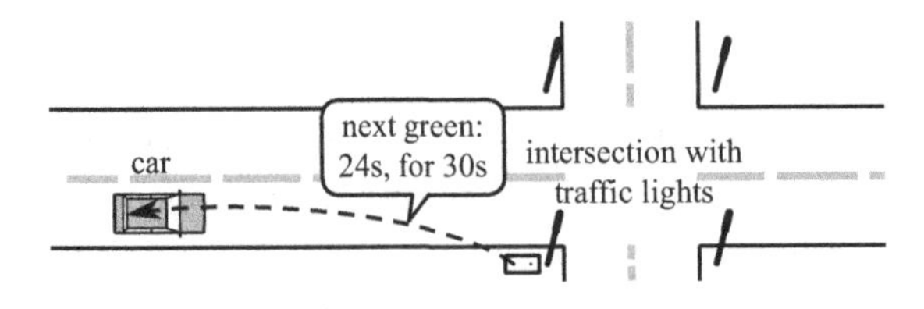

Figure 3.5 Data flow between traffic lights and approaching vehicles.

long as possible. The results obtained during the Travolution project have shown that reducing waiting times at traffic signals substantially cuts fuel consumption.

Adaptive traffic lights

So far, we have considered a unidirectional communication channel from the traffic lights to the vehicles on the streets. This is already a very powerful approach. Yet, assuming a return channel to the traffic lights would provide means for even more powerful applications that are based on influencing the traffic-light operation.

Since traffic-light control is a very sensitive task, this access must be very carefully controlled. In order to eventually improve road traffic safety with traffic lights, we must ensure that no security problem opens the door for unauthorized manipulation of the traffic lights. So, the first application using this return channel – and an application that is still under investigation (Noori 2013) – is focusing on a very limited number of systems having access to the control interface: emergency vehicles. The idea is quite simple: Assuming that the traffic light can be controlled by means of wireless radio communication, emergency vehicles such as fire engines or an ambulance can send specific control messages that, for example, configure the traffic lights in such a way as to show red for all directions. The emergency vehicle can then pass all intersections safely without having to maneuver carefully to pass due to a green light for crossing traffic.

We will discuss the resulting security issues later, in Chapter 7. Assuming all the security questions to have been addressed, we can finally also think about even more powerful applications. We can make the traffic lights adaptive (Gradinescu *et al.* 2007). This is actually not a new idea. Many of today's traffic lights already use sensors in the streets to learn about the presence of a car or the density of traffic flows. This information is then incorporated into the switching cycle of the traffic lights. Now, assuming we have IVC available, we could inform the traffic lights not only about our wish to pass an intersection but also about the expected number of vehicles in a certain direction. Thinking further ahead, we could even inform the traffic lights in advance about the entire path we aim to take. Of course, this brings up privacy concerns that need to be addressed (see Section 7.1.4 on page 311).

The remaining question is, what is a good communication channel for this application? There is no clear answer. The communication channel from the traffic light to the vehicles will most likely be a broadcast channel. This can easily be provided by WiFi or IEEE 802.11p. On the other hand, a cellular broadcast channel can also be used – this opens up many questions such as inter-mobile network operator (MNO) information sharing. For the return channel, the answer is clearly more complicated. The same broadcast channel seems to be an obvious candidate. Unfortunately, security issues are not that easy to solve in this case. Also, the traffic lights would have to have a backbone network connection to inform others about the switching decisions.

Virtual traffic lights

If we consequently continue towards the idea of implementing intelligent traffic lights, we will logically end up replacing all the physical traffic lights by virtual ones managed

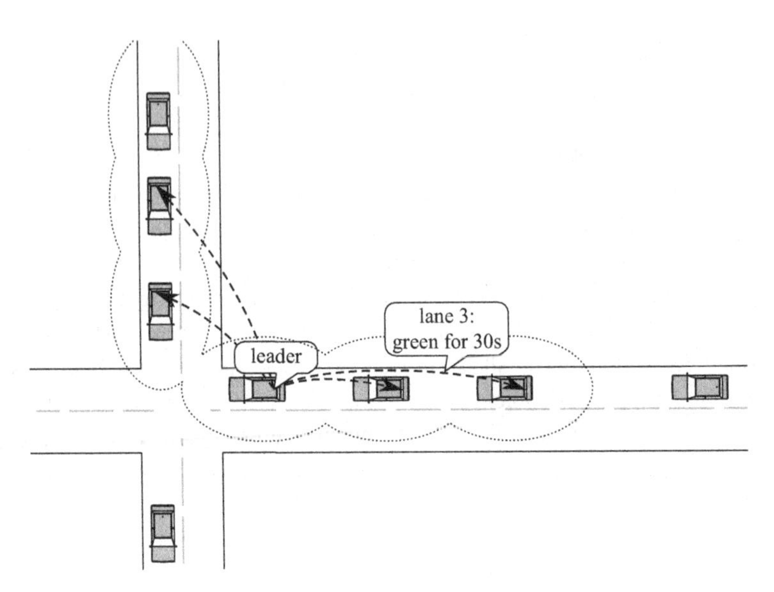

Figure 3.6 The concept of VTL: All cars approaching an intersection are managed in the form of a cluster coordinated by a VTL leader. The leader becomes a virtual traffic light.

and controlled by the cloud of vehicles on the road. Ferreira *et al.* (2010) investigated this concept in more detail. The resulting approach was named virtual traffic light (VTL) or self-organizing traffic lights.

The concept is depicted in Figure 3.6. All the cars approaching an intersection are managed in the form of a cluster. The cluster is coordinated by a VTL leader. The leader then becomes the (virtual) traffic light for all the other vehicles in its cluster. Basically, this leader internally operates a normal light-switching algorithm. Obviously, this must be coordinated with the other vehicles approaching the same intersection. It has been shown that, with the VTL approach, both the traveling time of commuters and the amount of carbon emission can be reduced substantially (Ferreira & d'Orey 2012). The main advantage is that now all intersections are automatically controlled by a (virtual) traffic light. This also reduces the costs for deploying and maintaining physical traffic lights. Cars not equipped with such a traffic light are, or course, difficult to integrate.

The concept described by Ferreira *et al.* (2010) does not go into details on how the cluster is to be established or how information is transmitted among the vehicles. Instead, it focuses more on the general applicability. Follow-up research investigated the feasibility of the approach when relying on IEEE 802.11p-based radio communication (Neudecker *et al.* 2012). It has been shown that, at least for beaconing frequencies of 10 Hz, the likelihood of receiving the needed control information in time is very high, resulting in reasonable deceleration rates at intersections.

The most critical issue is the cluster management. This can be done using algorithms proposed in the distributed systems context adapted for the very unreliable nature of broadcast-based systems. The following algorithm has been proposed by Sommer *et al.* (2014b) (the algorithm is based on Vasudevan *et al.* (2004)).

- If a car sees the need for an election it broadcasts an announcement message. This message contains its current distance and an election number. The car initially assumes that it is the one closest to the intersection, storing its own distance and ID. Additionally, a timeout defines how long the initiator waits for replies to the announcement.
- After a timeout, the initiator determines which node (i.e., which vehicle) won the election by virtue of having the shortest distance to the intersection and, to break ties, node ID) and informs this node. The winner sends out a 'leader' broadcast to the other cars. This message contains the ID of the new leader and also informs the other cars that the election has been concluded.
- If a car receives an announcement and is currently not participating in any election, the car gets engaged in the new election, replying with its own distance.
- If a car receives an announcement and an election is running that has higher precedence determined by the election number and, in the case of a tie, vehicle ID), the car participates in this election instead. Otherwise the node ignores the message.

In this way, the leader-election problem can be solved. Still, countermeasures for the potential case of a failing election (resulting in no leader or multiple leaders) must be incorporated.

3.1.5 Entertainment applications

A third class of applications benefiting from inter-vehicle communication is entertainment. We experience a seemingly endless growth of mobile entertainment applications these days, almost all focusing on mobile devices such as smart phones or tablets. Many of these applications rely on two options for accessing multimedia content: download and streaming. Both of these approaches have their advantages. Downloading makes content available for use in case the network connection is no longer stable or available at all. Streaming reduces the overall network utilization (saving download costs and energy on the mobile system), but requires a constant network quality. This can be further extended to real-time interactions over the network such as video conferencing or multiplayer gaming. In this section, we investigate concepts for making this kind of application also available in the vehicular context.

Content downloading and multimedia streaming

The need to download information from an external network such as the Internet to a vehicle is obvious. In a sense, this is exactly what we already studied in the context of TIS. In this section, we concentrate on the download of specific information and of larger data files. This can be, without any additional technical development, achieved using 3G or 4G networks. The technology is readily available (and many top-of-the-range cars are already equipped with the necessary communication technology), but with obvious restrictions on scalability.

Following the ideas described by Malandrino *et al.* (2011), we envision a network composed of vehicles dynamically connecting to other cars for exchanging information, as well as of infrastructure at the roadside such as APs or RSUs. The vehicles download content from the Internet through the APs. Each vehicle may be interested in different content. For downloading, the vehicle can either exploit direct connectivity with an AP, if available, or be assisted by other vehicles that act as relays between the downloader and the AP.

We can distinguish among the following three data-transfer paradigms (Malandrino *et al.* 2011).

- *Direct transfers* – If the downloader is in direct communication range of an AP, the vehicle can be assumed to have direct Internet access. This is similar to the typical Internet access using a mobile phone or a laptop computer via a WiFi AP.
- *Connected forwarding* – If the AP is not in direct communication range but a route could be established over a number of relay hops, we can use MANET routing concepts for downloading the content. Of course, the path must be stable for the duration of the download.
- *Store–carry–forward* – The remaining alternative is download using DTN concepts. The idea is that vehicles download (popular) content whenever they are within communication range of an AP. They can then act as content providers when encountering other vehicles.

We will focus on the first two approaches in this section. Both are outlined in Figure 3.7. On the right, a vehicle downloads while having a direct communication link to an AP. On the left-hand side, one vehicle acts as a relay for the transmission.

Malandrino *et al.* (2011) tried to optimize such a content-downloading system so as to maximize the aggregate throughput. In particular, they used the knowledge about AP locations and the estimated behavior of potential relay nodes to improve the overall content-downloading time. In order to enable sufficient data rates, the number of APs needs to be very large.

This problem can be addressed by dynamically introducing additional network infrastructure. The best way to do so is to make use of parked cars as information relays (Malandrino *et al.* 2012, Liu *et al.* 2012). Studies show that about 70% of all parked cars were parked on streets, i.e., having a direct communication link to passing vehicles (Morency & Trépanier 2008). On average the duration of one parking event was

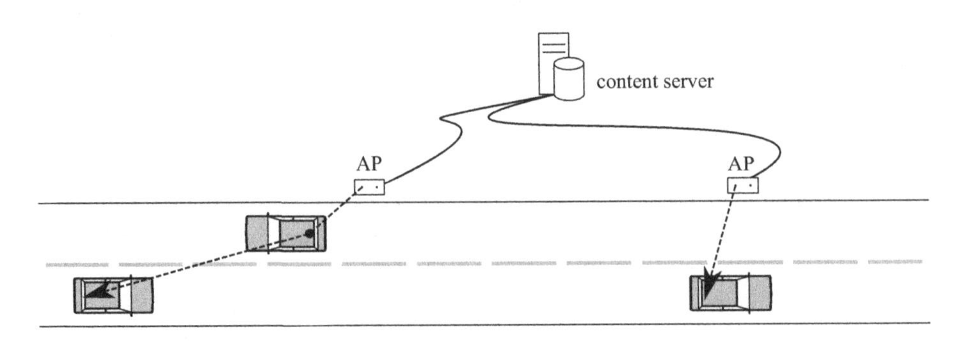

Figure 3.7 Content downloading in vehicular networks.

about 7 h. The study furthermore showed that parked vehicles were widely distributed throughout the whole city, which means there is a high probability that a parked node is within transmission range of a moving car. Authors of other studies have stated that, on average, a vehicle is parked for 23 h a day (Litman 2006).

The content-download application as studied so far, can seemingly rely on V2V communication using WiFi or IEEE 802.11p. Yet, communication through the link to the Internet will most likely not be stable enough for continuous uninterrupted transmission. This is, however, exactly what video streaming requires. Multimedia streaming was considered as one of the most interesting applications for vehicular networks in the beginning, and many approaches have been proposed. They range from new architectures (Guo *et al.* 2005) to cross-layer path-selection schemes (Asefi *et al.* 2010) and even to actually very interesting network coding concepts (Park *et al.* 2006).

After some years of research into this direction, the vehicular networking community agreed upon the fact that multimedia streaming over VANET is not just challenging but very unlikely to work in most scenarios. One option that remained was to abstract from the networking technology and to focus on peer-to-peer concepts instead (Zhou *et al.* 2011, Chu & Huang 2007). With these ideas, multiple different network technologies can be used in a complementary way to support uninterrupted multimedia streaming.

Multiplayer games

Assuming the ubiquitous availability of ad-hoc networking capabilities in vehicular networks, this technology will not only enable efficiency or safety-critical applications but also support the already-discussed multimedia streaming ideas. Going one step further, even multiplayer games can be supported by the same network infrastructure (Tonguz & Boban 2010, Palazzi *et al.* 2007).

Tonguz & Boban (2010) studied this idea in more detail. They distinguished between two different types of games: Internet multiplayer games and mobile multiplayer games. First of all, Internet multiplayer games are still the predominant focus of industry, since the player base is significantly larger than that of mobile multiplayer games. Furthermore, these Internet multiplayer games can make use of very high data rates to a central server. Data rates in mobile networks (here we are talking about 3G or 4G networks) are growing also, but have not yet reached that of a fixed Internet connection.

The main approach in the design of mobile multiplayer games has been to create games for small devices (e.g., mobile phones) that require players' movement in order to achieve the game objective. This approach did not lead to highly popular mobile multiplayer games. Tonguz & Boban (2010) envision a completely new era of IVC-based multiplayer games that offer the player the opportunity to engage in a location-aware, mixed-reality multiplayer game that takes advantage of inherent mobility; these features are not available in Internet multiplayer games. In this domain, energy is not the most critical resource, in contrast to mobile multiplayer games, which suffer from limited battery lifetime. The needed mixed-reality game engines will therefore have enough computational resources to present a high-quality gaming experience. The information needed to be communicated among vehicles, i.e., the playing parties, will be the context information of the player. Since location-awareness is needed and the player can participate only in that part of the game which is related to the current position of the vehicle, this seems to be feasible using all the available IVC technologies.

3.2 Requirements and components

In the last section, we studied a wide range of applications that can benefit from inter-vehicle communication (IVC). We even briefly touched on communication technologies that are suitable candidates for one or more of these applications. In this section, we want to explore and to characterize the application demands in a more holistic way. We start by presenting a taxonomy of applications and communication techniques, discuss communication principles, and derive application-based requirements on IVC protocols.

3.2.1 Application demands

Recapitulating the applications discussed in the previous section, we can distinguish several application categories. In the literature, one can find slightly different categorizations, some using three, some even more groups. For example, Willke *et al.* (2009) provide a nice overview of IVC protocols and applications and rely on three main classes.

In the course of this textbook, we will stick with the taxonomy depicted in Figure 3.8, i.e., defining two main classes of applications: *safety* and *non-safety*. Of course, selected applications might be categorized as both, i.e., exhibit overlapping demands and requirements. In the following, we study these two main categories in more detail, primarily referring to the applications already discussed in Section 3.1 on page 39. The discussion is then summarized in Section 3.2.2 on page 62 focusing on typical networking metrics.

Non-safety applications

The first domain, also motivating early research on vehicular networks, consists of so-called *non-safety* applications. As the name suggests, the category covers all those

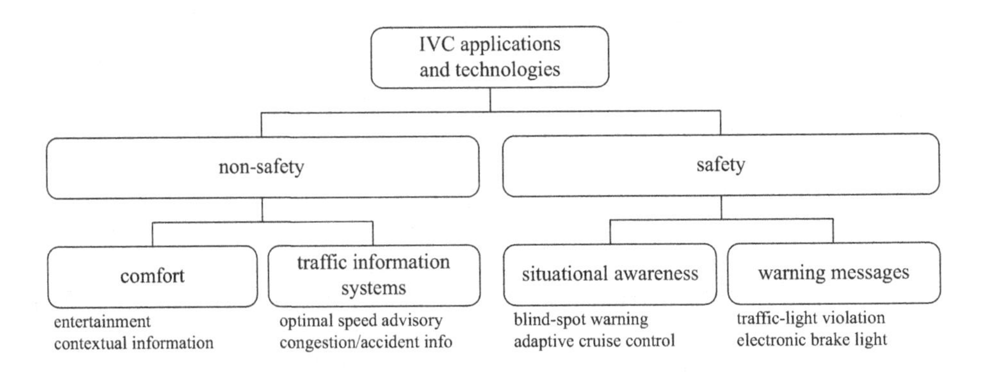

Figure 3.8 A taxonomy of IVC applications and technologies.

applications that, even though clearly enhancing the driving process, do not constitute any safety- or life-critical requirements. That is, if the application demands can be fulfilled by the network, one can drive more efficiently, or the driving experience improves, but loss of network connectivity is not critical. As shown in Figure 3.8, non-safety applications can further be split into those related to driving efficiency, i.e., all types of traffic information systems (TISs), and those related to improving the driver's comfort.

Traffic information systems

We have discussed the application field of TISs already in some detail. This was the first application field and still is one of the most widely discussed applications. Information about congestion and traffic jams obviously may help to reduce the travel time of individual vehicles.

TIS systems require continuous updates on the road traffic situation along the selected route. Vehicles may act not only as information sinks but also as sources generating traffic information related to the current position of the vehicle – so-called floating car data (FCD). Therefore, both directions of communication need to be considered.

Depending on the communication principles used (we discuss this aspect later in Section 3.2.4 on page 68), the downlink may be organized as efficiently as the uplink. On the uplink, just a few events within a period lasting on the order of minutes are generated and need to be communicated to other vehicles either directly or via some communication infrastructure, e.g., a TIC. On the downlink, only events related to the planned routes of the vehicles are of interest to the vehicles. If this can be filtered within the network, again just a few events within a period lasting on the order of minutes need to be downloaded. If, however, such filtering is not possible, the downlink may become a bottleneck due to the high amount of events in the whole network.

Another aspect that needs to be discussed is the granularity of traffic information data needed. Current systems like TMC do not consider this aspect, but try to provide events in as fine granularity as possible. Given the limitations of the downlink channel – whatever channel is used, it will be restricted in capacity – there is an upper bound on the number of events to be transmitted per time unit. Thus, even assuming the data channel will be fully available for the vehicle under consideration, either we have to accept higher latencies (increasing the time unit) or lower granularity (decreasing the number of events). These conflicting metrics can, however, be constrained according to a third dimension: space. Usually, fine-grained information about micro-jams is of interest only in the local vicinity and along the planned route of the vehicle. On the other hand, aggregated information also describing major and more critical traffic jams farther away is absolutely sufficient for planning longer routes. Aggregation techniques help to provide these capabilities.

Another application scenario that belongs to the broad field of TIS is communication between the vehicles and traffic lights. In order to optimize traffic flows on our streets, information not only about congestions and traffic jams, but also about the presence and most importantly the switching cycles of installed traffic lights, is needed. The aforementioned Travolution project investigated the applicability of a so-called green-light optimal speed advisory (GLOSA). Here, traffic lights inform approaching vehicles about the current switching state, i.e., the color of the traffic light, as well as about the switching time, i.e., when the traffic light will switch to the next color. From this information, the vehicle's OBU can calculate the best speed for approaching the traffic light in order to directly pass without stopping at a red light. The communication channel needed is unidirectional from the traffic light to the approaching vehicles, but the information needs to be delivered faster than is the case with general TIS data, i.e., on the order of seconds.

This concept can be extended to coordinated information delivery of traffic lights from subsequent intersections to help create a "green wave". The complexity of information processing and forwarding exponentially increases with the number of traffic lights involved. Also, if there is no direct communication between the traffic lights, moving vehicles may act as relays. This also makes the communication channel more complex and contributes to the load in the vehicular network.

When enabling bidirectional communication between the vehicles and the traffic lights, a next-generation traffic light control that is known in the literature as adaptive traffic lights can be realized. Even today's traffic lights can adapt their switching cycles by using sensors on the roads. Frequently, inductive sensors are used to determine the presence of vehicles and, thus, to prefer selected lanes or directions. Assuming vehicles can inform traffic lights not only about their presence but also about their intended routes, traffic lights can take more intelligent decisions, helping to optimize the road traffic flows on a larger scale. For this, a communication channel from the vehicles to the traffic light is needed. Again, a granularity in time on the order of seconds is needed, but there are far more sources in this uplink (all the vehicles) than in the downlink (one or multiple traffic lights).

In summary, the application demands for TIS range from the transmission of several events per minute for detecting coarse-grained congestion or traffic jams, (or a very high number of events, still within a time on the order of minutes, for fine-grained information) to several events per second for traffic-light control or GLOSA applications. Vehicles are involved as traffic sources and sinks but there are also infrastructure elements, i.e., traffic lights, generating but also receiving data for improving travel times of individual vehicles but also entire traffic flows. Table 3.1 summarizes the application demands.

Comfort applications

Comfort applications include all entertainment and contextual information systems that help improve the driving experience but have no direct influence on the route selection.

Table 3.1 Application demands: a traffic information system (TIS)

Metric	Range
Amount of data	Tens of bytes to hundreds of kilobytes
Latency	Seconds to minutes
Throughput	Medium
Direction	Downlink-oriented, limited uplink
Range	Global (TIS) to local (traffic lights)

The basis for most of these applications is content downloaded from the Internet in general or the cloud, as has become very popular in recent years. Content can be any data downloaded while browsing the Internet, emails read by some of the passengers in the car, or audio or video files to be played later on during the drive.

Regarding content downloading, we talk about downloading single data segments or chunks, or even files as a whole. The size of each segment can range from a few bytes to some gigabytes when considering full HD videos. Of course, for convenience, the download of such segments should be as fast as possible, but, essentially, users do not expect better than best-effort communication qualities.

The story becomes slightly different when considering multimedia streaming. In this case, audio or video data is to be downloaded for immediate playback. As with similar applications in the Internet, buffering and caching techniques can be used to mitigate communication delays and temporal outages. The data rates needed range from some hundreds kilobits per second for audio to several megabits per second for video streaming.

If content to be downloaded or streamed depends on the current position of the vehicle (e.g., for position-based modifications or even augmented-reality applications), the broadband downlink needs to be complemented by a narrowband uplink for continuously informing the server or cloud systems about the position of the vehicle. In this case, buffering and caching becomes more complicated because the content to be downloaded or streamed heavily depends on the current position and will become outdated very quickly. It seems, however, reasonable that other vehicles might become interested in the same type of content in the course of driving, so sharing this information among nearby vehicles becomes an option.

The most interactive form is games, more precisely multiplayer games. Gaming became one of the killer applications for high-speed Internet access. This is due to their interactive nature. Not only does audio or video data need to be downloaded – this can in many cases perfectly well be preloaded and cached – but also user actions need to be exchanged, with very low latencies to prevent lags in the game. Thus, only a few bytes of control data need to be exchanged between participating users, but within a few hundred milliseconds. These numbers hold for single-player interactive games as well as for multiplayer games. The challenge in the latter type of game is that a potentially very high number of players will interact at the same time and all the control information may already constitute a substantial amount of data that is to be exchanged with very low delays.

Concluding the discussion, we can see huge differences in throughput requirements and communication delays. There are no hard latency requirements for content downloading in general, yet, streaming and even multiplayer games establish more challenging requirements. Control information for games may have to be delivered within a few hundred milliseconds. For content download, best-effort services are sufficient, but a high-throughput downlink is needed. Communication partners are either the vehicle and a centralized entity such as an Internet server or the cloud in general, or multiple vehicles traveling in a close proximity, considering gaming applications. Table 3.2 lists typical requirements from the applications' point of view.

Table 3.2 Application demands: comfort applications

Metric	Range
Amount of data	Tens of bytes (control) to megabytes (streaming)
Latency	Hundreds of milliseconds to seconds
Throughput	Medium to high
Direction	Very downlink-oriented, limited uplink
Range	Internet/cloud to vehicle or local (gaming)

Safety applications

The second application domain covers all *safety*-critical applications. These applications constitute the key motivation for regulatory bodies to support and foster the deployment of vehicular networks. Application demands are much more strict in this domain. Figure 3.8 lists two categories of safety applications, namely situation-awareness applications and warning messages.

Situation awareness

Making vehicles aware of each other, i.e., collecting information about the other vehicles' positions, is part of many recent technological advances in the automotive field. IVC is considered to provide better capabilities than local sensors such as radar or cameras. A simple first application could be blind-spot warning. Assuming the other vehicles' positions, speeds, and headings can be learned by means of IVC, this information can help the driver to identify blind spots even in the case of uncertainty of those positions by means of visual control. To make this work, periodic updating of these items of information is needed, on the order of several updates per second. If the vehicle is in a rather slow movement, e.g., in a stop-and-go situation, it might suffice to inform all direct neighbors. Driving at cruising speed on a freeway, however, may influence vehicles that are more than a hundred meters away. All these vehicles must be informed with each situation-awareness message.

Using these situation-awareness messages, one can build a multitude of applications ranging from simple information-presentation-only solutions to (semi-)automated reactions. One example is the very popular ACC. Such systems currently operate using radar distance measurements. IVC may help by improving ACC to include even vehicles that are quickly changing lanes. The logical next step is CACC enabling the establishment and control of so-called road trains or platoons. Such platoons need to be managed by one of the vehicles, e.g., the leading one, in order to safely support acceleration, deceleration to a full stop, and other actions of the platoon. Talking about such management and control operations, the state (again, position, speed, heading, acceleration) of a vehicle must periodically be exchanged among vehicles following each other, as well as with those on a side lane. In addition, the leader of the platoon must be able to send control information to all following cars. Such control data needs to be sent periodically as well as during specific critical operations. According to the control theory behind platooning applications, the vehicles in the platoon must be able to exchange the aforementioned messages at intervals of less than one hundred milliseconds. If the reliability of the data transmissions is an issue, the interval further reduces.

Table 3.3 Application demands: situation awareness

Metric	Range
Amount of data	Tens of bytes to kilobytes
Latency	Fractions of a second
Throughput	Low
Direction	Broadcast to all neighbors
Range	Local neighborhood (within some hundred meters)

In between situation awareness and the traffic information category is the use of virtual traffic lights. Algorithms such as VTL always provide a fall-back solution, i.e., they do not constitute hard safety-critical applications. Nevertheless, the coordination of virtual traffic lights also relies on real-time messaging among the vehicles involved, with high reliability demands.

We can thus summarize the application demands in the field of situation awareness as follows. The current state the vehicle is in needs to be disseminated to all vehicles in its local vicinity in a periodic way at a rate no slower than once every few hundred milliseconds. Depending on the current speed and the situation, the vicinity of a vehicle can be defined as all directly neighboring vehicles but also all vehicles within a few hundred meters, meaning that a potentially high number of vehicles will have to receive this situation-awareness message. The demands in terms of reliability of the message transmission are clearly higher than for all non-safety communications. Table 3.3 summarizes the discussed demands of situation-awareness applications.

Warning messages

The last class of IVC-based applications is warning messages. This is the most safety-critical category of applications since these messages will typically directly trigger actions by the driver or even (semi-)automated reactions. The so-called electronic brake light is probably already well known from the media. A braking vehicle informs all following cars by means of IVC about this maneuver. Messages will thus be disseminated only when the braking action is initiated, but must be received within at most a few tens of milliseconds.

The most challenging variant of this application is an automated emergency brake. This can be caused for example by a car that is just becoming involved in an accident. The notification about this accident needs to be disseminated to following cars even faster than the information about a braking maneuver. Only automated actions are possible in this case, since the human reaction time will likely be too slow to prevent crashing into the cars involved in the accident.

Traffic-light violation is another application in this domain. So-called intersection collision warning systems (ICWSs) are expected to help in the case of intersections that are either very difficult to pass, or in the case of vehicles denying the right of way. Depending on the speed of the vehicles, the notification times can be relaxed compared with those for emergency brakes, yet not only vehicles following the car need to be notified about an intersection maneuver but so also do all vehicles approaching the intersection, even if (and especially in this case) buildings block the direct line of sight.

Table 3.4 Application demands: warning messages

Metric	Range
Amount of data	Less than a kilobyte
Latency	Tens of milliseconds
Throughput	Low
Direction	Broadcast to following vehicles / vehicles approaching the intersection
Range	Local neighborhood (less than one hundred meters)

Concluding the application demands, we notice rather local communication and little volume but strict real-time requirements to support this very safety-critical application field. Acceptable communication delays are well below one hundred milliseconds and messages might be very targeted, e.g., all following vehicles or all those approaching a particular intersection. Table 3.4 summarizes these findings.

3.2.2 Metrics to assess IVC solutions

In the previous section, we identified quite significant differences in application demands in terms of range of information spread, data volume per time unit, and hard latency requirements. We now bring this discussion more to the point related to typical networking metrics. In addition to our previous argumentation, we also introduce new metrics that are unique to vehicular networks and that need to be assessed in order to evaluate IVC concepts in a holistic way. We step through all these metrics in the following and recap the demands of the two main application categories.

Data rate and volume

The first and for many the most obvious and important metric is throughput. It is necessary to distinguish between the *data rates* required for particular applications and the overall *volume* of data to be transmitted. Both data rates and volume are very different for non-safety and safety applications.

In general, it can be said that all the safety-critical applications rely on little data volume per message. Yet, given that very high densities of vehicles on the streets need to be handled, this may translate into rather high data rates. Doing a simple "back of the envelope" calculation, we could expect several hundred cars driving in close proximity to a particular car on major freeways. If each vehicle is about to send situation-awareness messages multiple times per second, the network needs to handle thousands of messages per second, which may already translate into significant data rates. For example, if a message is on the order of 1000–10 000 bits, this results in 1–10 Mbit/s.

Considering non-safety applications such as video streaming or content download in general, the volume of messages to be transmitted is obviously larger. This also holds for the data rates, which for streaming applications must be very constant and, e.g., for supporting video, very high. For content download, high data rates are preferred, but the normal best-effort communication is acceptable. Thus, we talk about a data volume of a few bytes to several megabytes and data rates of up to several megabits per second.

Dissemination range or spread

A second criterion is the needed *dissemination range* or the *spread* of generated messages. The dissemination range can be used to describe the distance between the sender and the receiver(s) of a message by means of a geographic distance in meters or the distance in the network in terms of forwarding hops.

This discussion can be complemented by arguing about the number of receivers. Some applications need a simple unicast connection to a specific receiver, whereas others rely on broadcast, multicast, or flooding in a certain geographic area. Here, we follow the classic description of these terms in networking. Broadcast refers to a single message to all nodes within communication range, i.e., all nodes able to receive the same wireless transmission. Multicast describes the communication from a single sender to a group of receivers that are interested in this message. This can be described by means of a multicast group (and potentially to receivers in quite a large geographic area) or by means of all nodes sharing certain characteristics (likely in a smaller geographic area). Finally, flooding means the transmission to all nodes in the network. This can be (and usually has to be) restricted to a geographic or topological area (this is called a geocast, which we will discuss later in this book).

In the safety domain, the dissemination range or spread needed is rather small. Messages sent by vehicles in the local vicinity are of interest, meaning all the vehicles following a car (and maybe those ahead) or all vehicles approaching an intersection, etc. The communication will likely be following a broadcast or flooding approach. The dissemination range for selected safety-critical warning messages is outlined in Figure 3.9.

In the non-safety application domain, there is a wide range of different application demands. Content downloading or multimedia streaming relies on unicast connections over a potentially large geographic distance, whereas multiplayer gaming will likely require multicasts to all vehicles where passengers are participating in the game. It is worth looking at traffic information systems in particular, because this application can be designed in quite different ways as discussed in Section 3.1.1 on page 39. Centralized TIS will use unicasts for the uplink and may use multicasts on the downlink to the vehicle. In contrast, distributed TIS will have to inform all neighbors in a certain area

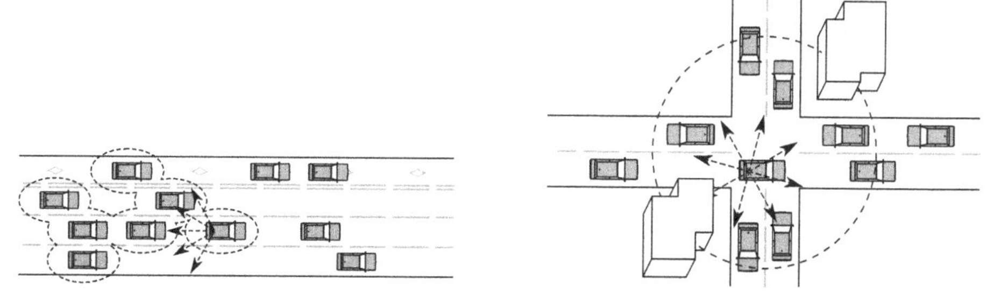

Figure 3.9 Dissemination ranges for safety applications: backwards on a freeway (left) or in all directions within a safety range at intersections (right).

(or even all participating vehicles in the entire network) about observed traffic jams and congestion. Dedicated flooding techniques over large areas will be the basis, which can best be measured by means of the information spread in the network.

Delay and reliability

Latency is for some applications most probably the most critical metric. Since delays cannot be measured if messages get lost, we discuss this metric in direct relation to *reliability*, which describes the likelihood of successful transmission. When looking at a textbook on computer networking, the delay and the variance of the delay are usually discussed in the context of the quality-of-service (QoS) of a network service or protocol. We do not use this term in this book, even though there is at least one application with high demands on the classical definition of QoS: multimedia streaming.

Stepping through the categories of vehicular networking applications, we start with safety-critical systems. In this field, the delay of a message transmission is the most critical metric. As discussed before, requirements range from a few hundred milliseconds down to tens of milliseconds. At the same time, high reliability is needed, especially for warning messages, in order to achieve very short inter-packet arrival times. Closely tied to this is the number of collisions on the channel, which substantially impact reliability. In the situation-awareness application domains, weaker requirements on reliability can be accepted, assuming a high enough message rate. In this case, newer messages can be used to tolerate the loss of previous announcements.

Non-safety applications do not impose high requirements on delay and reliability. Designed for best-effort communication, such applications are usually very robust concerning high delays and packet loss. The only exception is the aforementioned multimedia streaming class. Here, the absolute delay is not that critical but the variance of the delay matters. Depending on the encoding algorithms used, one may allow forward error correction or even completely tolerate the loss of single packets; demands in terms of reliability are comparably weak as well.

Application-specific metrics

When one is assessing specific vehicular networking applications, the networking metrics discussed above might not reveal the full picture of the capabilities of the protocols used. From the networking perspective, however, these are the metrics that may be used even in running systems, providing feedback for runtime protocol adjustments. Let's discuss two examples for application-specific metrics that are mainly used for performance evaluation of protocols and applications (see Section 6.3.4 on page 290).

Vehicle collision probability

For safety-critical applications, one can be interested in networking measures but one must demonstrate that safety as a metric, represented for example by a vehicle collision probability, is improved. This is the case for an ICWS. Here, vehicles approaching an intersection are supposed to be informed by means of IVC about other vehicles that might be on a collision course. If the system performs well, many crashes can be prevented. This metric can then be used to compare different ICWS approaches and the networking protocols used.

Travel time and CO_2 footprint
Looking at traffic information systems, the overall performance can be measured not by means of computer networking metrics but by the improvements experienced by the drivers. This is, unfortunately, not as straightforward as it seems. For the individual driver, the overall travel time is what matters most. Thus, improvements according to this metric are a great selling point. Looking at the problem from a more macroscopic perspective, the improvements on the microscopic level, i.e., for the individual drivers, might not result in solving the problem of congestion on our streets. Instead, entire traffic flows need to be considered. A metric partly showing improvements on this level is the emission of CO_2. If traffic flows smoothly, gas consumption and CO_2 emission are reduced, even though individual vehicles might have to be slowed down to compensate for capacity bottlenecks in the road networks.

3.2.3 Communicating entities

In the course of discussing IVC applications, we repeatedly mentioned so-called vehicle-to-vehicle (V2V) communication, i.e., referring to IVC as a means to exchange information among vehicles driving on our streets. For most of the applications discussed, this is exactly the overall objective, even though specific communication protocols might have to rely on additional components. These communication principles will be discussed in Section 3.2.4 on page 68.

A second entity involved in vehicular networks is the Internet – servers of cloud systems may be used, for example, as a central TIC or general storage to download content from. This is often referred to as vehicle-to-infrastructure (V2I) communication.

Besides driving *vehicles* and the *Internet*, there are several communication entities that might be involved in vehicular networking applications. Figure 3.10 outlines some of those entities. We will briefly introduce some of the most important ones in the following.

Roadside units

Roadside units (RSUs) have been introduced in the area of vehicular networking to bridge communication gaps between vehicles and to provide access to other networks and the Internet (Lochert *et al.* 2008; Salvo *et al.* 2012). This concept has been borrowed from what is being used very successfully in the WiFi world. Here, access points (APs) connect mobile nodes such as laptops or smart phones to the wired Internet. The communication to WiFi APs was experimentally validated as long ago as in 2005 (Ott & Kutscher 2005).

RSUs can be used in combination with almost any wireless communication technology to access backbone networks connected to them. The most obvious question to be answered is about the benefits of RSUs in relation to deployment and operational costs (Zheng *et al.* 2010).

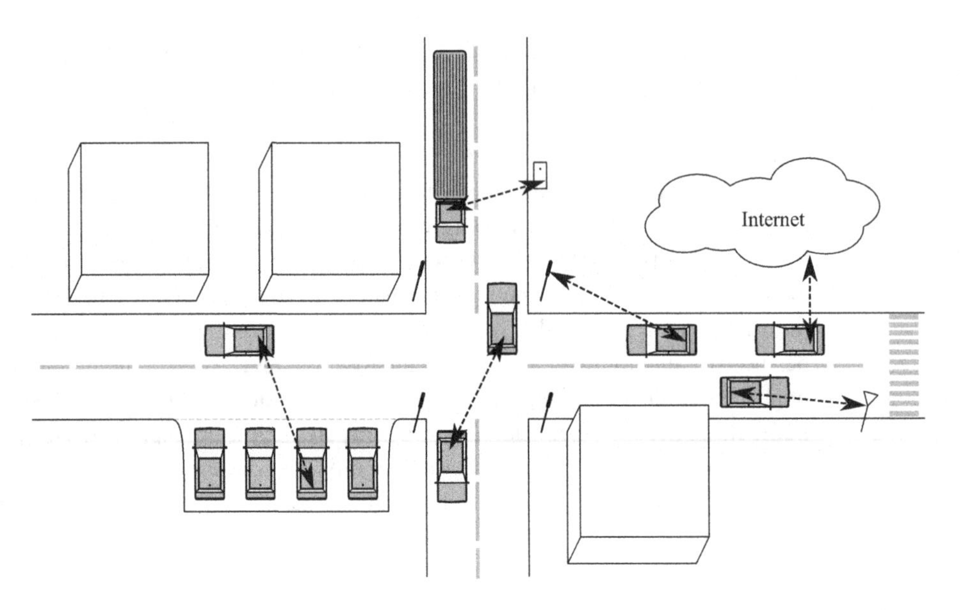

Figure 3.10 Communicating entities in a vehicular network.

Such RSUs are assumed to be connected to a backbone network and to be externally powered. There is, however, also the concept of so-called stationary support units (SSUs) (Lochert *et al.* 2008; Sommer *et al.* 2010e; Ahn *et al.* 2012). SSUs can participate in the vehicular network like any other vehicles but are deployed at a fixed position with no backbone connection. This reduces both deployment and operational costs. An SSU can, for example, be powered by an attached solar panel. Still, for some protocols SSUs have been shown to substantially improve the communication among the driving vehicles.

Cellular networks work in a similar way. Being completely infrastructure oriented, base stations are deployed in a cell structure in 3G/4G networks to provide optimized network coverage. Mobile phones and vehicles with embedded end systems can connect to these base stations, which, in turn, provide connectivity to a core network and eventually to the Internet. Base stations in cellular networks can also be used to broadcast information to all nodes that are currently connected.

Infrastructure elements like RSUs or SSUs come with a non-negligible operational cost. This cost is even higher for cellular networks such as UMTS and LTE, but they are already well enough deployed to support vehicular networks, at least during the early phase of market introduction. Especially in this early phase, infrastructure will be the main if not only possibility to establish network connections. V2V connections are feasible only given some minimum penetration rate – depending on the application, this can be 10%–90%.

Parked vehicles

If pre-deployed infrastructure is not (yet) available, alternatives have to be found. It has been discovered that parked vehicles that have already installed the vehicular communication technology may be an excellent alternative.

A detailed study of parking behavior in the area of Montreal, Canada offers interesting insights (Morency & Trépanier 2008): in 2003, out of 61 000 daily parking events, 69.2 % of all parked cars were parked on streets while only 27.1 % were parked on outside parking lots. A minority of 3.7 % was parked in interior parking facilities. On average the duration of one parking event was about 7 h. The study furthermore shows that parking vehicles were distributed throughout the whole city, which means there is a high possibility that a parked car is within transmission range of a moving car. Other studies found that, on average, a vehicle is parked for 23 h a day (Litman 2006).

Additionally, parked vehicles will likely be available at times when traffic is low, i.e., the probability of not having other driving vehicles in communication range that could act as relays themselves will be high – even if the penetration rate is already high. We therefore conclude that the use of parked cars as relays in vehicular networks can prove to be very helpful in supporting message exchange – at any given time, most cars are parked; of these, most are parked on streets. This concept has successfully been used for cooperative awareness applications (Eckhoff *et al.* 2011c; Sommer *et al.* 2014a) and for distributed content download (Liu *et al.* 2011; Malandrino *et al.* 2012).

Traffic lights and traffic signs

When talking about RSUs, one could think of not installing those devices as standalone units but integrating RSUs with available road network infrastructure such as traffic lights. The advantage of traffic lights is that they are located at positions perfectly suited to have lines of sight to all incoming streets and, therefore, to all possibly approaching vehicles.

Besides the suitable position for communication between approaching vehicles and the traffic light, vehicles are also interested in traffic-light cycles for the aforementioned GLOSA applications. We also discussed the possibility of using vehicles as sensors identifying and disseminating traffic flow characteristics to the traffic lights to enable adaptive changes of the traffic-light cycles. Thus, traffic lights will likely become a fundamental part of vehicular networking applications (Gradinescu *et al.* 2007; Tubaishat *et al.* 2007).

Similarly, we can try to extrapolate this concept to equipping even important traffic signs, in a first step likely the smart traffic signs available already on major freeways. The traffic signals can then directly communicate with passing vehicles to disseminate information displayed or even more relevant information to be processed by the OBU.

Third parties

Third-party participants in the vehicular network will usually provide relevant information by means of Internet or cloud services. Yet, this communication link might not be available all of the time and more local communication can also provide additional capabilities. Therefore, it is to be expected that also third-party information sources will join the vehicular network even in its early days. The most obvious example is information about available parking places. This information can either be collected and managed in a fully distributed manner by all participating vehicles (Caliskan *et al.* 2006) or can be provided as a service by a local company.

The latter option is an example of location-based services in general (Klimin *et al.* 2004; Lee *et al.* 2010c). We can think about many novel applications providing information that is perfectly well prepared according to the local position to passing vehicles. This ranges from parking place information via emergency information like directions to hospitals or police stations to special ads from local shops.

Other participants

Regarding all the non-safety applications, we could already conclude the discussion about entities participating in vehicular networks. Yet, looking at the safety domain, there are some that we have missed so far – and so has the broad field of vehicular networking for a while. Our streets are used not only by cars driving on the road but also by pedestrians, bicyclists, and others. Designing safety-critical applications for vehicles helps (and may solve the issue on freeways), but may lead to insecure situations if the other road users are not included in the picture.

Thus, in recent work also pedestrians and bicyclists have been added as participants in vehicular networks (Sugimoto *et al.* 2008; Riaz *et al.* 2006). This raises questions about the communication technology. Vehicles are energy autonomous – this even holds for electric cars, considering the little energy that is needed for wireless communication devices in comparison with all other systems. For pedestrians and bicyclists, however, the story is slightly different. Continuous communication might be prohibitive.

Thinking further along the lines of additional entities participating in vehicular networks, we will eventually arrive at a point at which vehicular networks disappear into a much broader type of network. This is what is often referred to as the Internet of things (IoT) (Atzori *et al.* 2010). In the scope of this textbook, we stop at discussing vehicular networks but partly also consider some of the listed entities such as RSUs, parked vehicles, and traffic lights.

3.2.4 Communication principles

Having a good understanding now about applications relying on inter-vehicle communications as well as their demands in terms of communication quality, dissemination range, and the communication entities involved, we will now spend some time looking at the major communication principles. In some aspects, vehicular networks can be regarded as a (very particular) class of mobile ad-hoc network (MANET), which is dominated by laptop computers and smart phones. They share many communication concepts: Interaction between multiple communicating entities can be realized *ad hoc*, using, for example, WiFi-like technologies. Alternatively, *infrastructure*-based access is possible, as realized by cellular networks. One major difference is that the differentiation between ad-hoc and infrastructure-based interaction is not that simple in vehicular networks, as we will show in the following.

The two traditional ways to disseminate data in wired networks are unicast and network broadcast dissemination of data, that is, the dissemination of data to exactly one individually addressed entity or the dissemination of data to all nodes participating in the network, respectively. Looking at the IVC applications, it also becomes apparent, however, that these dissemination primitives turn out to be the most exotic. Typical applications of vehicular networks require the dissemination of data not to one but to a number of participants, which are identified not by an address, but rather by fuzzy parameters such as their approximate position. Conversely, designing an application that would require data items to be disseminated to all participants (i.e., all cars) would yield a system that just would not scale (see Section 3.4.3 on page 104).

At this point, we aim to outline communication principles only on a very high level. Detailed discussions focusing on specific approaches and protocols follow in Section 3.3 on page 71.

Ad-hoc communication

Direct communication between neighboring vehicles within communication range can be established using different technologies. Early approaches have been based on WiFi, i.e., IEEE 802.11a/b/g/n. This has been conceptually quite successful, even though WiFi is very sensitive to high mobility. This concerns not only the short connection times (WiFi relies on a time-consuming association process) but also the potentially high speed of vehicles. Multi-path effects make radio signals difficult to decode, and signal fading is also hard to compensate.

Therefore, a new protocol standard has been developed, which is based on the IEEE 802.11a PHY, IEEE 802.11e MAC extensions, and new extensions introduced in IEEE 802.11p. The new protocol, which is often also referred to simply as IEEE 802.11p (or, colloquially, the new DSRC standard, named after the frequency band in which it operates) has been designed with all these limitations in mind. It is more robust and allows one to send and receive messages without any association process. We discuss WiFi-based access as well as IEEE 802.11p in detail in Chapter 4.

Technologies such as IEEE 802.11p provide direct communication only among those vehicles in communication range. This is certainly not sufficient for many of the applications discussed. Information needs to be spread over larger areas. Ad-hoc communication can be extended by means of relaying, i.e., using neighboring cars as helpers that forward the message to the intended recipient.

The concept of multi-hop ad-hoc communication is depicted in Figure 3.11. In this figure, a vehicle sends a message using multi-hop forwarding to all potentially interested cars. No additional infrastructure or helper is needed.

There are numerous alternatives for organizing this forwarding. A network topology can be established before sending a message, i.e., a network is established and routing algorithms are used to identify suitable paths to the destination. This requires some

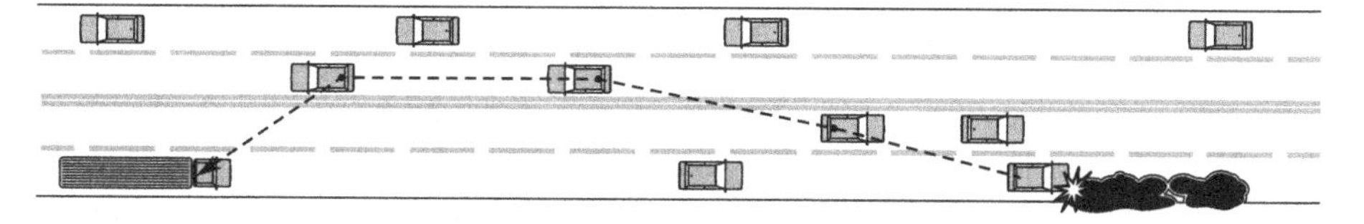

Figure 3.11 Ad-hoc communication between neighboring vehicles; multi-hop relaying helps disseminate information.

non-negligible overhead for topology maintenance. The other option is to flood the messages through the network. A common design paradigm uses only local (one-hop) broadcasts to transmit data and relegates the task of actually disseminating knowledge through the network to the application layer, lending the name *application-layer broadcast* to this class of dissemination strategies. Since packet loss and shifting network connectivity necessitate the continuous repetition of these broadcasts, this approach is also called *beaconing*: repeated undirected broadcasting of data. Beaconing can and should, of course, be made more intelligent, e.g., by exploiting geographic positions.

Infrastructure-based communication

Ad-hoc communication is possible only if a minimum vehicle density is available, i.e., if enough forwarders can be found. This will be the case in very dense road traffic situations but might be problematic if only a few cars are driving on the streets or if the penetration rate with this new technology is still marginal. Also, external network resources must be included in the network, as discussed before.

Therefore, network infrastructure can be deployed to provide connectivity and to connect the vehicular network to exterior networks. There are two options regarding how to design infrastructure-based vehicular networks. Like in WiFi networks, APs can be installed to provide network access. This can be done using WiFi directly or by deploying IEEE 802.11p access points. In the latter case, these are usually called RSUs. Such installations are being considered especially along major freeways or in urban environments. It is, however, not clear whether this is feasible on a large scale due to the associated costs.

The second option is to rely on using cellular networks. Such networks are available almost everywhere, providing ubiquitous access, and, if using 3G and 4G networks such as UMTS or LTE, also high data rates. The use of cellular networks has been evaluated in several projects. It seems to be obvious to use cellular networks for many of the vehicular networking applications, yet it is still not known whether low-latency safety applications can be supported and whether these networks scale better with increasing numbers of participating vehicles.

Figure 3.12 shows the conceptual approach of using cellular networks for IVC. Vehicles are connected to a base station in the local vicinity, and this base station connects to a backbone network. No additional routing or forwarding techniques are needed.

Hybrid approaches

Both ad-hoc networking and infrastructure-based access have their advantages and limitations. Therefore, ad-hoc and infrastructure-based communication can and must be combined in order to overcome all the challenges in vehicular networking. Ad-hoc communication is in general useful only in a rather local geographic region. We will discuss this in more detail later in this book. Infrastructure-based networks help to bridge long distances or to provide direct Internet access. On the other hand, latencies experienced while messages are routed through the backbone network or even the Internet will naturally be much higher.

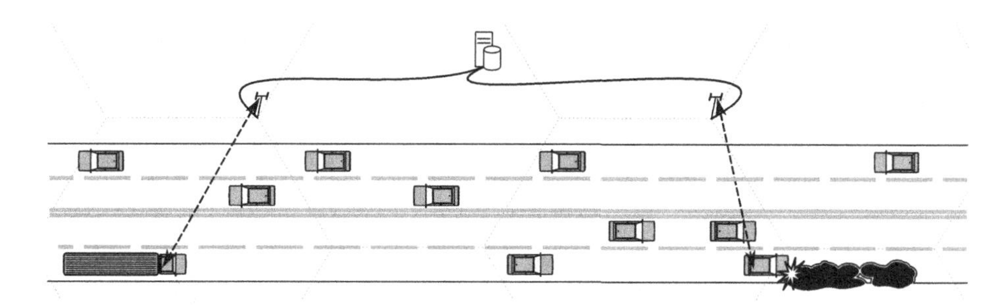

Figure 3.12 Infrastructure-based communication using, as an example, a cellular network.

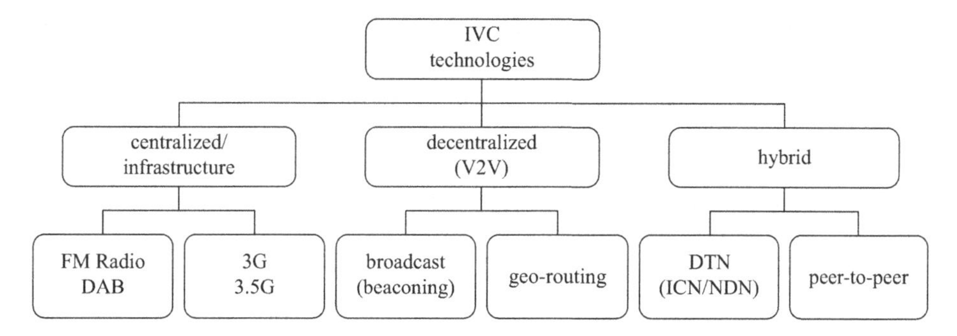

Figure 3.13 A taxonomy of IVC technologies that are being considered for real-world applications.

The idea is therefore to combine these two types of communication. Many of the examples we selected for investigating IVC protocols in detail make use of both concepts or, at least, allow one to integrate other types of communications to improve the overall protocol performance. This trend will become clear when we discuss IVC communication concepts in the following section.

3.3 Concepts for inter-vehicle communication

The focus of this section is to introduce the IVC concepts that have been considered for use in vehicular networking applications. The objective is to provide an overview mainly based on historical milestones rather than an in-depth discussion of specific concepts. We will return to selected approaches for medium access and information dissemination in Chapters 4 and 5, respectively.

Figure 3.13 shows an overview of the different concepts that have been considered for application in realistic IVC-based systems. In general, we can categorize IVC concepts as centralized or infrastructure-based, decentralized (i.e., pure V2V), and hybrid solutions.

In the following, we will slowly go through this variety of approaches, mainly following the historical course, i.e., first looking at solutions that have already been in use for a while, and then studying more visionary ideas. This includes the currently used

FM radio channels, the use of 3G or 4G networks to disseminate traffic information and other data, ad-hoc routing, geographic routing, and broadcasting-based approaches.

3.3.1 FM radio and DAB

The first approach towards a fully operational vehicular networking application was the use of FM radio and now DAB to broadcast traffic information to participating vehicles. In this context, participation means that the vehicles must have a unit installed that is able to tune into the respective radio channel, to receive and decode the messages. This system has become available and has been in use for a decade under the name traffic messaging channel (TMC). TMC establishes a simple unidirectional broadcast channel from the radio station to the vehicles using a digital channel in the analog FM radio transmission. We discuss TMC here.

The dedicated successor of TMC is the use of a more flexible (and more complex) approach named the Transport Protocol Expert Group (TPEG) protocol. It was developed for use in DAB radio, which provides higher data rates suitable for more fine-grained TIS information. The architecture and message format will be analyzed in more detail on page 74.

Direct digital broadcast of the TIS data via a satellite is also possible, but has not yet been achieved.

RDS/TMC

The term TMC is used both for an information-dissemination protocol over FM radio and for the overall TIS which couples central management of traffic information with this broadcast channel. The main architecture of TMC has been introduced already in Figure 3.1. Here, we do not concentrate on the information sources but discuss solely the broadcast protocol.

The transmission of digital data via FM radio is performed using the radio data system (RDS), using a dedicated channel of FM radio as depicted in Figure 3.14 (ISO 2003a). The transmission uses a simple binary phase-shift keying (BPSK) modulation on a 57-kHz subcarrier to separate the data channel from mono and/or stereo audio. It provides a data rate of 1.2 kbit/s.

RDS is responsible for, and was established mainly for, transmitting meta information alongside an FM radio channel. This typically includes information about the radio station or the song being played. Thus, TMC cannot use the full capacity of this wireless downlink channel. In particular, the RDS group identifier 8A, i.e., TMC, can send approximately 10 bulletins per minute. The resulting data rate is obviously very low.

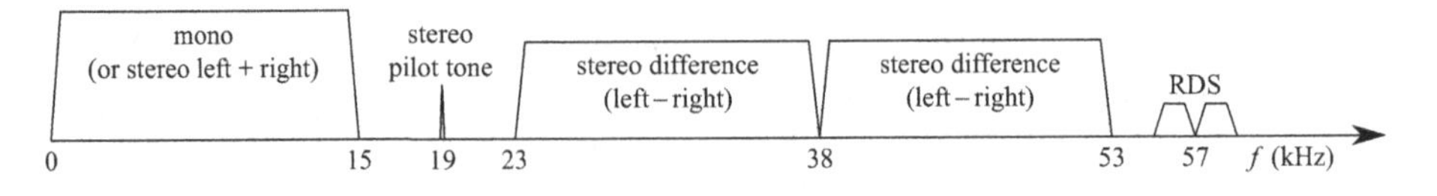

Figure 3.14 The location of the RDS channel for digital transmissions in analog FM radio.

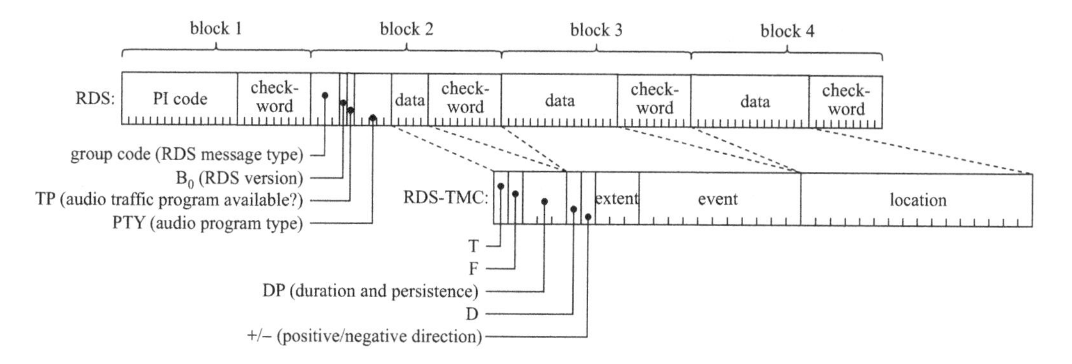

Figure 3.15 RDS/TMC message format.

Messages are prepared in the form of a carousel. Each TMC message is repeatedly broadcast as long as it is valid. The repetition time, i.e., the update frequency, is a function of the number of messages to be delivered, i.e., the granularity of TIS information, and the still-acceptable delay. The worst-case delay is the time between tuning into the TMC channel and the time at which the message is delivered. Thus, very few messages can be sent by TMC, since the messages need to be repeated at least every few minutes and the channel is bound to 10 messages per minute.

The coding protocol (ISO 2003b) and the event and information coding (ISO 2003c) are described in the respective standard documents. The message format of RDS blocks carrying a TMC message is shown in Figure 3.15. Each message has a length of 104 bit and contains essentially four 26-bit blocks, each of which is secured against error by a 10-bit checkword. The following information is contained:

- PI is the program identification code that identifies the radio station
- a group code encodes the RDS message type (here: `0x8a`, i.e., TPEG)
- B_0 encodes the RDS version
- TP and PTY give information on the audio programme; they are the last fields common to all RDS messages
- the TPEG-specific part starts with three flags, T, F, and D, which identify whether this is a first or subsequent group in a sequence
- F distinguishes between a single- and a multi-group message
- DP stands for duration and persistence, and contains information about the time span of the traffic event
- D is a flag if re-routing is required
- +/− identifies the direction of the road event (i.e., to the front or to the back of the position)
- Extent shows the extent of the current event
- Event contains an 11-bit event code, which is looked up on the local event code table
- Location refers to a location code, which is looked up against the location table database installed in the end system

Table 3.5 Selected TMC location codes

Country and Table	Code	Description
58:1 (Germany 1)	1	Country: Germany
	2	Country: France
	264	Area: Bavaria
	12579	Motorway junction: Irschenberg
25:1 (Italy 1)	1	Continent: Europe
	2	Country: Italy
	68	Area: Piedmont
	2046	Motorway junction: Bardonecchia

Table 3.6 Selected TMC event codes

Code	Description
101	Service level: standing traffic (generic)
102	Service level: 1 km of standing traffic
103	Service level: 2 km of standing traffic
104	Service level: 4 km of standing traffic
105	Service level: 6 km of standing traffic
106	Service level: 10 km of standing traffic
107	Expected service level: standing traffic
211	Incident: broken-down car(s)
212	Incident: broken-down truck(s)
393	Incident: broken-down car(s), danger
394	Incident: broken-down truck(s), danger
1478	Security alert: terrorist incident

Obviously, all the location and event codes need to be installed in the end device. Tables 3.5 and 3.6 list selected examples for location and event codes, respectively (note that the meaning of location codes is dependent on which country's code table is loaded; e.g., the German code table uses location code 1 for *Germany*, while the Italian code table uses the same location code to refer to *Europe*). TMC does not provide any (real) security measures. It has already been demonstrated that the system can be cheated by sending falsified messages.

TMC has, for example, been extended by Navteq to provide regional value-added services. The resulting system is financed by per-decoder license fees. The main difference is that data collection and processing are now fully automatic. This was the first system deployed that makes use of FCD. For event prediction, expert systems and neural networks are used for early warnings of predicted events. The system is restricted to major roads and allows one to upload and download data also using cellular networks.

TPEG

The Transport Protocol Expert Group (TPEG) protocol was planned as a successor to the RDS-based TMC. It was finally released in April 2000 (TPEG 2006a). It was intended for use with DAB, and the main objective was to develop a communication protocol that can make use of a faster communication medium. Also, in contrast to the very limited information that can be encoded in TMC, TPEG was designed with a strong focus on extensibility.

The design goals of TPEG can be summarized as follows

- A unidirectional, byte-oriented stream is provided for the downlink channel from an infrastructure-based TIC to the participating clients.
- It follows a modular concept that can easily be extended to upcoming needs of some future TIS.
- The encoding follows a hierarchical approach to enable compatibility with future extensions.
- An integrated security approach is used to detect malicious or falsified information.

Figure 3.16 outlines the TPEG architecture. As can be seen, TIS information can be generated automatically by sensors, can be imported from external databases, or can also be manually integrated. The information dissemination was initially planned to use DAB but was later extended to use arbitrary communication channels. Within the vehicles, the received data can be used by the OBU to automatically calculate re-routings or to present the information to the driver.

The TPEG message format (TPEG 2006b) is outlined in Figure 3.17. The TPEG transport stream is divided into sections, each containing data for a specific service. A summary of the service data in the stream is made available in the form of a service info header. In our example, the header announces two services, of which the second one is shown in more detail. Each service section is, again following the hierarchical concept, divided into several sections. The first section describes the service (in the example it is about travel information from the BBC) and provides information about the data

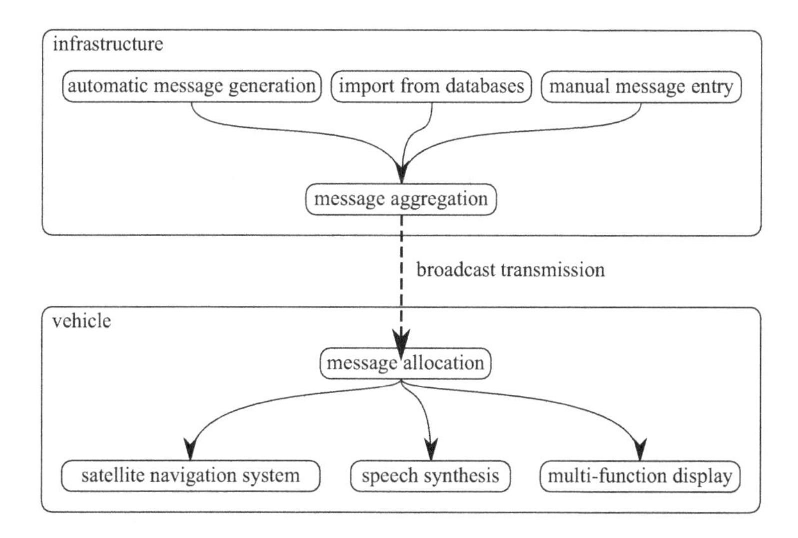

Figure 3.16 TPEG architecture.

Table 3.7 Selected TPEG information types

Code	Description
RTM	Road traffic message
PTI	Public transport information
PKI	Parking information
CTT	Congestion and travel time
TEC	Traffic event compact
WEA	Weather information for travelers

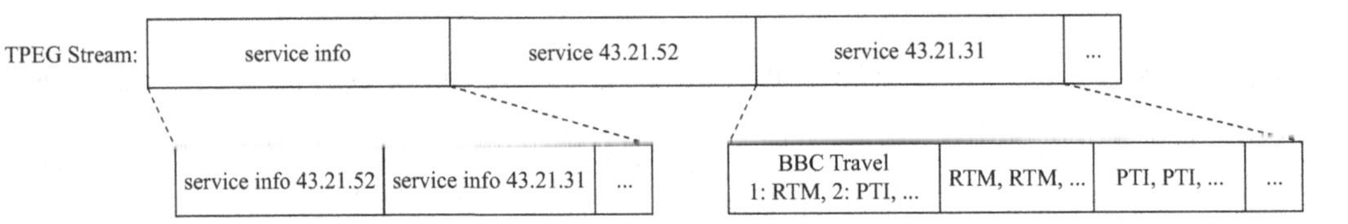

Figure 3.17 TPEG message format.

sections included. RTM and PTI stand for road traffic message and public transport information.

Thus, TPEG is conceptually very close to TMC but much more flexible. Given the extensibility, there cannot be a complete lookup table for all information types defined by TPEG applications. Several types have been specified already, and selected examples are listed in Table 3.7.

A variant for dissemination over technology other than DAB, such as the Internet, is to encode TPEG messages in XML (TPEG 2006c). A sample XML-encoded TPEG message is shown in Listing 3.1. This example has been taken from the BBC traffic service. It contains a single road traffic message announcing two traffic obstructions.

Geo-referencing can be much more accurate in TPEG than in TMC. The standard supports a hybrid approach to geo-referencing. Position information can be provided in a very fine-grained manner using exact geographic coordinates based on the WGS 84 definition. Alternatively, intersection location (ILOC) descriptions can be used in a normalized, shortened textual representation of street names intersecting at a desired point. Even human-readable plain text can be used; however, that is very difficult for a navigation unit to process automatically. Finally, location table codes are supported (cf. TMC).

3.3.2 Cellular networks

The idea of all cellular networks is to divide the world into cells, each of which is served by a base station. This allows, for example, frequency reuse in frequency-division multiple access (FDMA) protocols exploiting spatial diversity of the carrier used. Figure 3.18 outlines this concept. A car passing through the network will connect to the respective local base station. Thus, it needs to perform handovers at the cell borders.

```
1  <tpeg_message>
2    <originator country="UK" originator_name="BBC Travel News"/>
3    <summary xml:lang="en">M1 Bedfordshire - Broken down lorry southbound
         between J11, Dunstable and J10, Luton Airport, queueing
         traffic.</summary>
4
5    <road_traffic_message message_id="2259387"
         message_generation_time="2008-03-05T07:22:46+0" version_number="5"
         start_time="2008-03-05T06:55:49+0" stop_time="2008-03-05T08:22:46+0"
         severity_factor="&rtm31_4;">
6
7      <obstructions number_of="1">
8        <vehicles number_of="1">
9          <vehicle_info vehicle_type="&rtm1_3;"/>
10         <vehicle_problem vehicle_problem="&rtm3_8;"/>
11       </vehicles>
12     </obstructions>
13
14     <network_performance>
15       <performance network_performance="&rtm34_2;"/>
16     </network_performance>
17
18     <network_conditions>
19       <position position="&rtm10_1;"/>
20       <restriction restriction ="&rtm49_255;"/>
21     </network_conditions>
22
23     <location_container language="&loc41_30;">
24       <location_coordinates location_type="&loc1_3;">
25         <WGS84 latitude="51.893453" longitude="-0.469775" />
26         <location_descriptor descriptor_type="&loc3_7;" descriptor="M1;"/>
27         <location_descriptor descriptor_type="&loc3_24;"
               descriptor="Bedfordshire"/>
28         <location_descriptor descriptor_type="&loc3_10;"
               descriptor="Dunstable"/>
29         <location_descriptor descriptor_type="&loc3_32;" descriptor="M1; 11"/>
30         <location_descriptor descriptor_type="&loc3_8;" descriptor="A505;"/>
31
32         <WGS84 latitude="51.853818" longitude="-0.42358" />
33         <location_descriptor descriptor_type="&loc3_7;" descriptor="M1;"/>
34         <location_descriptor descriptor_type="&loc3_24;"
               descriptor="Bedfordshire"/>
35         <location_descriptor descriptor_type="&loc3_10;" descriptor="Luton
               Airport"/>
36         <location_descriptor descriptor_type="&loc3_32;" descriptor="M1; 10"/>
37         <location_descriptor descriptor_type="&loc3_8;" descriptor="M1;"/>
38         <direction direction_type="&loc2_7;"/>
39       </location_coordinates>
40     </location_container>
41
42   </road_traffic_message>
43 </tpeg_message>
```

Listing 3.1 A sample TPEG message (BBC traffic service).

The design of the network infrastructure depends on the generation of the cellular network. Figure 3.19 outlines the hierarchical structure of UMTS networks, i.e., cellular networks of the third generation. The end system (named UE in this context) is always connected to a specific cell. The cell, i.e., the antenna, is managed by a so-called NodeB, that is the base station for UMTS cells. Multiple NodeBs are connected to a radio network controller (RNC), which finally connects the cells to the IP-based core network.

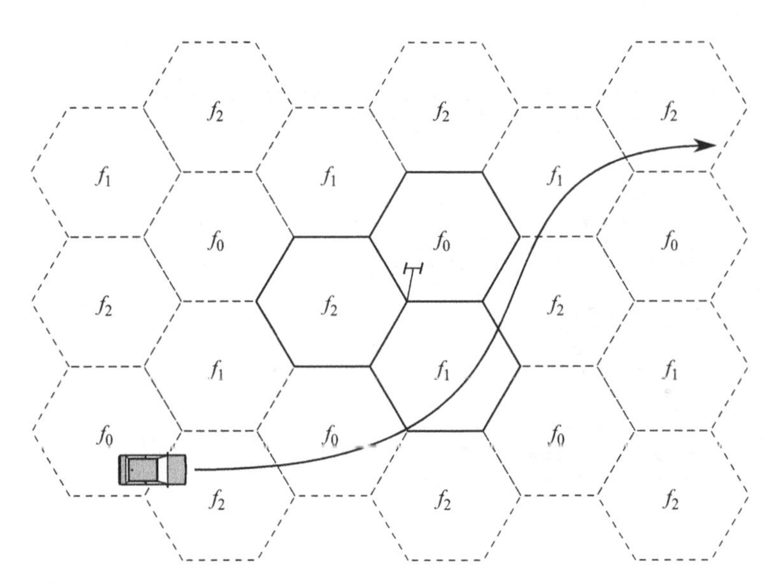

Figure 3.18 The layout of cellular networks; f_i represents the carrier frequency used, which is reused in non-neighboring cells thanks to spatial diversity.

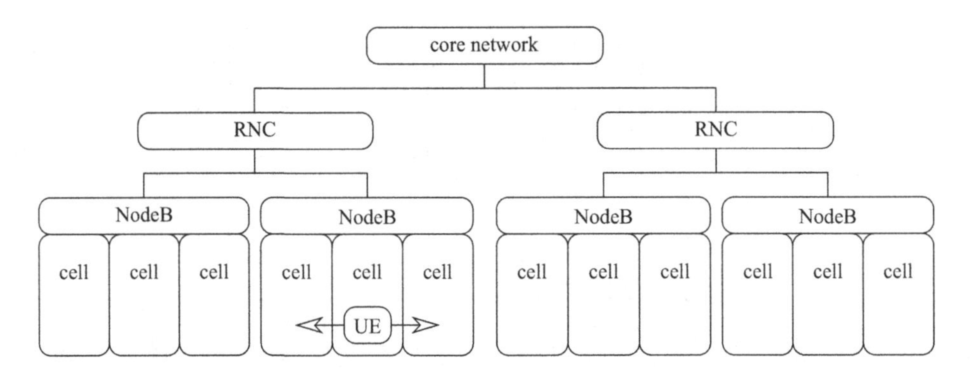

Figure 3.19 The hierarchical structure of UMTS networks.

At this point, we only very briefly introduce the feasibility of using cellular networks for information dissemination in vehicular networks. The technology is discussed in more detail in Section 4.1 on page 107. A full introduction of data transmission in cellular networks of the second (2G) to fourth (4G) generations is beyond the scope of this textbook. Very good tutorial-like presentations are available in the literature (Dahlman *et al.* 2008, 2011).

Table 3.8 provides an overview of the different generations of cellular networks and the protocols used. We are now heading towards 5G networks, which are currently being researched to identify suitable technologies and protocols. From 2G to 4G, the protocols evolved quite drastically. LTE Advanced networks of the fourth generation no longer provide dedicated voice channels. Instead, the networks have converged from circuit-switched networks, also providing packet-based transmission using dedicated circuits,

Table 3.8 An overview of the different generations of cellular networks and the peak data rates supported

Generation	Protocol	Downlink	Uplink
2G	GSM[a]	9.6 kbit/s	9.6 kbit/s
2.5G	GPRS[b]	9.6–50 kbit/s	9.6–50 kbit/s
	EDGE[c]	384 kbit/s	384 kbit/s
3G	UMTS[d]	384–2000 kbit/s	384–2000 kbit/s
	HSDPA/HSUPA[e]	10 Mbit/s	5 Mbit/s
	HSPA+[f]	100 Mbit/s	23 Mbit/s
3.9G	LTE[g]	300 Mbit/s	75 Mbit/s
4G	LTE Advanced	1 Gbit/s	1 Gbit/s

[a] Global System for Mobile Communications (GSM)
[b] General Packet Radio Service (GPRS)
[c] Enhanced Data Rates for GSM Evolution (EDGE)
[d] Universal Mobile Telecommunications System (UMTS)
[e] High-Speed Downlink Packet Access (HSDPA)/High-Speed Uplink Packet Access (HSUPA)
[f] High-Speed Packet Access (HSPA)
[g] Long Term Evolution (LTE)

to pure packet-switched networks. Voice is now provided as a side feature using, e.g., voice over IP (VoIP).

Also, the data rates supported have increased substantially. As listed in Table 3.8, early 2G networks just supported data rates of up to 9.6 kbit/s. In contrast, 4G networks support peak data rates of up to 1 Gbit/s. Since the channels used are shared among all participants, the data rates listed are peak values and will not be available all the time to all participating nodes. The end-to-end latency supported strongly depends on the capabilities of the core network and is not to be seen as an inherent parameter of the protocols used.

The use of 3G and 4G networks in vehicular networking applications has been investigated in several projects. The good news is that the speed of vehicles is not a limiting factor. Field operational tests (FOTs) at 290 km/h have shown that the signal drops only after sudden braking maneuvers. In these cases, the handover prediction fails. Open questions include delay and capacity guarantees.

From these FOTs, we can learn that both UMTS and LTE, i.e., cellular networks of the third and fourth generations, can support common vehicular networking applications in general. The main limitation is the available capacity, i.e., only low vehicle densities can be supported for time-sensitive or traffic-intensive applications. This situation can be improved by additional investments in much smaller cells, e.g., along freeways.

From other experiments (e.g., CoCar described in Section 4.1.5 on page 116), we already have some initial insights (Sommer *et al.* 2010d; Valerio *et al.* 2008) into the delay performance of UMTS and LTE networks. It turned out that direct V2V communication via the core network of the service provider is not able to support the latencies required for safety-critical applications. In high-load scenarios, delays may add up to several seconds. Yet, there is a technology that was defined already for UMTS

(ETSI 2012b) and then made available in LTE as well: multimedia broadcast/multicast service (MBMS) and eMBMS, respectively. MBMS/eMBMS implements cellular multicasting. The original idea was to use this technology for establishing centralized services such as a TIS, where data needs to be delivered to all participating nodes. This kind of cellular broadcast has been reused in CoCar for developing local reflectors to disseminate information from a vehicle via the reflector to all other vehicles in the local cell, i.e., all neighboring vehicles. In this way, delays of about 100 ms can be supported and, thus, so can safety-critical applications.

One of the most critical open questions concerns the communication paths when multiple MNOs are involved. In general, the communication in such a scenario, i.e., when one user is using MNO A's network and the other one is using B's network, respectively, is essentially performed via the Internet. That means that users will likely experience unpredictable and long delays, and also that data rates will vary. This problem has been identified, and the current trend towards using a shared physical infrastructure and mobile virtual network operators (MVNOs) may help to solve the issues.

Still, cellular networks are best suited for early applications even if the penetration rate is still very low. Good examples include all types of information download and even decentralized TIS applications. We discuss the PeerTIS concept (Rybicki *et al.* 2009) in Section 5.7.3 on page 225.

3.3.3 Ad-hoc routing

Following our historical overview of vehicular networking approaches, many of the early concepts for IVC tried to make use of protocols developed for mobile ad-hoc networks (MANETs). Also the term vehicular ad-hoc network (VANET) was coined in this context. We will study MANET routing protocols and their application for VANETs in detail in Section 5.1 on page 138. In the following, we will discuss the main concepts only, focusing on the general ideas.

Mobile ad-hoc networks

Data dissemination in MANETs has been investigated for almost two decades (Basagni *et al.* 2004), the most prominent research question being routing in dynamically established network topologies. Much more so than wired networks, MANETs have to cope with dynamic node topologies, allowing nodes to join and leave the network at any time. A nice but slightly outdated overview of MANET routing is provided by Abolhasan *et al.* (2004).

Figure 3.20 outlines the main ideas of ad-hoc routing. Nodes are considered to be uniformly distributed over a geographic area. As direct wireless communication is not feasible between all nodes, multi-hop routing is used to exchange information between arbitrary nodes. In Figure 3.20, lines between the nodes characterize possible communication links. If a node wants to send a message to an arbitrary other node, the topology of the network needs to be known, at least between the two nodes. In the given example, node Ⓐ sends a message to node Ⓙ. An alternate path is depicted as well.

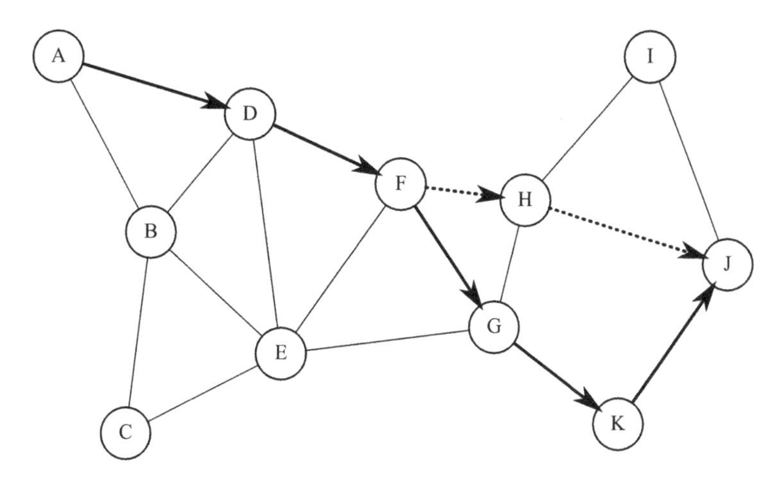

Figure 3.20 The MANET routing approach for a typical network topology consisting of randomly distributed nodes.

Different concepts for ad-hoc routing have been developed, being based either on continuously maintaining the network topology in advance of sending messages (this is known as *proactive* routing) or discovering the network topology only if a message is to be sent (*on-demand* routing). Many of the early approaches to MANET routing converged to a few protocols that have also been standardized by the Internet Engineering Task Force (IETF). In particular, the *IETF MANET (Mobile Ad-hoc Networks)* working group[2] focuses on general-purpose ad-hoc routing protocols. Based on this, the new *IETF ROLL (Routing Over Low Power and Lossy Networks)* working group[3] investigated MANET routing in cases in which energy is a major resource. This paradigm shift was nicely formulated by Watteyne *et al.* (2011). In vehicular networks, energy is considered not to be a main issue. Even considering electric cars, the energy consumed by wireless communication is negligible compared with other energy consumers.

The most critical issue in MANETs, however, is node mobility. Although the name *mobile* ad-hoc networks suggests that this issue has been dealt with appropriately, it has not. Indeed, almost all of the MANET routing protocols can handle a little node mobility but fail if the network topology changes too quickly and too often. We shall discuss both protocols and mobility issues shortly.

Vehicular ad-hoc networks

In general, VANETs are quite similar to MANETs (Hartenstein & Laberteaux 2008; Li & Wang 2007, Lee *et al.* 2010a). Information needs to be transported between nodes in a network that is dynamically created by nodes moving in a geographic area. This similarity motivated a rich and deep set of investigations, many of the early works even suggesting that the same set of algorithms can be used in VANETs again. Yet, it turned out that there are some quite significant differences between a generic MANET and a typical VANET.

[2] IETF MANET Charter, see http://datatracker.ietf.org/wg/manet/charter/.

[3] IETF ROLL Charter, see http://datatracker.ietf.org/wg/roll/charter/.

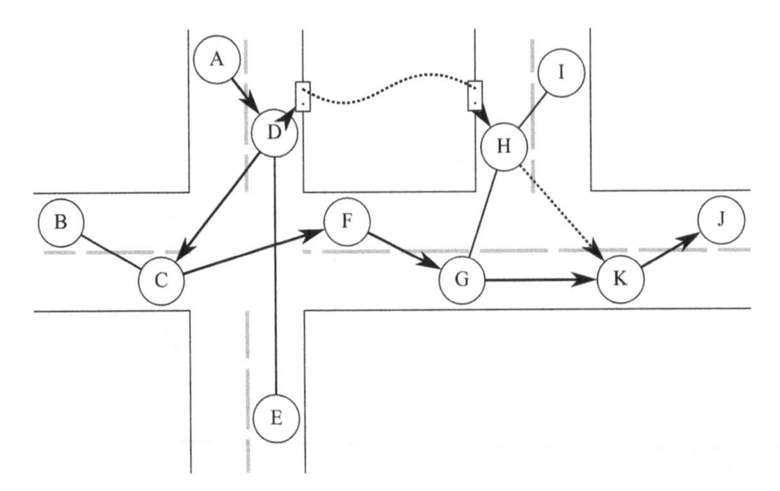

Figure 3.21 Routing in VANETs: the topology of the network is strongly constrained by the road network.

Figure 3.21 shows a typical example of what was originally considered to be a VANET. The vehicles on the road form a dynamic ad-hoc network to exchange information among the participating vehicles. In addition to the vehicles, RSUs that may connect the VANET to some backbone network or the Internet are shown. Obviously, the vehicles are located on streets and cannot occupy the space in between these streets. This changes one of the main assumptions behind the development of MANET routing protocols. Again, we highlight the communication path between node Ⓐ and node Ⓙ. As an alternative, a path via RSUs is shown. In contrast to the MANET case, the communication links are constrained by the road topology.

Three more aspects need to be mentioned in our discussion of using MANET concepts for establishing VANETs. First, the node density has always been assumed to be constant in MANETs. Even though protocols have to deal with different node densities, there is no need to handle fast changes. Second, mobility is much more predictable. Third, the connectivity is much more dynamic in vehicular networks, most prominently in urban environments, than in other ad-hoc networks. This is mainly due to quick changes of network connectivity due to buildings and other obstacles blocking the radio communication.

The effect of connectivity in urban scenarios has been discussed by Viriyasitavat *et al.* (2009), among others. We will check the applicability of MANET routing protocols in the following, referring to a study investigating the use of the DYMO routing protocol in vehicular networks (Sommer & Dressler 2007; Sommer *et al.* 2008). The results clearly show that generic ad-hoc routing will fail in vehicular environments due to the wide spread of possible network conditions and the high dynamics in the network topology.

Ad-hoc routing protocols

The field of ad-hoc routing protocols has become big and partly confusing. There is not only a huge number of protocols that have been developed in the last two decades but also a wide spread of different application scenarios for which these protocols have

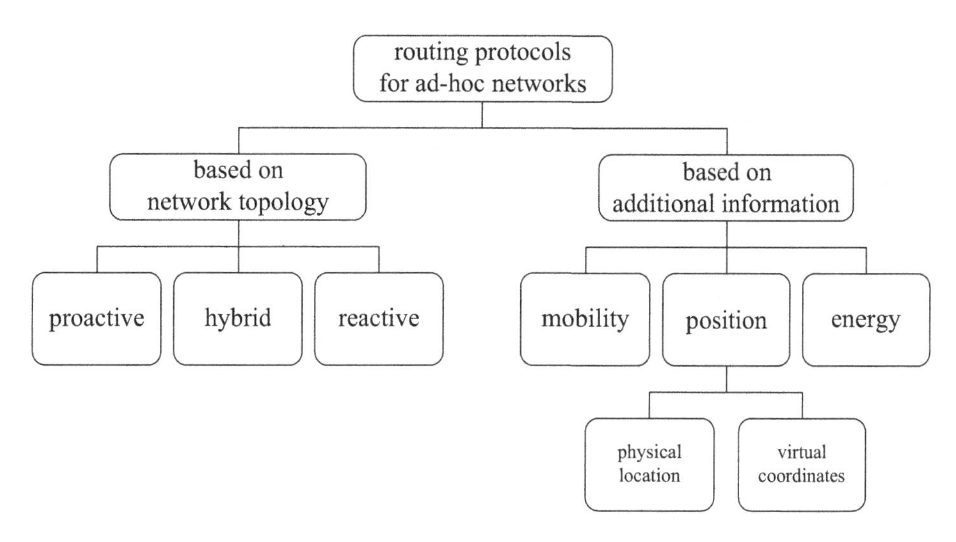

Figure 3.22 A taxonomy of ad-hoc routing protocols.

been optimized (Basagni *et al.* 2004). We will study selected protocols in greater detail in Section 5.1 on page 138. At this point, we aim to acquire a broad understanding of ad-hoc routing protocols, their different concepts, and their applicability in VANETs.

A general taxonomy of ad-hoc routing protocols is depicted in Figure 3.22 (Basagni *et al.* 2004; Dressler 2007). We can distinguish two main classes of routing protocols: those based on topology information (most commonly the k-hop neighborhood of nodes) and those based on utilizing additional information available to the nodes. A good overview of ad-hoc routing solutions as used for VANETs is provided by Lee *et al.* (2010a).

Network topology based on routing updates

The idea of topology-based routing concepts is to determine the network topology either in advance or on demand. The subclasses are therefore called *proactive* and *reactive* routing, respectively. These concepts, also including hybrid versions, constitute the basis for most of the generic MANET routing approaches. Well-known examples include the proactive routing protocols Destination Sequenced Distance Vector (DSDV) and Optimized Link State Routing Protocol (OLSR), as well as the reactive protocols Dynamic Source Routing (DSR), Ad-hoc On-demand Distance Vector (AODV), and DYMO (Basagni *et al.* 2004; Dressler 2007). We will study these examples in detail in Section 5.1 on page 138. The general idea of all proactive routing protocols is to determine the network topology before sending any data packets. Thus, as soon as a node system starts, it initiates its routing protocol instance and starts exchanging routing protocol messages with neighboring nodes. This is the same concept as what is used in the Internet. Depending on the degree of dynamics in the topology and the frequency of message exchanges among the nodes, the network might be quite busy just updating the network topology. Reactive routing protocols, instead, remain silent as long as no data exchange is taking place. If a data packet is about to be sent, the routing protocol instance starts looking for a suitable path to the destination node. Usually, this involves flooding a so-called route request through the entire network until the destination node, i.e., a path to that node, has been discovered. Thus, the overhead caused by the routing protocol is marginal as long as no information has to be sent. The route discovery itself can be comparatively expensive (in terms of routing protocol overhead).

Almost all of these protocols have been experimented with also in the scope of vehicular networks. This led to several improved versions. Yet, as reported, for example, by Sommer *et al.* (2008), even the most recent protocols such as DYMO are not able to handle the dynamics of vehicular networks and to scale with the needs of the highly varying node densities.

Location-specific resources-based routing

On the other hand, routing concepts have been investigated that are based on other, very specific characteristics and resources. We can distinguish several subclasses. First, since high node mobility is a weak spot of all classical MANET routing solutions, approaches have been developed to explicitly exploit high node mobility to forward messages according the *store–carry–forward* principle. One of the best-known examples is last-encounter routing (LER) (Sarafijanovic-Djukic & Grossglauser 2004, Grossglauser & Vetterli 2006). Depending on the mobility patterns of the participating nodes, this approach can be quite successful. Unfortunately, there is no published literature on LER in the context of a typical vehicular networking scenario. Second, energy can be used as the key resource that needs to be optimized. This has been studied in the context of wireless sensor networks (WSNs), but is not that relevant to VANETs. Third, geographic coordinates may replace the classical address- or ID-based routing (Mauve *et al.* 2001). This is very helpful in the context of VANETs. We will investigate geographic routing concepts later in this section.

Limiting factors

Without going into the details, the most limiting factor for using MANET routing in vehicular networks is *mobility*. There are multiple published studies on this issue. In particular, the typical mobility patterns of vehicles have been investigated and studied in terms of their impact on routing protocols (Camp *et al.* 2002, Bai & Helmy, 2004).

Since vehicles driving on road networks follow quite specific patterns, partly predefined by the road network itself, partly defined by more social behavioral rules affecting the actions of drivers, these patterns will have a strong influence on the resulting network topology and, therefore, on the practicability of ad-hoc routing concepts (Sommer & Dressler 2008). Knowledge of this has even led to the development of specific routing approaches that are aware of the mobility pattern of the vehicles involved (Hung *et al.* 2008).

Scalability is an issue of all MANET solutions, of course, including ad-hoc routing protocols. This is a result of the limited capacity of the wireless network. The groundbreaking work by Gupta & Kumar (2000) led to follow-up studies such as that by Hong *et al.* (2002), who investigated this theoretical upper bound in more detail for practical solutions. Scheuermann *et al.* (2009) investigated the scalability of information dissemination in VANETs. Using their model (see Section 3.4.3

on page 104 for details), it is possible to determine scalability criteria for granularity vs. dissemination distance of messages in multi-hop vehicular networks. Of course, the vehicle density and the aggregation techniques used vary with the range and amount of information transfered, yet the upper bound just gets slightly shifted.

Finally, *delay-bounded* operation is required by many vehicular networking applications. This concerns not only hard safety-critical operations but also any real-time transmissions that need to be completed within a fixed time interval. This problem has been investigated in part already, with selected solutions being presented in the literature (Skordylis & Trigoni 2008, Chigan & Li 2007). In general, it is simply impossible to guarantee certain performance measures such as the maximum latency for transmissions over multiple hops in a MANET or VANET.

Many of the limits discussed could be mitigated if one were able to add capacity. Actually, this is indeed possible. There are two ways to achieve this objective. First, increasing the capacity of IEEE 802.11 wireless communication is possible through cooperative coded retransmissions (Loyola *et al.* 2008). Furthermore, multi-channel operation is an option by adding additional spectrum. The selection of the communication channel becomes a research issue for itself that has been addressed already for MANETs (Kyasanur & Vaidya 2005, Bahl *et al.* 2004). First approaches also try to improve VANET operation by providing access to multiple channels for increased spatial diversity (Klingler *et al.* 2013).

3.3.4 Broadcasting

Since ad-hoc routing will not be feasible in many or even most of the IVC scenarios, alternatives that rely on the basic operation of the wireless communication between nodes have been studied. Essentially, each unicast between two neighboring nodes is technically a broadcast that can be overheard by all nodes that are within communication range. So, the simple idea is to make use of broadcast communication only and to consider re-broadcasting the information if necessary. In the extreme case, this results in network-wide flooding of information – obviously overloading the network if more than a few nodes are participating. Of course, some application-layer intelligence can be used to restrict broadcasts to certain areas.

The concept of broadcasting has become known as *beaconing* in the scope of vehicular networks. In the following, we will study the general concepts of beaconing as well as the ideas behind more sophisticated adaptive beaconing approaches.

Dedicated short-range communication

Before studying broadcasting techniques, we need to spend a moment on lower layers in the communication protocol stack, in particular discussing medium-access technologies. Broadcasting builds upon short-range radio communication, which we know quite well from the WiFi protocol suite. In particular, carrier-sense multiple-access (CSMA)-based medium access is used as defined for WiFi. All the little details of WiFi and the vehicular-networking-related variants will be studied in detail in Section 4.2 on page 118.

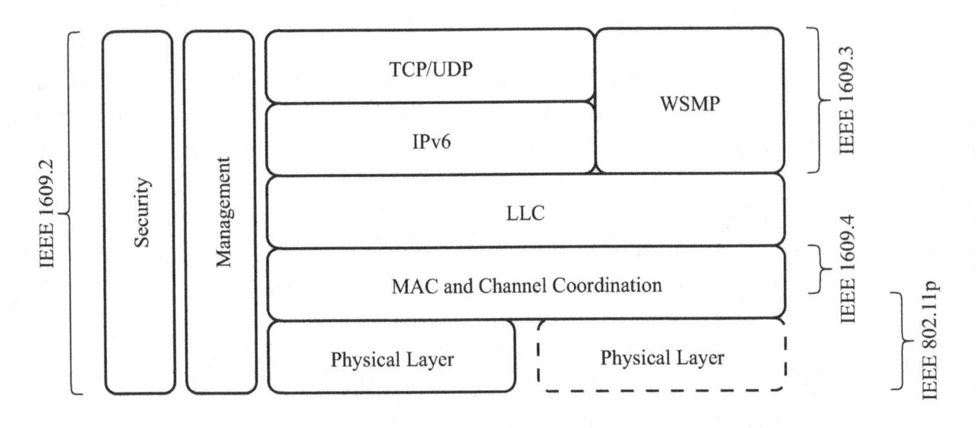

Figure 3.23 The IEEE 1609 WAVE protocol architecture.

Using WiFi

In fact, most of the aforementioned VANET routing approaches build upon WiFi. The IEEE 802.11 protocols provide all the functionality needed not only to allow infrastructure-based networking, i.e., using an access point (AP), but also for ad-hoc communication between arbitrary WiFi devices. The usability of WiFi (IEEE 802.11a/b/g) for vehicular networking has been investigated in several projects and experiments. One of the more detailed studies looked at what has been called a "drive-thru Internet" (Ott & Kutscher 2005). In this context, the authors performed many experiments driving a car at varying speeds along a street equipped with an AP. The main result was that communication is indeed possible, but the association procedure (i.e., the initial connection to a basic service set (BSS) provided by some AP or by another node (see Section 4.2.1 on page 119 for details)) takes too much time. Even when one is operating in ad-hoc mode, this association procedure needs to be performed in order to exchange messages. Besides this, there are other problems related to the use of WiFi in vehicular environments, including reliability issues, problems related to multi-path effects, and speed-dependent signal fading.

After identifying the root causes limiting the usability of WiFi, new protocol variants have been developed. Actually, this caused quite substantial investigations of different approaches to medium access in the vehicular networking domain. Eventually, the search for a more suitable approach resulted in the definition of inter-vehicle communication (IVC) communication protocol stacks, which have been standardized partly by the Institute of Electrical and Electronics Engineers (IEEE), the European Telecommunications Standards Institute (ETSI), and the Society of Automotive Engineers (SAE).

IEEE 802.11p and IEEE 1609 WAVE

The WAVE protocol stack is defined in several standards. An overview is depicted in Figure 3.23. IEEE 802.11p is usually used to describe the lower protocol layers, whereas WAVE encompasses the full protocol stack. WAVE is defined in the IEEE 1609 suite (IEEE 2006, Uzcátegui & Acosta-Marum 2009). From the networking perspective, the IEEE 1609.3 (IEEE 2010a) and IEEE 1609.4 (IEEE 2011) are most important, describing network services and channel management, respectively. Also included are security services in IEEE 1609.2 (IEEE 2013), which we will study in the context of security and privacy in Chapter 7.

The basis of the WAVE protocol stack is IEEE 802.11p (IEEE 2010b). IEEE 802.11p defines the PHY and the lower layer MAC.

The PHY is almost identical to IEEE 802.11a. IEEE 802.11p also operates in the 5-GHz band using orthogonal frequency-division multiplexing (OFDM). Compared to IEEE 802.11a, it ensures reduced inter-symbol interference to mitigate multi-path effects. This can be achieved by doubling all timing parameters and reducing the channel bandwidth to 10 MHz compared with 20 MHz for IEEE 802.11a. In this way, the needed reliability of transmissions can be guaranteed even for vehicles driving at more than 200 km/h. However, it also results in a reduced throughput of 3–27 Mbit/s instead of 6–54 Mbit/s. The communication range of IEEE 802.11p is up to 1000 m, given a maximum transmission power of 800 mW.

The MAC layer comes with extensions to IEEE 802.11 that include randomized MAC addresses, QoS support, and a new ad-hoc mode. In order to eliminate the rather long time for the initial association process, a new outside-the-context-of-a-BSS (OCB) mode defines the use of a *wildcard BSS identifier* in transmitted packets.

A high QoS is provided by integrating the IEEE 802.11e enhanced distributed channel access (EDCA) mechanism into the WAVE protocol definition. EDCA was developed as a replacement of the IEEE 802.11 distributed coordination function (DCF), with the main motivation being to provide support for high- and low-priority data flows. First and foremost, EDCA allows earlier access to the wireless channel by modifying first the interframe spacing, i.e., supporting shorter interframe spaces for high-priority messages, and secondly the contention window (CW), i.e., reducing the CW for those higher-priority packets. This is managed by classifying messages into one of four access categories (ACs) (AC0 to AC3). We will study the EDCA mechanism in detail in Section 4.2.1 on page 121.

WAVE channel management

The WAVE standard also defines channels and channel access in general. Following the high expectations of vehicular networking to provide road-safety solutions and to help prevent crashes and, most importantly, fatalities, the responsible regulatory bodies worldwide tried to allocate spectrum for IVC. This resulted in successful allocation of wireless spectrum in the USA, Europe, and Japan.

Seven 10-MHz channels have been dedicated to IVC in the USA and five in Europe, by the US Federal Communications Commission (FCC) and the Electronic Communications Committee (ECC), respectively. For these activities, WAVE uses a dedicated frequency range in the 5.9-GHz band, which is reserved exclusively for V2V and V2I communication. Even though it is strictly regulated, no license costs are involved. In Japan, only a single channel has been allocated, in the 700-MHz band.

The different channels have been pre-defined with different roles. Most importantly, we can distinguish a single control channel (CCH) and multiple service channels (SCHs), which can be used by IVC applications. The CCH is to be used for management and safety information using broadcasts. In particular, so-called WAVE service advertisements (WSAs) will be used on the CCH to announce the transmissions of following WAVE short messages (WSMs) on one of the SCHs. Single radio systems will obviously have to switch between the channels, periodically tuning to the common CCH and then to one of the SCHs depending on their interest in specific messages. The slots are synchronized using GPS as a reference clock. The resulting performance has been studied in several works (Wang *et al.* 2008, Eckhoff *et al.* 2012).

This operation has the advantage of being simple and allows one to work with a single radio system. On the negative side, it obviously introduces additional delays. Let's assume the following situation: A time-critical message to be sent over an SCH is created close to the end of the CCH period. The earliest time it can be advertised is in the next CCH, then it can be sent in the following SCH – if this will be allowed at all. The standard defines 50-ms slots for CCH and SCH; thus, there will be a delay of at least 100 ms until the message can be sent, not counting the uncertainty due to CSMA.

This can be changed if multi-radio systems are used. In this case, one radio can constantly tune to the CCH, being able to almost instantly send safety-critical warning messages. Other messages still have to be scheduled for transmission in the following time slot and, most importantly, on a specific SCH. This constitutes a complex scheduling problem that needs to be solved in a fully distributed manner (Klingler *et al.* 2013).

Beaconing

The concept of beaconing was introduced first in the scope of the aforementioned SOTIS idea (Wischhof *et al.* 2003, 2005). SOTIS was the first approach towards a fully decentralized TIS. The idea is that all participating vehicles periodically broadcast their local knowledge base to neighboring vehicles. These, in turn, incorporate the received information into their local knowledge bases and start broadcasting the resulting superset. The result is a store–carry–forward approach built on broadcasting, i.e., beaconing, as a basic building block.

Figure 3.24 illustrates how, in this way, knowledge will be gradually disseminated among participating cars, even if the network sporadically becomes disconnected. Even isolated clusters of cars will infrequently receive beacons transmitted, e.g., by vehicles driving in the opposite lane. Using the accumulated contents of their local knowledge bases, cars will be able to perform road-traffic analysis or warn the user about oncoming hazards.

Each vehicle maintains a local table containing information about currently known traffic problems. In each communication, the table is transmitted as a whole to the neighboring vehicles. That, first of all, means that the table must fit a single broadcast packet's payload. SOTIS suggests data aggregation to handle the table size. Aggregation is performed according to the distance to an event. The rule of thumb is to aggregate more aggressively the farther away events are located, i.e., only details of major traffic jams from remote locations will be kept, but every item of fine-grained data about local micro-jams will be retained.

But let us discuss the communication aspects in more detail (Wischhof *et al.* 2005). The approach is to *periodically* broadcast a message to all neighboring vehicles.

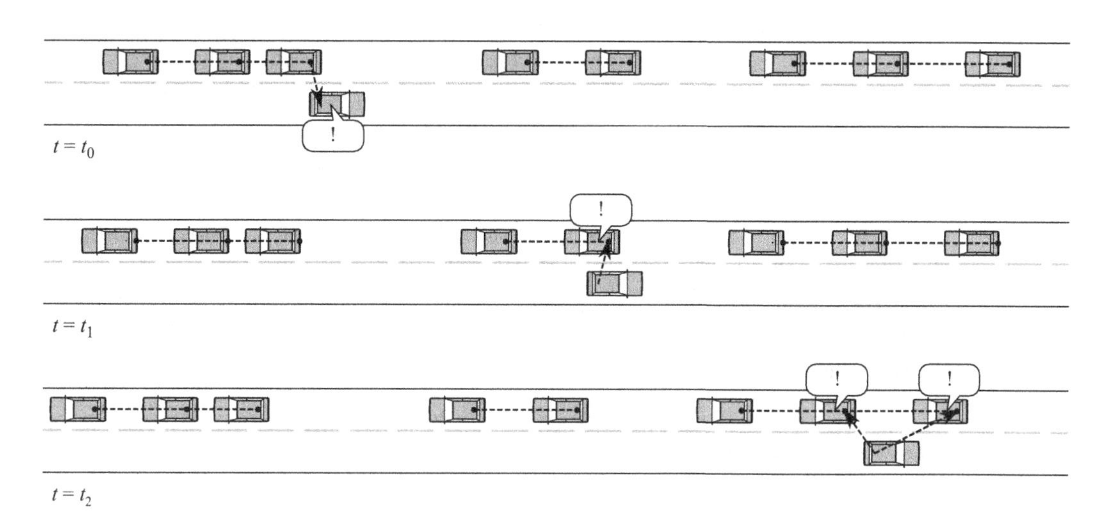

Figure 3.24 Message dissemination via beaconing, as envisioned in the SOTIS architecture.

The periodicity has been specified to be on the order of 1–10 Hz. Now, we have to consider the two extreme cases.

- 1 Hz: At this frequency, vehicles in a sparse environment might pass each other's communication range without being able to receive a beacon. This will be especially important with low penetration rates at early deployment stages.
- 10 Hz: Assuming higher beacon frequencies, we have to look at quite dense scenarios such as occur on a major freeway (five to ten lanes during rush hour). In such a scenario, we will easily encounter more than a thousand cars within communication range. Broadcasting at 10 Hz will fully overload the wireless channel.

The beaconing concept has also been incorporated into early standardization efforts for IVC by standardization bodies such as the European Telecommunications Standards Institute (ETSI) and the Society of Automotive Engineers (SAE) in the European Union and in the USA, respectively. The resulting protocols known as cooperative awareness message (CAM) (ETSI 2010a) and basic safety message (BSM) (SAE 2011) are based on the IEEE 802.11p standard. The idea of CAM and BSM is to periodically broadcast so-called awareness messages at frequencies in the range 1–10 Hz. This was later extended to the range of 1–40 Hz. The protocols need to be configured to a fixed beaconing frequency, i.e., there is no mechanism available that dynamically adapts the frequency according to the current situation. Obviously, the same restrictions apply for sparse and dense scenarios. More details about CAM messages can be found in Section 5.3.2 on page 172.

This simple broadcasting has been extended to two-hop-broadcasting and even to multi-hop broadcasting solutions (Korkmaz *et al.* 2006; Amoroso *et al.* 2011). The main idea is to extend the communication range while keeping the use of the channel as low as possible. We will study selected *static beaconing* approaches in greater detail in Section 5.3 on page 167.

Before discussing more recent beaconing concepts that do indeed adapt the beaconing interval, we have to briefly mention one of the most critical problems in wireless networks using broadcast-based communication, namely the broadcast storm problem (Ni *et al.* 1999). Even if low beaconing frequencies are being used, re-broadcasting by all nodes that received the message might instantly cause their neighbors to re-broadcast again. As one can easily imagine, this may quickly lead to a substantial number of messages on the wireless communication channel that exhausts all the available resources. We study the problem in detail in Section 3.4.2 on page 102.

Adaptive beaconing

Since the static beaconing approach clearly has several weaknesses, the vehicular networking research community focused on making beaconing situation-aware. This resulted in a set of proposals for *adaptive beaconing* that aimed at mitigating the problems of not reaching all vehicles in sparse scenarios and overloading the wireless channel in dense scenarios, respectively. The common idea of all these approaches is to make the beaconing adaptive either in the time domain, i.e., changing the beaconing interval, or in space, i.e., exploiting spatial diversity by changing the transmit power of the radios.

Adaptive traffic beacon

One of the first approaches is the adaptive traffic beacon (ATB) protocol (Sommer *et al.* 2010b, 2011c) (cf. Section 5.4.1 on page 175 for all protocol details). The main objective of ATB is to exchange information in knowledge bases by sending beacons as frequently as possible, but to maintain a congestion-free wireless channel. ATB achieves this by employing two different metrics, the *channel quality* C and the *message utility* P, to calculate the beacon interval I at which to disseminate messages.

The conceptual architecture of ATB is outlined in Figure 3.25. The system was designed for direct V2V communication using one-hop broadcasts. Yet, ATB also allows one to incorporate infrastructure elements such as RSUs to overcome low-density or low-penetration-rate problems. By adapting to channel quality and message utility, ATB leaves the channel uncongested and allows communication opportunities for higher-priority messages or other protocols.

In the following, we briefly introduce the different metrics ATB uses to assess channel quality and message utility. Each metric is derived by considering one particular measure of either channel quality or message utility and calculating its value relative to a fixed maximum value.

The channel quality C is estimated by means of three metrics, which are indicative of network conditions in the past, present, and future, respectively. Looking into the past, ATB counts the number of collisions on the channel, actually estimating the degree of overload of the wireless channel. The current channel quality can be estimated by assessing the signal-to-noise ratio (SNR) for the most recently received

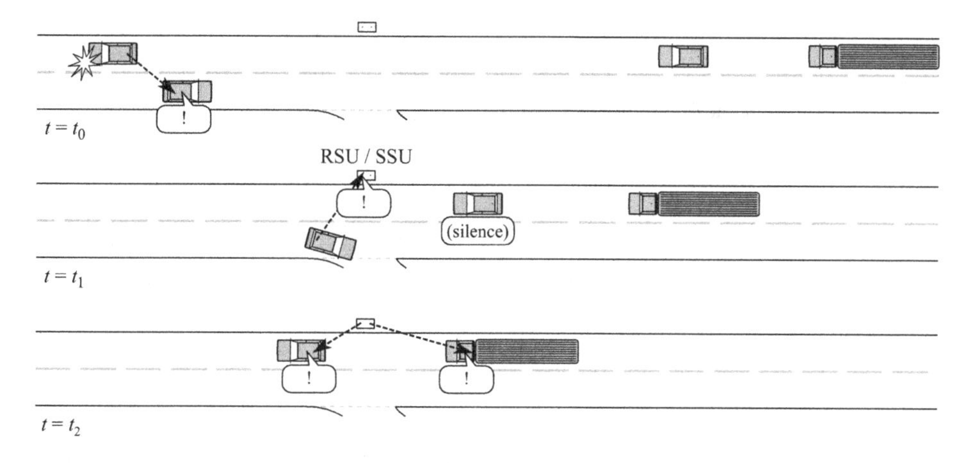

Figure 3.25 The conceptual architecture of ATB.

messages. If a neighboring car is very close, this indicator allows one to prefer short-range transmissions that will likely be more relevant and have a higher success probability. This is in line with findings published for example by Tonguz *et al.* (2010). Finally, the load on the channel in the next time interval can be estimated by looking at the number of neighbors. Since the channel quality metric C in turn depends on the value of I that was chosen by nearby vehicles, ATB exhibits properties of a *self-organizing system* (Dressler 2007): On a macroscopic scale, vehicles participating in the VANET will independently arrive at beacon intervals that enable them to use the shared channel commensurately with their own and other nodes' needs.

The message utility P is derived from two metrics. First, a node accounts for the distance of a vehicle from an event, which is the most direct indication of message utility. Second, it accounts for message age, thus allowing newer information to spread faster.

ATB adjusts I such that it becomes minimal only for the highest message utility and the best channel quality; in all other cases, channel use is reduced drastically, allowing uninterrupted use of the channel by other applications:

$$I = I_{\min} + \left(I_{\max} - I_{\min}\right) \times \left(w_I C^2 + (1 - w_I) P^2\right) \tag{3.1}$$

The relative impact of both parameters w_I is designed to be configurable, e.g., in order to calibrate ATB for different MAC protocols. Sommer *et al.* (2010b) reported that an empirically derived value of $w_I = 0.75$ worked very efficiently in most scenarios.

The ATB protocol is assumed to behave in a fair way for all participating nodes. Yet, this is not guaranteed by protocol design. Follow-up works extended ATB to support fair distribution of the available channel capacity to all nodes (Schwartz *et al.* 2012b, 2014).

Adaptive transmission power

Besides the beaconing interval, a prime candidate for adaptivity in beaconing is the dynamic adjustment of transmit power. In the scope of vehicular networks this concept was popularized by Artimy *et al.* (2005), who proposed adjusting the transmit power of nodes as a function of locally measured road traffic density, so as to keep the network connected without "wasting" channel capacity. A statistical approach to adaptive transmit power control has been investigated by Egea-Lopez *et al.* (2013), who showed that congestion control in general can successfully be addressed using power management only.

Torrent-Moreno *et al.* (2006) use transmit power control to minimize packet collisions while still maintaining a fair allocation of channel capacity to participating nodes. Their approach, distributed fair power adjustment for vehicular networks (D-FPAV), is designed to work in tandem with others, e.g., adaptive beacon intervals, allocating transmit power in a fully distributed fashion so that free channel capacity can be reserved for event-driven messages. For this, they define a threshold of maximum tolerable channel load caused by beaconing (called *MBL*) and try to keep the beaconing load below that. More formally, they determine a power assignment to vehicles so that the minimum of their transmit powers is maximized and at the same time the channel load caused by beaconing remains under the defined threshold. In order to obtain information on the position of nodes in their vicinity, nodes piggyback n-hop information about their neighbors on periodic beacons (with n large enough to gather information about all nodes in the node's maximum transmission range). The gathered position information about all potential neighbors is then used by each node to locally forecast how chosen transmit powers of its neighbors might impact channel load and, thus, which transmit power the node itself should best use. It could be shown that the D-FPAV algorithm manages to reserve channel capacity for event-driven messages even in highly mobile and high-density networks, successfully trading a higher probability of receiving event-driven messages for a lower probability of receiving beacons.

An interesting approach to transmit power scheduling was investigated by Kloiber *et al.* (2012). It was shown that the use of a random transmit power substantially reduces the load on the wireless channel but, compared with perfectly scheduled options, with less impact on the average transmission latency. This idea opens up new possibilities for advanced beaconing systems. Yet, as shown by Tielert *et al.* (2013), the average latency does not reflect the quality of the algorithms for safety-critical applications.

Decentralized congestion control

The idea of adaptive beaconing – its advantages having been confirmed in several projects (Schmidt *et al.* 2010) – was finally picked up by the ETSI in its standardization process leading towards cooperative awareness protocols. The resulting framework is called decentralized congestion control (DCC) (ETSI 2011). DCC is a rather complex compilation of different concepts, in the best case complementing each other. Two of them are relevant to our discussion: transmit rate control (TRC) and transmit power control (TPC).

TRC is based on the very successful approaches for adaptive beaconing (Sommer *et al.* 2011c; Schmidt *et al.* 2010; Schwartz *et al.* 2014). Since TRC controls the rate of the beacons, it essentially controls the load on the wireless channel. The main metric used by TRC is the so-called busy ratio b_r of the channel. This metric can be measured by each node individually by overhearing the channel and accumulating the time the channel was busy, i.e., other nodes were transmitting in these time slots. The beaconing frequency is then set using a simple state machine that switches between a minimum frequency (maximum interval I_{max}) if a certain threshold of b_r has been exceeded. A second threshold is used to differentiate between a maximum frequency (minimum interval I_{min}) and some medium value:

$$I_{min} \overset{\text{channel congested}}{<} I \overset{\text{channel free}}{<} I_{max} \tag{3.2}$$

TPC, on the other hand, focuses on gaining spatial diversity by configuring the transmit power for each transmission. Initially, this was considered the main approach for making beaconing for cooperative awareness applications feasible in very crowded environments. TRC and TPC can also be used in combination. It turned out, however, that transmit power control is not as successful as had initially been expected (Tielert *et al.* 2013). The main weakness is that power control may quickly result in the need to re-transmit a message in order to deliver the data to a specific destination. For example, consider two vehicles approaching an intersection. If a building is blocking the line of sight between the two vehicles, it might be possible to establish a connection using the highest transmission power, but the attempt will most likely fail for reduced-power transmissions. The additional re-transmission simply takes time and makes it hard to achieve the required real-time behavior.

Still, transmit power control techniques make a lot of sense for specific applications. One example is the control and management of platoons of multiple automatically controlled cars, i.e., so-called road trains (Segata *et al.* 2012; Segata 2013). Here, the communication between a vehicle and its successor or predecessor can be very well controlled within its transmission range.

Dynamic beaconing

The most problematic aspect of TRC as it has been standardized is that it reacts rather slowly to fast changes of the network topology. Hence, more reactive protocols have been investigated. So, the question is whether and, if so, why network topologies change that quickly on our streets even though there is a fixed upper speed limit.

Figure 3.26 outlines one of the scenarios in which the topology changes extremely quickly. A single car is driving on a secondary street towards a major road. Initially, it has (effectively) no neighbors at all and can send beacons at the highest possible rate. On entering the intersection, it spontaneously comes within communication range of hundreds of other vehicles. The currently selected beaconing rate therefore overloads the wireless channel at this new position.

Dynamic beaconing (DynB), which also relies on the busy ratio b_r on the wireless channel but uses a more reactive scheme for calculating an adequate beaconing interval, addresses this issue. Essentially, DynB tries to always maintain a desired busy ratio

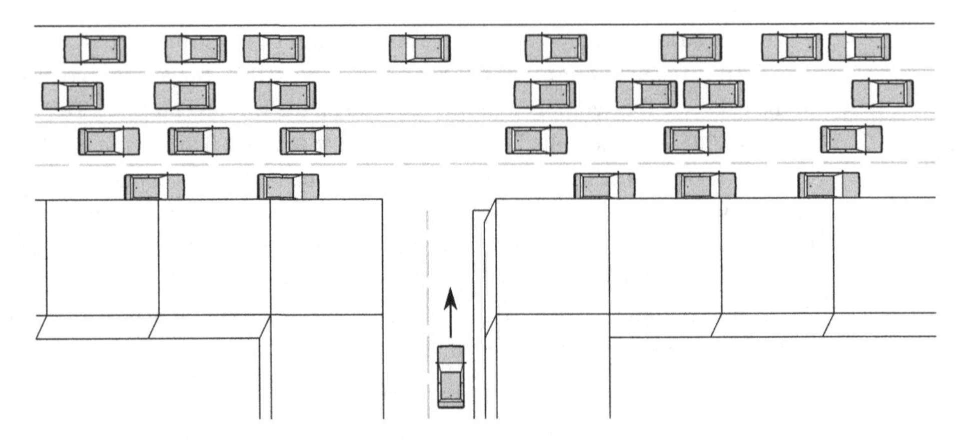

Figure 3.26 An example of a rapid topology change.

b_{desired} by adapting the beacon interval using the current busy ratio and the estimated number of neighbors as key parameters.

We only touched on some of the internal aspects of adaptive beaconing concepts in our discussion of selected approaches. We study these ideas in depth in Section 5.4 on page 174.

Infrastructure support

All the beaconing solutions discussed will deliver the expected performance only if a certain penetration rate of vehicles equipped with the IVC communication devices has been achieved. This, of course, also holds for the ad-hoc routing concepts. But what penetration rate is to be considered an absolute minimum? For traffic information applications such as SOTIS, even quite low penetration rates on the order of 10% have been shown to be sufficient. For other solutions, the minimum needs to be much higher.

This initial deployment problem could be substantially alleviated if additional infrastructure were available and if the protocols were to fully make use of this infrastructure. In our discussion, we have already mentioned the concept of roadside units (RSUs). RSUs can be placed along major freeways or at dedicated positions in urban environments. Typically, that would be intersections, since the RSUs would have perfect line-of-sight communication to cars on all the intersecting streets. At these positions, the RSUs could even be integrated with the traffic lights to integrate these into the vehicular network.

The optimal positioning and placement of the RSUs is a challenging problem in itself (Lochert *et al.* 2008; Aslam *et al.* 2012; Barrachina *et al.* 2012). Their placement is essentially determined by the costs involved (i.e., the fewer RSUs the better) but one must also consider the performance gain achieved. This gain is not linear with the number of RSUs because some positions are more relevant or provide better radio coverage. Thus, coverage is usually used as the optimization criterion (Salvo *et al.* 2012).

RSUs are usually considered to be connected to some backbone network. In practice, this connection is the major cost factor during deployment. Lightweight systems without such a backbone connection would therefore be preferred. In the literature, such systems became known as stationary support units (SSUs) (Lochert *et al.* 2008; Sommer *et al.* 2011c).

Finally, one might reduce the infrastructure support to a fully self-organizing and dynamic approach when considering only vehicles equipped with radio communication devices. As discussed before, parked cars can be used to close the gap and provide relay functionality at critical positions such as intersections (Eckhoff *et al.* 2011c; Sommer *et al.* 2014a; Liu *et al.* 2011; Malandrino *et al.* 2012).

Integration of store–carry–forward

If no infrastructure is available and the penetration rate is still too low, the concept of delay/disruption-tolerant networks (DTNs) can be used (Fall 2003). The idea is to explicitly assume that there will be communication gaps between the nodes in the network and to rely on the *store–carry–forward* technique for transmitting a message to a certain destination. The most critical point is how nodes have been assumed to move when designing the DTN approach. Most of the DTN concepts have been investigated for human motion (Hui *et al.* 2005). Compared with this, the motion of vehicles is rather easy to predict, especially considering the given road network topology.

For communication on freeways, one of the best-known approaches is distributed vehicular broadcast (DV-CAST) (Tonguz *et al.* 2010). Since the motion of vehicles can easily be predicted in this scenario, one simply has to distinguish between vehicles driving in the same direction and those driving in the opposite direction. The idea of DV-CAST is depicted in Figure 3.27. If the network is connected, i.e., the density of vehicles equipped with the IVC technology used is high enough, DV-CAST uses the already-discussed broadcast-based communication (see Figure 3.27(a)). If, however, communication gaps can be detected, DV-CAST switches to the *store–carry–forward* operation mode. Now, messages are no longer disseminated using multi-hop broadcasts, but vehicles driving in the opposite direction are exploited as information ferries (see Figure 3.27(b)). The remaining problem is to decide which operation mode is to be used in each time step.

Internally, DV-CAST manages local topology updates by means of periodic hello messages. These hellos contain exact GPS information of the vehicle together with the current heading. Even though these messages are called hellos in DV-CAST, we can easily assume that we can rely on the CAM or BSM messages discussed before, which already include the needed information.

All participating vehicles maintain three neighbor tables:

- all vehicles driving in the same direction, driving ahead;
- all vehicles in the same direction, driving behind;
- all vehicles on the opposite lane.

Now, messages that are to be sent contain the source position and the intended destination region of interest (ROI). Using the algorithm outlined in Figure 3.28, DV-CAST

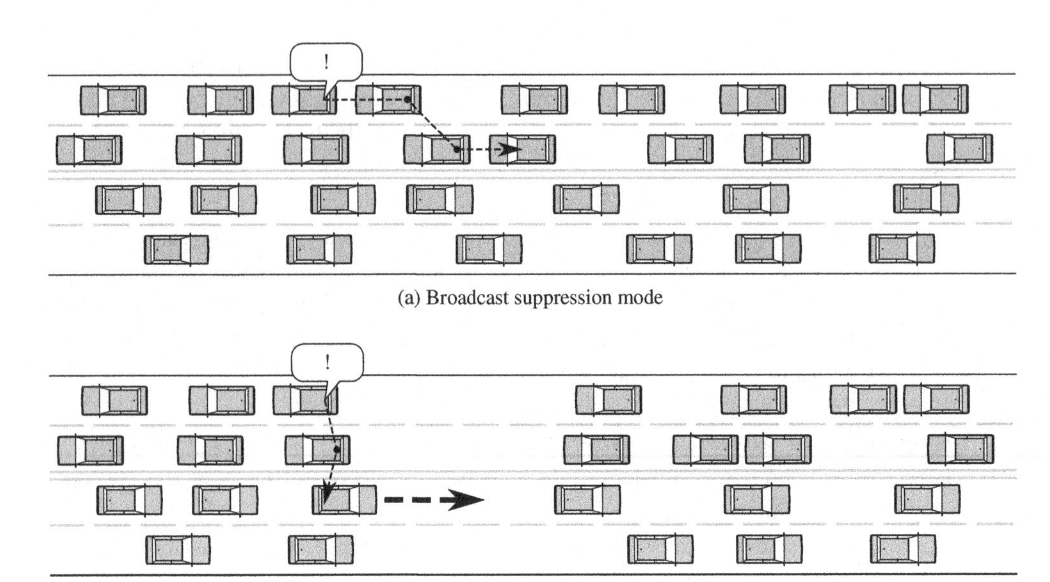

Figure 3.27 The mode of operation of DV-CAST.

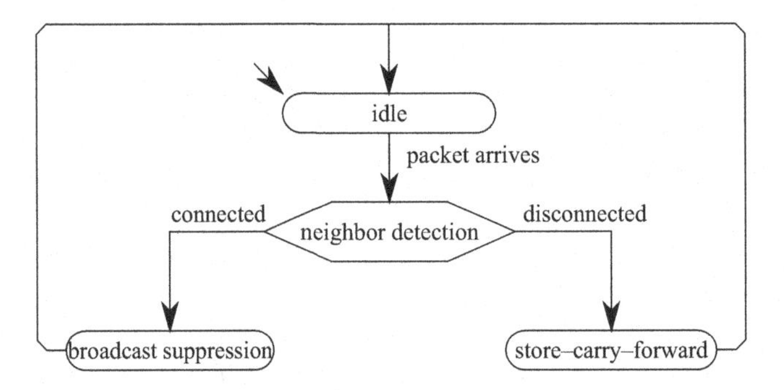

Figure 3.28 The conceptual work flow of DV-CAST.

switches between the two operation modes *broadcast suppression*, i.e., beaconing, and *store–carry–forward* depending on whether the neighbor tables contain entries in the designated direction. We discuss DV-CAST in detail in Section 5.7.1 on page 219.

In urban environments, the mobility is less restricted than it is on freeways. Therefore, more intelligent predictions are needed. One example is the urban vehicular broadcast (UV-CAST) algorithm that was designed explicitly taking these considerations into account (Viriyasitavat *et al.* 2010).

3.3.5 Geographic routing

As we have discussed, ID- or address-based routing using MANET protocols is not suitable for VANETs. Instead of using something like IP addresses, vehicles can rely on another type of address information, namely geographic coordinates. Geographic routing (or geo-routing) has been investigated by our research community in the context

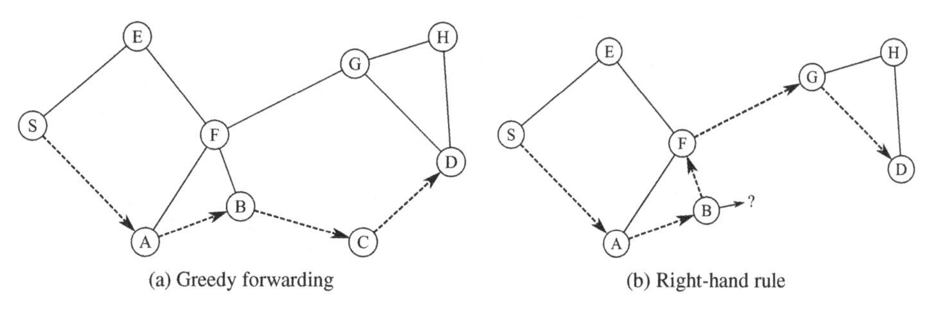

Figure 3.29 The basic operations of geo-routing.

of MANETs. Depending on the application, this approach has some clear advantages, there being only one major concern: the availability of accurate geographic position information. In the vehicular domain, however, satellite navigation systems are such a basic instrument that this information can be assumed to be available with very high accuracy. Thus, position-based routing in vehicular networks does indeed seem to be a very promising option (Mauve *et al.* 2001). We will discuss the different approaches to geographic routing and *geocasting* in detail in Sections 5.2.1 and 5.5 on page 153 and 196, respectively.

Greedy routing

The concept of geographic routing is quite simple: Messages are using a geographic location as a destination address and, in each forwarding step, routing is performed not according to some routing table entries but rather on the basis of neighborhood information. This neighborhood data is used to identify the node which is closest to the destination in terms of geographic distance. This approach is called *greedy forwarding*.

Figure 3.29 outlines the basic operation of geo-routing. As long as possible, greedy forwarding is used to get to the destination using as few hops as possible (see Figure 3.29(a)). Yet, greedy forwarding is very sensitive to the so-called "dead-end" problem. At some point, there might be no further node in the right direction to achieve progress towards the destination. The simplest solution to overcome these situations is to change the forwarding technique to what is called the "right-hand rule" (see Figure 3.29(b)), which is used to get out of an unknown labyrinth. A node that detects that no further progress to the destination can be achieved relays the packet backwards but using the "rightmost" node in that direction. This procedure continues until one of the nodes involved can switch to greedy forwarding again.

One of the first approaches implementing geo-routing is greedy perimeter stateless routing (GPSR) (Karp & Kung 2000). It uses greedy forwarding for as long as possible. If no progress is possible, it switches to "face" routing, which is similar to but more efficient than the right-hand rule. A face is the largest possible region of the plane that is not cut by any edge of the underlying graph. It can be exterior or interior. During face routing, each packet is sent around the face using the right-hand rule. This approach requires a planar graph, which needs to be established beforehand using topology control techniques.

Geo-routing has been used and extended by many approaches. Practical issues have been discussed in depth by Kim *et al.* (2005).

Virtual coordinates

The problem of dead ends and the possible inefficiency of face routing to overcome this problem have been investigated in a number of approaches. The main idea of most solutions is a hierarchical solution that can be further extended to completely switch to virtual coordinates on the overlay by performing some coordinate transformation (Flury *et al.* 2009; Tsai *et al.* 2009). This transformation allows one to completely overcome the dead-end problem at the cost of quite substantial efforts for topology management.

Instead of starting with geographic coordinates and performing a centralized coordinate transformation, virtual coordinate solutions are no longer based on geographic positions. One of the first virtual-coordinate-based protocols was geographic routing without location information (GRWLI) (Rao *et al.* 2003). Instead of using real node locations, it constructs an n-dimensional virtual coordinate system, which is based on finding the perimeter nodes and their locations. Then, a relaxation algorithm is used to find the virtual location of all nodes. However, the drawback of having many dimensions resulting from a large n is that the formation of virtual coordinates requires a long time to converge (Liu & Abu-Ghazaleh 2006).

Routing protocols like virtual ring routing (VRR) and virtual coordinate protocol (VCP) are more efficient and provide another benefit, namely an integrated DHT. DHTs have been used for a long time in Internet-based peer-to-peer (P2P) protocols for maintaining, updating, and downloading content in a fully distributed manner. If it is integrated with a (virtual-coordinate-based) routing protocol, the system provides not only routing functionality but also the ability to store and retrieve arbitrary information.

VRR (Caesar *et al.* 2006) is a routing protocol inspired by overlay DHTs. It uses a unique, location-independent key to identify nodes. VRR organizes the nodes into a virtual ring in order of increasing identifiers. For routing purposes, each node maintains a set of cardinality r of virtual neighbors that are nearest to their node identifier in the virtual ring. Each node also maintains a physical neighbor set with the identifiers of nodes it can communicate with directly. The forwarding algorithm used by VRR is quite simple. VRR picks the node with the identifier closest to the destination from the routing table and forwards the message towards that node. The problem of such protocols is that the adjacent nodes in the virtual ring can be far away in the real network. As a result, forwarding to the nearest node can result in a very long path.

In contrast, VCP (Awad *et al.* 2008, 2011) integrates underlay routing and overlay DHT management. This is accomplished by placing all nodes on a virtual cord, which is closely correlated with the network topology. As a side effect, the cord can also be used to associate data items with nodes. A hash function is used to create values in a predefined range $[S, E]$, and each node in the network maintains a part of the entire range. The routing mechanism relies on two concepts. First, the virtual cord can be used as a path to each destination in the network. Additionally, locally available neighborhood information is exploited for greedy routing towards the destination. The principle is outlined in Figure 3.30. In our example, node Ⓒ with the virtual identifier 0.25 sends

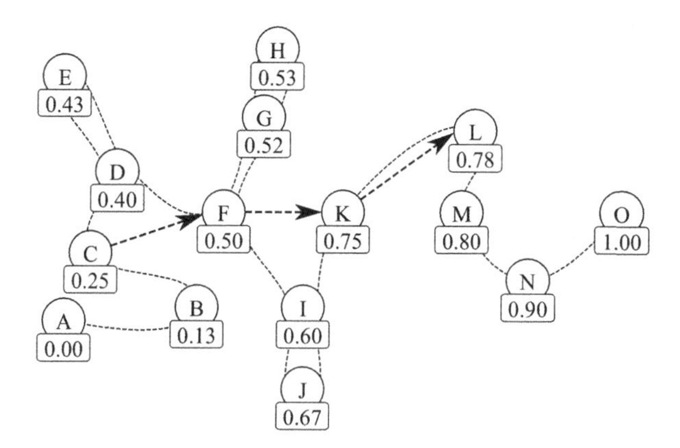

Figure 3.30 Greedy forwarding using the VCP coordinates.

a message to node Ⓛ, which has identifier 0.78. Forwarding is performed using the greedy approach. Since the virtual identifiers have been created in a smart way, the protocol guarantees dead-end-free forwarding. We will discuss the VCP protocol in detail in Section 5.2.2 on page 157.

Geo-assisted forwarding

Even if geographic routing or the use of virtual coordinates is not an option, so-called geo-assisted forwarding can be used. In this case, geographic coordinates are not used as the prime metric for taking routing decisions, but they are used to optimize message forwarding.

One of the best-known examples in the vehicular networking domain is topology-assisted geo-opportunistic routing (TO-GO) (Lee *et al.* 2009, 2010b) (we discuss this approach in detail in Section 5.5 on page 196). The idea here is to make vehicles able to route packets "around a corner", if required. Especially in urban environments, the line-of-sight communication is frequently blocked by large buildings. Yet, routing using a relay node that is located exactly at the intersection will be successful in delivering the message to all of the vehicles involved. However, mandating that messages are never forwarded further than to the next junction will limit progress. Thus, it would be beneficial to predict whether a message that is forwarded toward a junction will have to "turn a corner" or continue straight down the road. TO-GO realized this idea by maintaining two-hop neighborhood information.

Nodes periodically send hello beacons like in the CAM approach. In addition to the awareness information, each beacon contains the number of neighbors of a node, the neighbors' IDs summarized by a Bloom filter (Bloom 1970), and the IDs of the neighbors furthest down the road. In this way, all nodes learn about all two-hop neighbors.

Message forwarding using TO-GO is depicted in Figure 3.31. Depending on whether a message from node Ⓢ should go to node Ⓒ or node Ⓓ, TO-GO chooses one of two behaviors. If the message should reach node Ⓒ, TO-GO forwards the message no further than node Ⓑ to exploit its position on the junction – instead of greedily forwarding the message to node Ⓔ. The exact opposite applies if the message is addressed

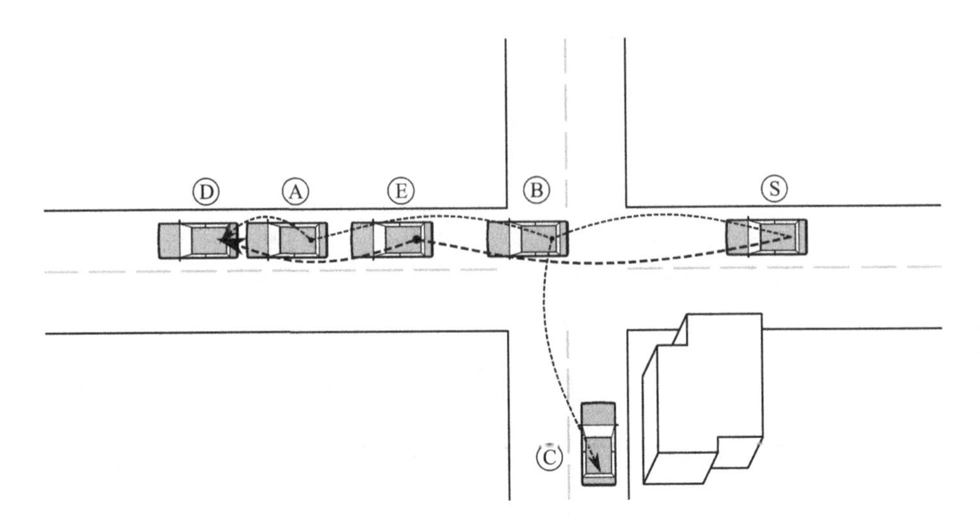

Figure 3.31 Geo-assisted forwarding using TO-GO.

to node Ⓓ. Here, it can choose to greedily forward the message to node Ⓔ without "stopping" at the intersection.

Geo-assisted forwarding has been subjected to standardization as well. The resulting approach has been named GeoNetworking (ETSI 2010b). Essentially, it works similarly to geographic routing but also adds broadcast-suppression techniques (see Section 3.4.2 on page 102). We discuss selected geo-assisted routing concepts in more detail in Section 5.5 on page 196.

3.4 Fundamental limits

After introducing vehicular networking applications and IVC protocols, we must also talk about the limits and scalability of IVC. As we have seen, there are very different approaches to IVC, each with its own constraints. We also talk briefly about business models, too. This is still one of the most unclear but also most important questions.

Whatever technology will eventually be used and deployed at scale, there will be initial bootstrapping problems until a certain minimum penetration rate has been reached. Also, given that each technology has its own advantages and drawbacks, concepts for using multiple technologies complementing each other are needed.

In this section, we discuss fundamental scalability problems of IVC and we also outline how heterogeneous networking approaches can be used to overcome some of the issues mentioned. Also, we will discuss the *broadcast storm problem* and how it can be mitigated, as well as fundamental scalability limits of VANETs.

3.4.1 Towards heterogeneous networks

At the beginning of this chapter, we discussed the demands of vehicular networking applications (Section 3.2.1 on page 56). At this point, we need to look back at this

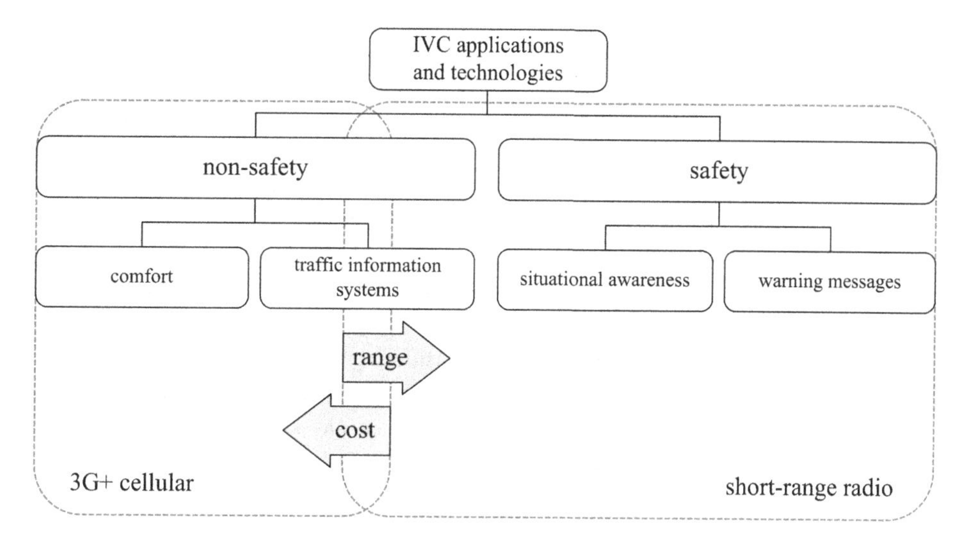

Figure 3.32 A taxonomy of IVC applications, with candidate technologies.

discussion and the provided taxonomy differentiating vehicular networking applications into safety and non-safety (see Figure 3.8). In particular, with our knowledge about the different concepts and technologies to realize IVC, we can investigate which of these technologies might be most appropriate for which applications.

We extend our previous figure to also show candidate communication technologies in Figure 3.32. Mainly 3G or 4G networks will help to realize entertainment applications such as multimedia streaming. Ad-hoc networks simply do not provide the needed degree of reliability and the required data rates. On the other hand, warning messages that need to be delivered in real time within at most 50–100 ms can, at the moment, be supported only using short-range radio broadcast. The big question is about the respective limits for 3G or 4G networks and short-range radio broadcasting.

Let's discuss the two most controversial properties: cost and range. Cost is the argument invoked by those preferring short-range radio broadcast solutions. Since there is as yet no clear business model regarding how 3G or 4G solutions can be realized for really large-scale TISs and even for safety applications, it is frequently assumed that only IEEE 802.11p-like solutions will have a chance in the market. On the other hand, MNOs have a strong incentive to get into this new market. The frequencies for 3G and 4G networks have been extremely expensive, and new income from vehicular networks is considered one option as a way to break even. At the moment, substantial work towards fully distributed TIS applications based on cellular networks and in combination with IEEE 802.11p for neighbor detection has been done in the PeerTIS context (Rybicki *et al.* 2009) (see Section 5.7.3 on page 225).

Range, on the other hand, is the most limiting factor for ad-hoc networking solutions. It is unlikely that distances beyond a few kilometers can be bridged using ad-hoc solutions only. Infrastructure elements like RSUs extend the range but establish another costs question. So, it remains unclear which of the technologies will be able to replace the other, if at all.

A future trend of vehicular networks is the move away from focusing on just a single technology and towards designing systems that can make use of multiple different technologies, creating *heterogeneous vehicular networks* (Tung *et al.* 2013). On looking into the literature, however, one sees that the underlying assumptions, concepts, and even goals of such approaches are very fuzzy. In an effort to move this research area forward by clarifying the foundations, identifying commonalities and differences of existing approaches, and outlining future research directions, a working group was formed in the context of an international seminar of experts at Schloss Dagstuhl to tackle these questions (Dressler *et al.* 2014).

One of the key motivations for considering such heterogeneous vehicular networks is the widespread availability of multiple technologies – both on today's portable devices like smart phones and in modern cars' sat nav systems or multimedia units. While cellular networks such as LTE will be a big helper during any initial roll-out of short-range communication technology, cellular networks will, in the medium term, not be able to offer sufficient network capacity without a drastic increase in deployment density and/or price (Vinel 2012; Rakouth *et al.* 2012). They might, in the long term, even be unable to offer sufficient capacity altogether.

Heterogeneous vehicular networking is further motivated by the fact that each of the currently available wireless technologies offers unique benefits, but also unique drawbacks. It was argued that the reasons to have WiFi lie in the downloading of added-value content and in the creation of a truly integrated environment, which would not be limited to cars as the only road users: Indeed, WiFi would foster the integration of bicycles and pedestrians into the network. Further, because of its tailored physical layer, dedicated channel(s), and tight locality, IEEE 802.11p can offer unique benefits in safety- and cooperation-awareness applications, due to their tight latency requirements. At the other end of the spectrum, cellular technologies are widely available, and designed for delivering large amounts of data over arbitrary distances. On the down side, they could face further hurdles when multicasting or local broadcasting is a strong requirement. Indeed, the lack of specific multicast support even in current 4G networks, coupled with problems of inter-MNO communication, is a critical limitation (Sommer *et al.* 2010d).

Casetti *et al.* (2013) identified two basic, opposing trends in heterogeneous vehicular networking that can be classified as follows.

- One pushes for a generalized network stack that abstracts away from lower layers to decouple applications from the technology employed, aiming to provide *data-offloading* services, or an *always-best-connected* experience to upper layers.
- One follows a *best-of-both-worlds* approach, exposing information and control of lower layers to applications, enabling them to selectively use the best-fitting technology for a particular task.

3.4.2 The broadcast storm problem

We now want to have a closer look at one of the most critical problems in wireless networks using broadcast-based communication, the broadcast storm problem (Ni *et al.*

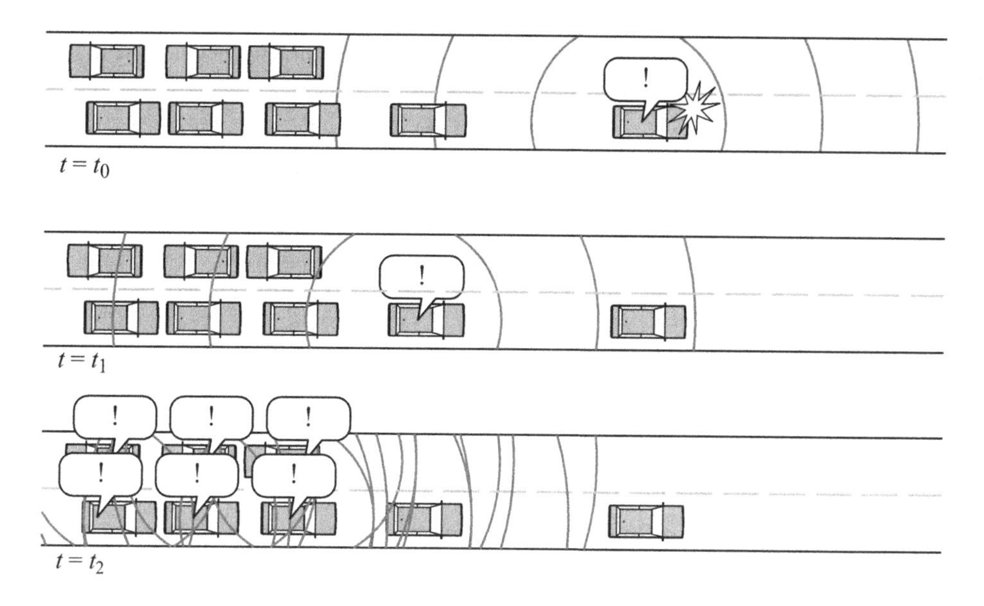

Figure 3.33 The broadcast storm problem for beaconing-based information exchange.

1999). Re-broadcasting data with the objective of achieving multi-hop forwarding of data (whether directed or undirected, i.e., using flooding) always leads to the question of which node should be used as a forwarder. In general, this question cannot be answered if no additional information is available.

Figure 3.33 outlines the problem. In our example, the rightmost vehicle creates a new message, which is to be delivered to all other nodes in the local neighborhood, i.e., an n-hop-broadcast is to be realized. In our example, no additional information is available. Thus, each node receiving the message must also re-broadcast if the message age (usually, a time-to-live value is used to count the number of hops) has not yet expired. As can be seen in Figure 3.33, first, a single car re-broadcasts the message, which is received in turn by another six vehicles. All of them re-broadcast again. Please note that this re-broadcast will happen at all the six cars almost simultaneously, leading to severe congestion situations in the wireless network. If more nodes are around, the channel will become completely overloaded (Tonguz *et al.* 2006).

In order to overcome this problem, several *broadcast-suppression* techniques have been proposed. This includes lightweight solutions such as probabilistic flooding, the maintenance of neighbor information, and the management of topology information. Of course, all these ideas come with limitations, namely either blind guessing without reasoning or control overhead for the examples mentioned.

In the following, we discuss techniques for broadcast suppression that were first introduced by Wisitpongphan *et al.* (2007b). In VANETs, geographic position information or information about the signal strength can be exploited with no additional cost. The distances between the vehicles involved can be estimated in the range $0 \leq \rho_{ij} \leq 1$.

For GPS-based measures, the distance D_{ij} can be directly calculated and then calibrated to the approximate transmission range R:

$$\rho_{ij} = \begin{cases} 0 & \text{if } D_{ij} < 0 \\ D_{ij}/R & \text{if } 0 \leq D_{ij} < R \\ 1 & \text{otherwise} \end{cases} \tag{3.3}$$

Alternatively, the received signal strength (RSS) measure of previously received messages can be used. This will, of course, provide only a rough estimate:

$$\rho_{ij} = \begin{cases} 0 & \text{if } \mathrm{RSS}_x \geq \mathrm{RSS}_{\max} \\ (\mathrm{RSS}_{\max} - \mathrm{RSS}_x)/(\mathrm{RSS}_{\max} - \mathrm{RSS}_{\min}) & \text{if } \mathrm{RSS}_{\min} \leq \mathrm{RSS}_x < \mathrm{RSS}_{\max} \\ 1 & \text{otherwise} \end{cases} \tag{3.4}$$

From this distance metric, different broadcast-suppression mechanisms can be derived. One of the simplest approaches is called *weighted p-persistence*. The idea is to make a probabilistic decision regarding whether to re-broadcast with a variable p_{ij}. Now, we just have to ensure that the larger the distance from the source of the message broadcast, the higher the probability of re-broadcasting. In this case, we can set $p = \min(p_{ij}) = \min(\rho_{ij})$. Each node now re-broadcasts with this probability. If the node did not choose to transmit, it overhears the wireless channel until a timeout expires. If no other node re-transmitted the message, the node will eventually re-broadcast itself. The advantage is simplicity; the disadvantage is that synchronized re-broadcasts may still occur, resulting in collisions on the wireless channel.

Instead, it would be better to make the farthest node re-broadcast first, followed by all other nodes ranked by their distance from the original transmitter. This model is realized in the *slotted 1-persistence* approach. Here, the distance is divided into slots and each slot is associated with a certain timeout, with the farthest slot being assigned the shortest timeout. Now, only vehicles in a certain slot (this relation is again calculated using ρ_{ij}) are allowed to re-broadcast. Nodes closer to the transmitter send later, and only if more distant nodes have not yet re-broadcast the message.

These two techniques can also be combined, resulting in *slotted p-persistence*. This also provides means for de-synchronizing all nodes allocated to the same slot.

We have already discussed quite a few applications of broadcast suppression without mentioning the underlying problem. Adaptive beaconing concepts such as ATB, DCC, and DynB inherently provide broadcast suppression, and algorithms like DV-CAST explicitly implement the slotted p-persistence approach.

3.4.3 Scalability of VANETs

The scalability of vehicular networking approaches is one of the most critical limitations. In the last decade, substantial progress towards understanding this fundamental characteristic has been made. As one of the first steps, fundamental capacity limits in wireless networks were identified by Gupta & Kumar (2000). In their milestone paper, the authors identified for the first time the upper capacity limits for unicast

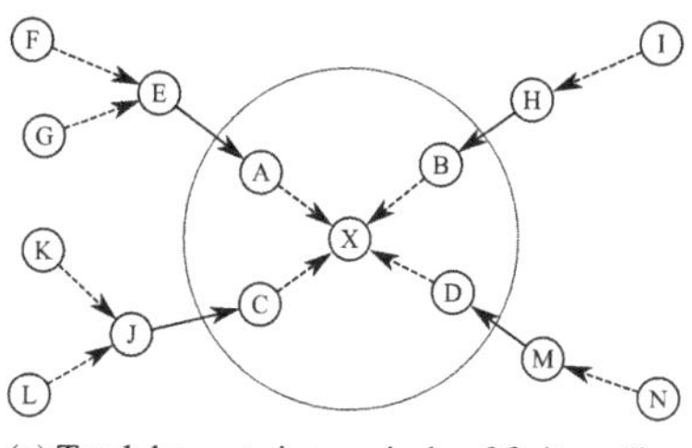

(a) Total data rate into a circle of finite radius

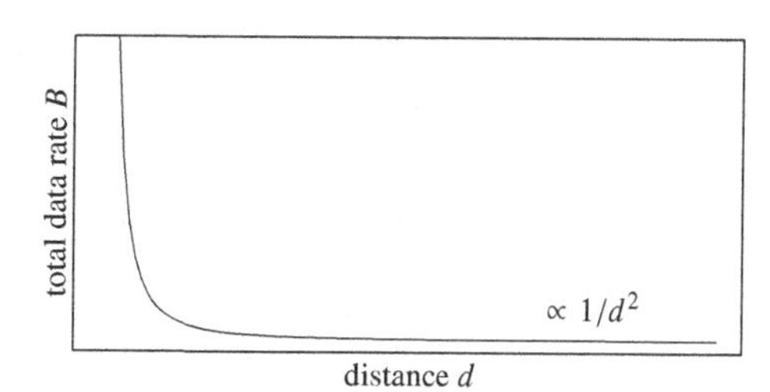

(b) Upper bound of total data rate for distance d

Figure 3.34 The scalability of information dissemination in vehicular networks.

communication in ad-hoc networks. They identified a fixed upper bound, which cannot be alleviated using other routing or dissemination strategies. Their result was later generalized to include also broadcast and multicast communication.

With reference to the subsequent research activities in the mobile ad-hoc networking domain, Grossglauser & Tse (2002) showed that – theoretically – the capacity of wireless networks increases if node mobility is taken into account. This is a major change compared with the initial results for fixed wireless networks. It has also been shown how mobility can be exploited to realize efficient data transmissions in highly mobile networks (Grossglauser & Vetterli 2006). Unfortunately, this approach does not work for arbitrary mobility models.

Still, for information dissemination in VANETs it was not fully understood whether the same upper limits can be assumed. One of the most important works looking at the scalability of VANETs was published by Scheuermann *et al.* (2009). The authors concentrated on the dissemination range in a vehicular networking environment and then assessed the upper capacity bounds.

As illustrated in Figure 3.34, the amount of information transported across any (arbitrary) border needs to be finite. Thus, the total data rate going into or out of a circle of radius d has to decrease exponentially for the system to be scalable. In their paper, Scheuermann *et al.* (2009) determined the specific upper bound, which is illustrated in Figure 3.34. Without going into all the details, the results point to general options for designing scalable dissemination protocols:

- A maximum dissemination range d for any information is defined;
- the update frequency is reduced with increasing distance; or
- information is aggregated as distance increases.

4 Access technologies

Radio access technologies like WiFi, IEEE 802.11p, and LTE form the basis of any communication stack, and the choice of technology heavily influences application performance. They are the topic of this chapter.

In general, two families of radio access technologies can be differentiated: those based on cellular networks and those based on short-range radio. Traditionally, these two families were conceptually vastly different. Cellular networks relied on central coordination, whereas short-range radio operated in a fully distributed fashion. Cellular networks used licensed spectrum, whereas short-range radio had to make do with unlicensed spectrum. These differentiations are no longer strictly true. Cellular networks are slowly moving towards (some) distributed control, while short-range radio, particularly for inter-vehicle communication (IVC), is profitting from infrastructure support and central services. Further, short-range radio for IVC can now rely on allocated dedicated spectrum. A new trend further blends licensed and unlicensed spectrum into spectrum that has primary users, which can access the spectrum with absolute priority, but one also allows its white spaces to be filled by non-primary users.

We will take a detailed look at the concepts and the underlying principles of representative radio access technologies from both families, always with a focus on their use in vehicular networks.

This chapter is organized as follows.

- Cellular networks (Section 4.1) – We start by following the evolution of the Third Generation Partnership Project (3GPP) family of cellular networks: from GSM, via UMTS, to LTE, with a perspective on future technologies. We will always be considering both halves of a cellular network, that is, the air interface and the radio access network (RAN), as well as the core network.
- Short-range radio technologies (Section 4.2) – We then turn towards a classical short-range radio technology, following the evolution of IEEE 802.11 wireless LAN (WLAN) and its many extensions. We discuss one extension in particular detail: IEEE 802.11p, which extended WLAN for use in vehicular networks. Lastly, we discuss efforts building on WLAN to provide a complete IVC protocol suite.
- White spaces and cog radio (Section 4.3) – In the final section of this chapter, we take a look at the use of white space as a novel physical medium for radio access technologies and its pros and cons for IVC.

4.1 Cellular networks

From small beginnings in the 1990s, when GSM networks started replacing first-generation local installations, to an estimated 6 800 000 000 contracts, that is, 0.96 per citizen worldwide in 2013 (ITU 2013), cellular networking has become an almost ubiquitous technology. This makes cellular networks a worthwhile technology to investigate also in the context of vehicular networking.

Three major standardization bodies govern the specification of cellular technology. These are the 3GPP (short for *Third Generation Partnership Project*), 3GPP2 (short for *Third Generation Partnership Project 2*), and – to some degree – the IEEE (short for *Institute of Electrical and Electronics Engineers*). In this section, we will focus on the 3GPP family of standards and their evolution, that is, GSM, UMTS, and LTE. Competing standards are defined by the 3GPP2 (called *CDMAone*, *CDMA2000*, and UMB, short for *ultra mobile broadband*, now discontinued in favor of LTE) and the IEEE (in particular, mobile WiMAX, short for *worldwide interoperability for microwave access*).

However, the core concept is always the same: A large area of land is covered by a network of base stations, each serving one part of this area, a cell. A mobile phone (or any authorized end device, such as a car) connects to the base station serving the cell it is in, always switching to the most appropriate base station as it moves across cell boundaries.

These networks have continuously evolved. This is visible in the 3GPP numbering released standards consecutively.[1] Traditionally, GSM has been called a *second-generation (2G)* network, because it succeeded analog cellular systems, and UMTS a 3G network. Because of marketing pressures, later distinctions like *fourth generation (4G)* do not mean much, and have been further diluted by terminology like *3G+*, *3.5G*, and *true 4G*. In the following, we will follow the historical evolution of GSM, via UMTS, to LTE without referring to these terms. Very good tutorial-style presentations of this topic are also available in the literature (Dahlman *et al.* 2008, 2011).

In talking about the evolution of cellular networks, we will always be considering two parts – although the choice of technology in either often influenced choices in the other. First, we consider the technology used on the air interface between the mobile phone and the base station. Closely linked to that is the RAN, over entities among which the tasks of coordinating transmissions, allocating channels, coding, modulating, and actually transmitting data are distributed. Distinct from that is the second part, the core network, which provides global services (such as authorization, authentication, billing, and roaming) as well as interconnecting everything and connecting to the landline telephony network or the Internet.

In discussing radio access technologies, we will refer to duplexing and multiple-access schemes. Duplexing refers to how uplink and downlink communication are sharing available resources. In frequency-division duplex (FDD), two separate frequencies are used; one for the uplink (to the base station) and one for the downlink (to the mobile

[1] The exception being in the year 2000, where the versioning scheme changed to continue from *Release 99* to *Release 4*.

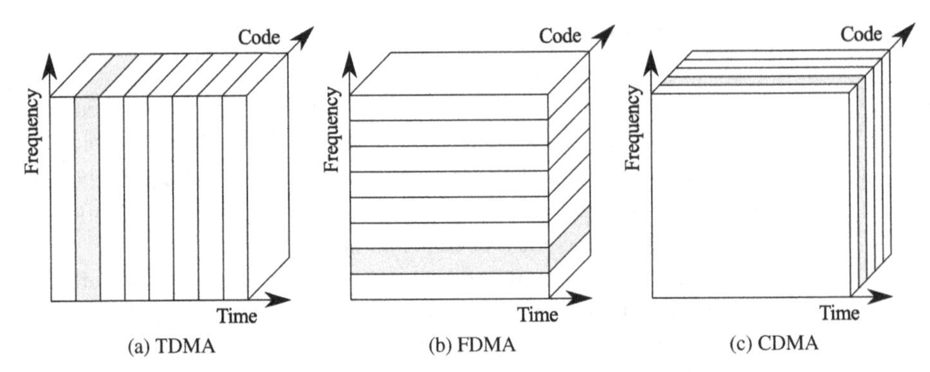

Figure 4.1 An illustration of the orthogonality of TDMA, FDMA, and CDMA.

phone). In time-division duplex (TDD), the same frequency is used, alternating between uplink and downlink. Multiple-access schemes further subdivide available resources into independent channels, e.g., for serving multiple users. They can be categorized along three complementary axes, as illustrated in Figure 4.1; that is, a system can use one, two, or all three of them.

Time-division multiple-access (TDMA) might be the most straightforward scheme. Here, the network serves different channels at different times. This relates closely to how most people would likely organize discussions involving multiple participants: by making them take turns. Time is commonly subdivided into equal-sized slots, which can then be allocated to, e.g., different users. This scheme is highly robust against frequency shifts and can tolerate arbitrary power levels. However, there is, of course, some overhead involved for keeping tight time synchronization among all devices and making sure that a signal sent in one time slot that is delayed (e.g., by arriving via multi-path reflections) cannot interfere with later slots.

Frequency-division multiple access (FDMA) is just as simple to implement. Here, different channels are carried on different subcarriers, that is, on different frequency bands within the allocated spectrum. This completely eliminates the need for time synchronization, but puts heavy requirements on transmitter hardware to make sure that, despite imperfections and high communication speeds, it sends only within a small frequency band – or large unused guard bands must be introduced to further separate the frequency bands used.

An advanced variant, OFDMA (short for *orthogonal frequency-division multiple access*) combines FDMA with orthogonal frequency-division multiplexing (OFDM) to alleviate this problem: Instead of allocating individual subcarriers on non-overlapping frequency bands separated by unused guard bands, overlapping frequency bands are allocated, but distributed so that all subcarriers are orthogonal to one another and thus cannot interfere. This allows much better use of the spectrum: It directly increases the spectral efficiency, that is, the amount of data that can be transferred per second and per hertz of bandwidth. Further, the use of OFDM allows the use of transmitter macrodiversity, that is, if multiple base stations need to transmit the same data, they can choose a non-interfering set of subcarriers and all broadcast on the same frequencies – resulting in what is called a single-frequency network (SFN). Indeed, mobile phones receiving such data from more than one base station can then benefit from diversity gain. A drawback of OFDMA is its high peak-to-average power ratio (PAPR), that is, its wastefulness in terms of energy expended for the transmission of data.

Code-division multiple access (CDMA) is the most involved of the schemes. It unifies many of the advantages of TDMA and FDMA, such as robustness against time and frequency shifts. In fact, it actually enables rake receivers to make use of multi-path signals that arrive with different time delays. The major downside is that, as we will see in the following, CDMA requires that all signals arrive at a receiver with roughly the same power level despite, e.g., devices sending at vastly different distances from the base station. Thus, closed-loop power control is required, that is, a dedicated channel needs to be used to continuously give feedback to participating devices on how much higher or lower their transmit power should be.

Figures 4.2 and 4.3 illustrate the concept of CDMA in a simplified way: Each channel gets assigned an orthogonal spreading code that is multiplied by each data bit to be sent. The resulting signal is then transmitted over the air. Note that the signal that results from overlaying all transmitted signals has multiple distinct power levels that need to be preserved for the despreading to work. Each receiver essentially performs the same process, multiplying the received mixed signal by its assigned spreading code and summing the resulting signal for the duration of a bit. In this ideal example, this allows each receiver to perfectly recover its part of the original data.

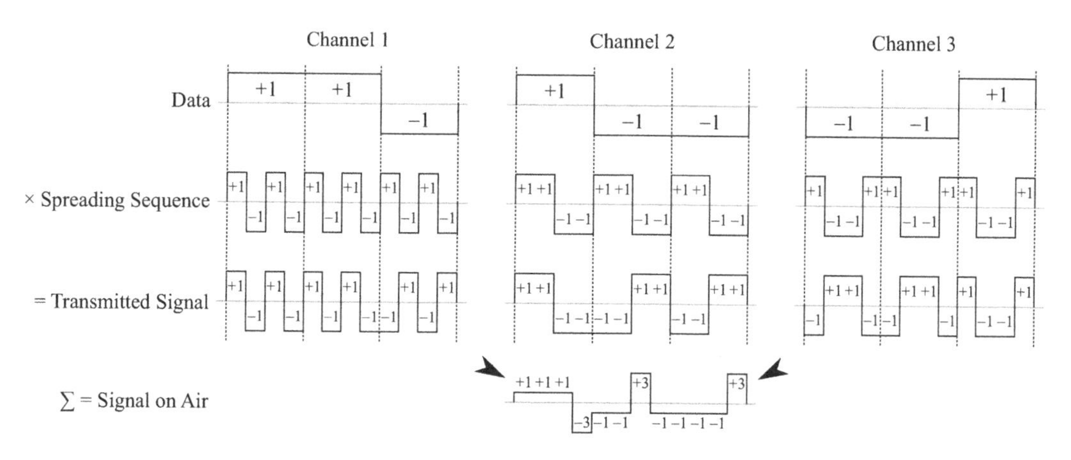

Figure 4.2 Sample data transmission in a CDMA system: spreading.

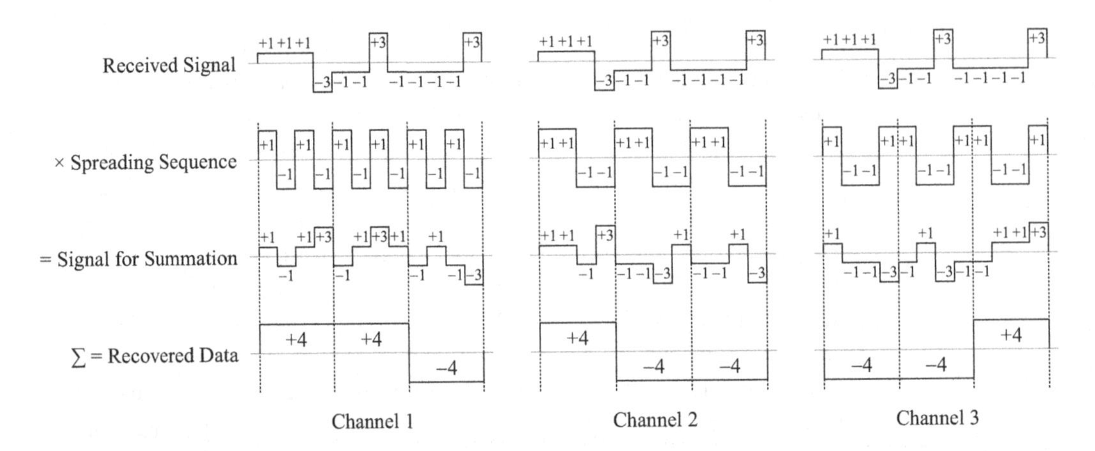

Figure 4.3 Sample data recovery in a CDMA system: despreading.

With these basic building blocks in mind, we can now take a closer look at three types of cellular networks, each representative of one evolutionary step: GSM, UMTS, and LTE.

4.1.1 GSM

GSM was conceived in the early 1980s, by a European Conference of Postal and Telecommunications Administrations (Conférence Européenne des Postes et Télécommunications, CEPT) working group called *Groupe Spéciale Mobile*, or GSM for short. In 1992, the developed system was launched commercially in seven European countries; by mid 1995, a total of 86 countries were operating GSM networks; and in late 1995 the first commercial US service launched.

The main novelty of what became known as the *Global System for Mobile Communications* (GSM) was its ability to connect mobile telephones wirelessly to each other and to regular landline telephones. To achieve this while keeping investment costs low, GSM reused landline technology, enhancing existing landline switching centers (which used to be responsible for establishing electrical circuits between each caller and callee) by adding functionality for mobile users, thus creating mobile switching centers (MSCs). These MSCs could then be connected to one another and, via a gateway MSC, to those of the fixed-line telephony network. Together they formed the core network for carrying voice data in a GSM architecture.

Mobile telephones would thus need merely a way to set up a connection to their nearest MSC, which would complete the circuit to the caller or callee. Initially, this did indeed involve a circuit-switched connection – only much later did technology move to virtual circuits established in a packet-switched network. Implementation of the access technology to allow telephones to connect to MSCs is split between two components, as illustrated in Figure 4.4. *Base station controllers* (BSCs) serve all higher functions, such as channel allocation, and are connected to anything between one and several hundred

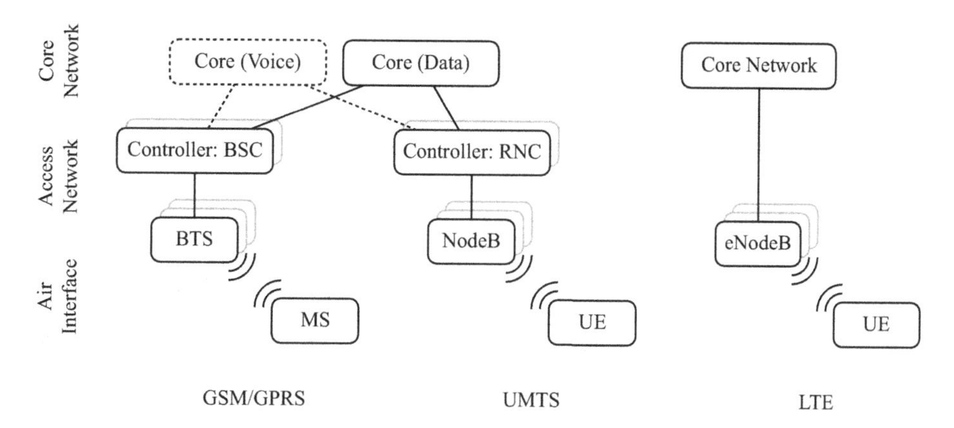

Figure 4.4 Simplified architecture of GSM/GPRS, UMTS, and LTE, the three generations of cellular networks discussed.

base transceiver stations (BTSs) that implement the actual physical layer of the air interface.

Incoming calls to a mobile phone require the network to know in which base station's service area the mobile currently is. One option would be to have mobile phones actively push updates to the core network whenever their location changes. This would, however, incur massive traffic in the network. The other extreme would be to not keep track of mobile devices' locations, but instead broadcast *paging* messages in the whole network whenever a call arrives, which the mobile phone responsible can then answer. This, however, would also incur unacceptable load. GSM therefore standardized a compromise between these two approaches: The network is divided into coarse location areas served by MSCs. Mobile devices update the core network only when their current location area changes, or after a pre-configured time interval has elapsed. Fine-grained localization of the mobile device is performed only when needed, via a paging message broadcast in all cells in the location area.

On the physical layer, GSM employs FDD, that is, separate frequencies for uplink and downlink, and FDMA and TDMA for providing multiple channels. For the popular GSM-900 standard, 124 frequency bands per direction, each 100 kHz wide, are used in the 900-MHz band and each is subdivided into eight time slots, affording a total of 992 channels.

Initially, data could be transmitted over GSM only by means of an acoustic modem. Later, with the introduction of CSD (short for *circuit-switched data*) technology, mobile telephones could reserve a circuit via the MSC to directly carry data on a channel at 9.6 kbit/s (or, with HSCSD, short for *high-speed circuit-switched data*, use multiple channels with more efficient coding, and thus at higher speeds). This meant that any data connection would tie up resources by allocating a circuit, irrespective of whether it was being used or not.

This changed in 1997 with the specification of a GPRS (short for *general packet radio service*) (Cai & Goodman 1997) that would allow packet-oriented data transport over flexibly allocated channels on the air interface. No circuit needed to be established in the core network. Instead, transport of this data in the core network was done by directly routing it to a newly specified *serving GPRS support node* (SGSN) in the core network and, further on, to a *gateway GPRS support node* (GGSN) that would route data to the Internet (or any other packet-based network in operation). These network nodes operated in parallel with the old nodes used for voice communication, as illustrated in Figure 4.4. Analogously to the concept of location areas (and location updates) to facilitate paging mobile phones on incoming voice calls, GPRS also introduced the parallel concept of routing areas served by SGSNs (and routing-area updates) for paging mobile phones on incoming data connections.

Later, GPRS was further augmented by EDGE (short for *Enhanced Data Rates for GSM Evolution*), most prominently, by allowing more efficient coding to provide typical data rates of 100 kbit/s and typical round-trip times below 500 ms (Negreira *et al.* 2007).

However, the use of GSM as a backbone for one-to-many transport of data is still infeasible. Because of its lack of multicast data dissemination, leading to many channels being established and torn down for such transmissions, common delays for the dissemination of data to 100 vehicles are in the range of 3000 ms (ETSI 2012a).

4.1.2 UMTS

Seeing how the GSM network was growing fast and was increasingly serving more and more short-lived data connections, it was recognized that, despite the rapid introduction of more flexible scheduling and more efficient coding, it would not be able to support the growing resource demands of mobile data connections for much longer. To meet these demands, a new system, UMTS (short for *Universal Mobile Telecommunications System*) was developed using different technology in the access network, but sharing the same core network as GSM, as illustrated in Figure 4.4. The first commercial deployments of this system started in 2001, operating on a new set of frequencies in parallel with established GSM networks.

On the air interface, UMTS now uses pure CDMA, dedicating the whole frequency band to all users simultaneously. It can use either static FDD or flexible TDD for duplexing the uplink and downlink. Taken together, the resulting physical layer is called W-CDMA or TD-CDMA, respectively. The user-facing radio access network was slightly restructured to be more in line with the packet-switched data transport, but remained unmodified in its basic structure. In particular, this meant that data transmissions would still incur high overheads in the core network, and establishing data sessions with other mobile devices would still be costly in terms of paging and establishing channels. This system made it possible to reach typical speeds of 384 kbit/s.

Later extensions of UMTS include HSDPA (short for *High-Speed Downlink Packet Access*), which moved scheduling decisions from the radio network controller (RNC) into the base station (called NodeB) to reduce delays, and also allowed more efficient coding as well as data transmission on multiple logical channels in the downlink. A complementary extension in the uplink, HSUPA (short for *High-Speed Uplink Packet Access*), allowed more efficient coding in the uplink as well. These extensions led to data speeds being pushed to 7.2 Mbit/s.

As discussed, the use of CDMA always necessitates that mobile devices closely control their transmission power. This is realized in UMTS by performing *closed-loop power control*, the continuous exchange of power-control messages between the mobile and infrastructure, for as long as a dedicated channel (DCH) is established for a mobile. Because of the continued cost associated with keeping a channel established, small amounts of data were allowed to be transmitted over a shared random-access channel (RACH) in the upstream and a shared forward-access channel (FACH) in the downstream. Since the required power level for transmitting on the RACH cannot be known to a mobile device, a procedure known as *power ramping* is employed: A mobile device will first try transmitting with the smallest possible power level and then keep repeating this transmission at increasing power levels until it receives an acknowledgement by the base station. It can then transmit a small chunk of data, safe in the assumption that (for a short time) its signal will be guaranteed to arrive at an appropriate power level. Transmitting data over the RACH is thus both time- and energy-intensive, and has the potential to interfere with any other mobile device trying to use the RACH for data transport or, e.g., establishing a DCH. This required that mobile network operators (MNOs) trade off the cost of forcing mobile devices to use the RACH versus the overhead of keeping a DCH established.

This is commonly realized via a simple timeout mechanism: After acquiring a DCH channel, a mobile is allowed to use this channel until no data has been transmitted for, e.g., 2 s. Afterwards, the mobile device is still known to the network and can receive data via the shared FACH channel (incurring one-way delays of 150 ms) – but the RNC can handle only a limited number of devices in this state. After, e.g., 30 s of no transmission the mobile device is therefore forced to enter a sleep state, from which it requires an additional 500 ms to wake if new data is to be transmitted. This makes the realization of vehicular network applications that require the exchange of sporadic time-sensitive data infeasible.

A further extension was specified to support broadcast or multicast transmission of data, aptly called the MBMS (short for *multimedia broadcast/multicast service*) extension. With the help of new dedicated control nodes in the core network, the radio access network would be able to receive multicast transmissions and send them to connected devices over a common shared channel. Since multicast transmission of data is a core requirement of most vehicular network applications, MBMS can thus make cellular networks a viable platform for wide-area distribution of information, eliminating the need to establish millions of unicast connections to disseminate the same piece of information. Unfortunately the MBMS extension has not seen widespread deployment in UMTS networks because of the high investment costs for changes of the core and radio access network for which there is low demand by consumers.

4.1.3 LTE

LTE (short for *Long Term Evolution*) is representative of a new generation of cellular networks that were launched commercially in the year 2009. It is based on a redesigned air interface, redesigned radio access network, and redesigned core network.

The underlying technology of all network components in LTE is IP. As illustrated in Figure 4.4, while it can be (and is) operated in parallel with existing cellular networks, it does not (and cannot) use most of the existing infrastructure, instead requiring separate network components – unlike UMTS, which can use (and relies on) existing infrastructure for voice calls. This has the drawback that there is no simple way to provide voice services to mobile phones, necessitating technologies like *voice over LTE* that can encapsulate such traffic for transport to new core network components, or voice over IP (VoIP) applications on the mobile device, or simply providing a legacy fallback network like GSM or UMTS that a device may connect to for performing voice calls.

Instead of a complex radio access network (RAN) of multiple components and functionality split across different physical devices, LTE integrates all radio access into a single entity called an *eNodeB*, which is directly networked with other eNodeBs as well as connected to the (new) core network. The main benefit of this approach lies in the fact that there is much less signaling across different entities and, thus, smaller round trip times and connection setup delays.

Just like the core network and the RAN, the air interface of LTE has evolved, too. At a basic level, LTE moves away from earlier standards' goal of combating fluctuations in radio channel quality and traffic patterns, instead trying to adapt to and exploit these fluctuations. For this it is now based on FDMA and TDMA. With regard to FDMA, in the downlink direction, OFDMA is used since the high PAPR associated with OFDMA is no problem for base stations – they have virtually limitless energy reserves. In the uplink direction, because of the high PAPR of OFDMA, a more appropriate variant called SC-FDMA (short for *single-carrier FDMA*) is therefore employed. The LTE air interface can make flexible use of a wide variety of different bandwidths and frequencies, supporting bands as narrow as 1.4 MHz and as wide as 20 MHz, as well as frequencies as low as (currently) 700 MHz and as high as 2.7 GHz. The frequency band used can be split between uplink and downlink in either an FDD or a TDD fashion. Cell sizes of up to 100 km can be supported (though at extremely low performance), and mobile devices may move as fast as at 500 km/h (for some frequency bands). Data modulation of up to 64-QAM is supported and spatial multiplexing (that is, MIMO) of up to 4×4 is standardized for the downlink (although mobile devices commonly have no more than two antennas). This results in a theoretical peak download rate of 300 Mbit/s and a latency of 5 ms (assuming best-case channel conditions and no load on a 20-MHz-wide band, as well as a 4×4 MIMO mobile device).

Downlink scheduling and modulation/coding decisions are taken directly at the eNodeB every 1 ms, depending on the channel characteristics from and to each connected mobile device, and quality-of-service (QoS) considerations. For scheduling, the available frequency band is split into 0.5-ms slots. A combination of twelve subcarriers and six or seven slots is called a resource block (RB) and constitutes one scheduling unit. Depending on the available frequency band, up to 100 RBs (for 20 MHz) are available for scheduling. One or multiple RBs are allocated per direction and user, and are used for transporting frames of length 10 ms, which are further subdivided into 10 subframes. Each subframe carries one *transport block*, which in turn carries the payload data. This means that, even in a 5-MHz cell, up to 200 active data connections can be managed.

LTE Advanced introduces further features, like allowing MIMO also on the uplink, introducing beamforming techniques, allowing multiple 20-MHz bands to be bundled (if available), and standardizing private femtocells. Some of these features are feeding back to UMTS (in the context of HSPA evolution).

Finally, and to the benefit of many potential vehicular networking applications, LTE introduces an evolved Multimedia Broadcast/Multicast Service (eMBMS): This eMBMS (sometimes also called *LTE Broadcast*) can also exploit a new multicast-broadcast single-frequency network (MBSFN) channel for broadcasting the same data in multiple cells on the same frequency, which is made possible by increasing the cyclic prefix length. This allows the MBSFN to deliver an experience similar to perfect coverage and mobile devices to benefit from diversity gain. The spectral efficiency of eMBMS can reach up to 0.5 bit/(s Hz) while still maintaining almost 99% coverage (ETSI 2012a). Further, unlike UMTS, the commercial adoption of this eMBMS service is picking up, and the first service was launched in early 2014.

One of the most recent additions to LTE is LTE Direct, a fundamentally new concept that introduces the possibility of (centrally coordinated) device-to-device data transmission and service discovery. Here, devices can petition their eNodeB for part of its resources for sending service-discovery beacons. Likewise, if a direct device-to-device data transfer should be initiated, these devices can petition the eNodeB for resources for such a direct channel as well. This channel can then be used like any other LTE channel, with the difference that data no longer needs to be relayed before it reaches its destination.

4.1.4 Future developments

Looking back at the evolution of cellular networks from GSM into LTE, one trend is immediately clear: More involved modulation and coding schemes and new technologies like MIMO and beamforming have improved throughput. More and more spectrum has been dedicated to individual channels to satisfy bandwidth demand that could not be covered with better technology alone. Current proposals are looking towards millimeter-wave (mmW) technology beyond 28 GHz, which would have up to 1 GHz of bandwidth available.

Further, from the strict hierarchies prevalent in GSM and UMTS to the almost mesh-like topology of self-governed LTE base stations, networks have become increasingly more flexible to accommodate new use cases, and their structure has been simplified to reduce latency. These trends are forecast by many to continue, and researchers are now looking towards making multi-hop networks and machine-to-machine communication first-class use cases of cellular networks.

The European Union FP7 research project METIS (short for *Mobile and Wireless Communications Enablers for the 2020 Information Society*) is looking to address the challenges of the next generation of cellular networks. It targets a system concept that will support the expected three-orders-of-magnitude-higher data volume that future cellular networks will need to be able to handle, along with two orders of magnitude more devices connected to cells, while further increasing throughput and decreasing latency.

Noteworthy is the composition of the project partners: Aside from major vendors and MNOs and many academic partners, the project has been joined by the automotive industry.

4.1.5 Use of cellular networks for IVC

Many vehicular networking research projects nowadays are based to some degree on the use of cellular networks. In 2006, project CoCar (short for *Cooperative Cars*) kick-started these developments. Researchers from industry (vendors, MNOs, automotive firms) joined academic partners to investigate both the technical and the commercial feasibility of a system that would rely solely on cellular networks. The technology of choice at that time was UMTS.

Use cases ranged from a general traffic information service to a time-sensitive cellular hazard-warning service. Cars would generate reports of traffic conditions or of perceived hazards to other vehicles, and transmit them via custom protocols (called *TPDP* and *FTAP*, respectively) to CoCar services, which would take care of their redistribution.

CoCar services were envisioned to be provided by three distinct logical entities in the network: *CoCar Reflectors* were employed for low-latency replay of received messages into the cell of origin and passed messages on to a central *CoCar Aggregator*. This entity could consolidate and verify incoming reports by cars and external sources, and then set up long-running dissemination of Transport Protocol Expert Group (TPEG) bulletins (see Section 3.3.1) via *CoCar Geocast Managers*. These logical entities were designed to operate independently of where they would be located, either on servers in the Internet, attached to the GGSN, as services running on nodes in the core network, or (for the CoCar Reflector) as extensions of software in the RAN.

Sommer *et al.* (2010d) were able to show the general feasibility of such a system, but had to conclude that, while a UMTS-based system would be ideal for a market-introduction scenario of IVC technology, results for MBMS-incapable systems did not appear very promising for full-scale deployment of services. Assuming MBMS to be available, though, the system might be a feasible basis for a traffic information system (TIS).

Figure 4.5 illustrates the results graphically: Shown are the empirical cumulative density functions (eCDFs) of delays between a message being sent and it being received by other vehicles on the road. Two different service levels have been configured for wide-area dissemination of TPEG bulletins: repeating messages every 30 s or every 60 s (labeled *double interval*). Figure 4.5(a) shows delays for those messages received via an immediate multicast back to the originating cell by a *CoCar Reflector*. It highlights that good parametrization of the system could bring delays down to below 125 ms for as much as 95% of all cars – assuming all cars are serviced by the same MNO. Figure 4.5(b) displays the same metric, but for messages received via multicast bulletins in other cells (that is, via the *CoCar Aggregator* and *CoCar Geocast Managers*).

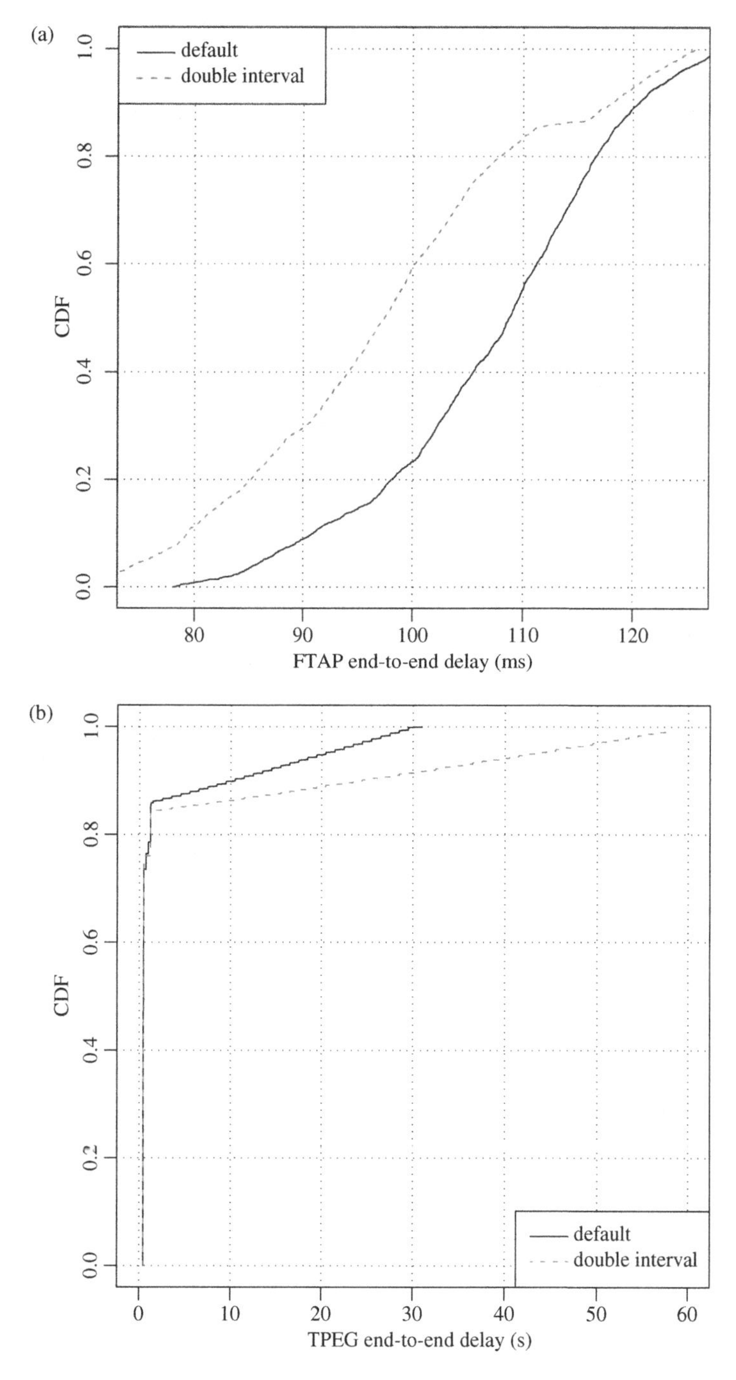

Figure 4.5 Round trip delays of messages in the CoCar system, differentiated between cars in the same cell (a) and cars in other cells (b).

Here, the delays are on the order of seconds. While these delays are certainly above any reasonable limit for safety messages, the system is still very adequate for, e.g., a TIS.

In 2009, an immediate follow-up project, CoCarX (short for *Cooperative Cars Extended*), extended the research in two directions. First, the just-launched LTE which promised to include eMBMS was chosen as a basis technology. Second, the cellular network was envisioned to be supported by IEEE 802.11p short-range radio technology (see Section 4.2.2 on page 122), forming a heterogeneous vehicular network.

By first exchanging messages via IEEE 802.11p, aggregating them locally, and only then sending them via LTE, this system could much better cope with a high load. The results are detailed in a report by the ETSI (2012a): Sending an aggregate of 40 cooperative awareness message (CAM) multicasts at 2 Hz could be shown to be feasible in cells serving up to 140 vehicles (assuming good channel conditions in urban areas) or up to 100 vehicles (in rural areas). However, increasing the sending rate to 10 Hz for as few as 20 cars, as would be needed for cooperative awareness, increased delays to well over 500 ms, even in well-covered urban conditions. For hazard warnings, however, the system proved feasible, provided that the network offered enough capacity to keep all vehicles connected. Further, if eMBMS can be assumed to be deployed in a sufficiently dense configuration, downlink resource usage was envisioned to be negligible and uplink delays the limiting factor. The envisioned delays are below 35 ms for CAM uplink data transmissions of as much as 190 cars per cell driving in a rural scenario (when beaconing at 10 Hz) and 1200 cars per cell (when beaconing at 1 Hz). Multicast dissemination of these CAMs via eMBMS is expected to add only a little more in terms of overhead.

Similar results have been obtained in the sim^{TD} (see Section 6.1.2 on page 240) field operational test (FOT) that concluded in 2013. Although sim^{TD} was not primarily investigating cellular communications, its large-scale field trials included some data logging of connectivity metrics to the Internet. While executing the sim^{TD} field tests, vehicles logged between 80% and over 90% availability of Internet services via cellular networks, with delays of below 200 ms in 80% of the measurements conducted (sim^{TD} 2013).

The CoCar and CoCarX projects were succeeded by numerous projects aiming at the integration of IVC technologies – based (in part) on modern cellular networks – into complete systems. One example is the Converge (short for *COmmunication Network VEhicle Road Global Extension*) project that now seeks to provide a reference architecture for data and content exchange in scenarios that are both multi-operator and multi-technology.

4.2 Short-range radio technologies

A wide variety of short-range radio technologies[2] can be used for IVC, and a multitude of approaches in the past have investigated the use of Bluetooth and Zigbee, along

[2] *Short* here means typical distances of some hundred meters and up to a few kilometers – in contrast to cellular networks, which specify cell sizes of tens of kilometers and up to 100 km.

with millimeter-wave (mmW) or optical links. Short-range radio research in the context of IVC, however, has always been dominated by approaches that were based first on WiFi and then on a technology that is alternately called IEEE 802.11p, the *new* DSRC standard, or (very colloquially) *vehicular WiFi*.

Before we can take a detailed look at IEEE 802.11p and higher-layer standards, we will therefore briefly review its roots, the family of classic wireless LAN (WLAN) standards.

4.2.1 Wireless LAN

Wireless LAN (WLAN) is the name of IEEE working group 802.11, which was established in September 1990 to create a standardized way for wireless local-area networking compatible with other IEEE 802 LAN standards such as the widely successful Ethernet standard specified five years earlier as IEEE 802.3 (see Section 2.3 on page 32). The first standard, IEEE 802.11-1997, was released in 1997 and has been updated with numerous amendments (e.g., IEEE 802.11n), many of which have since been integrated into updated versions of the standard (e.g., IEEE 802.11-2007).

The original IEEE 802.11 defined multiple options for the physical layer, one of them using infrared light, others using the ISM bands (short for *industrial–scientific–medical radio bands*), either direct-sequence spread spectrum (DSSS) or frequency hopping, and proved hard to implement in a compatible manner. IEEE 802.11a and IEEE 802.11b were the first amendments to the standard that specified new physical layers, along with clear implementation rules. IEEE 802.11a specified OFDM operation on one of the U-NII bands (short for *unlicensed national information infrastructure bands*), that is, in the 5.2–5.8-GHz range; 20 MHz of bandwidth were used for 52 subcarriers, four of which served as pilot subcarriers. Depending on the desired robustness vs. the desired speed, any of BPSK, QPSK, 16-QAM, and 64-QAM modulation could be used with code rates from ½ to ¾, resulting in raw data rates from 6 Mbit/s to 54 Mbit/s. IEEE 802.11b specified extensions to the original DSSS technique using CCK modulation on a 22-MHz channel in the ISM bands to achieve raw data rates of up to 11 Mbit/s. More recent physical-layer amendments that have gained commercial acceptance are IEEE 802.11n and IEEE 802.11ac; they specify OFDM on channel bandwidths up to 160 MHz, modulations up to 256-QAM, code rates up to ⅚, MIMO and beamforming, and can achieve data rates of up to 866.7 Mbit/s per spatial stream (that is, up to a theoretical maximum of 7 Gbit/s for eight spatial streams) – and physical layers delivering even higher performance have been standardized as well.

Part of the commercial success of WLAN is due to the *Wi-Fi Alliance*, an association of manufacturers and related industry formed in 1999 (then called the *Wireless Ethernet Compatibility Alliance*) to push IEEE 802.11b and ensure interoperability of devices by offering testing and certification. Many Wi-Fi Alliance certifications correspond directly to IEEE standards, such as the physical-layer standards introduced in amendments IEEE 802.11a/b/g/n, security features introduced in amendment IEEE 802.11i (branded *Wi-Fi Protected Access 2*, WPA2), and QoS extensions introduced in IEEE 802.11e (re-branded as *wireless multimedia extensions*). Other certifications refer to novel uses of IEEE 802.11, such as allowing a mobile device to act as an access point

(AP) for direct connections with automatic key configuration (branded as *Wi-Fi Direct*). Other certifications again tried to introduce wholly new functionality, like *Wi-Fi Protected Setup* that was meant to allow more user-friendly key distribution. Wi-Fi certified IEEE 802.11-based radio access technology is commonly referred to simply as WiFi.

On the MAC layer, WLAN uses carrier-sense multiple access with collision avoidance (CSMA/CA) following what is called the WLAN distributed coordination function (DCF). While it is not transmitting, each device (called a *station*) has to continuously monitor the channel to determine whether it is idle. This procedure is called clear-channel assessment (CCA). A device may transmit only if the channel has been idle for the duration of an interframe space (IFS), which is measured in integer multiples of slots (e.g., 13 µs long, depending on the physical layer and channel bandwidth). Different lengths of the IFS are supported to realize different access priorities. In order to work around problems when not all stations can see one another (commonly referred to as *hidden terminal problems*), the MAC layer also specifies a ready-to-send (RTS) query frame and a CTS (short for *clear to send*) response, both sent at highest priority (utilizing a SIFS, short for *short interframe space*), that notify receiving stations of a *virtually* busy channel, even though they might not hear the original transmitter. A SIFS is also used for transmitting MAC-layer ACK frames, which are used to acknowledge correctly received unicast transmissions.

If a station wants to transmit a frame, but cannot yet transmit (because the channel is busy or virtually busy, or the applicable IFS has not elapsed yet), it queues the frame and triggers a *backoff* procedure, illustrated in Figure 4.6, to avoid synchronous access to the channel as soon as it becomes idle. It draws a random backoff time from a pre-configured interval (the *contention window*) and waits. This backoff time is then decremented during intervals during which the channel is free. If the channel becomes busy, the backoff time is frozen until after it has been free again for at least an IFS. When the backoff time reaches zero, the frame can be transmitted. If a transmit attempt fails, the contention window is increased (possibly multiple times), until a transmit attempt succeeds, at which point the contention window is reset. The backoff procedure is also invoked after a station has finished transmitting, to give other stations a chance to use the channel.

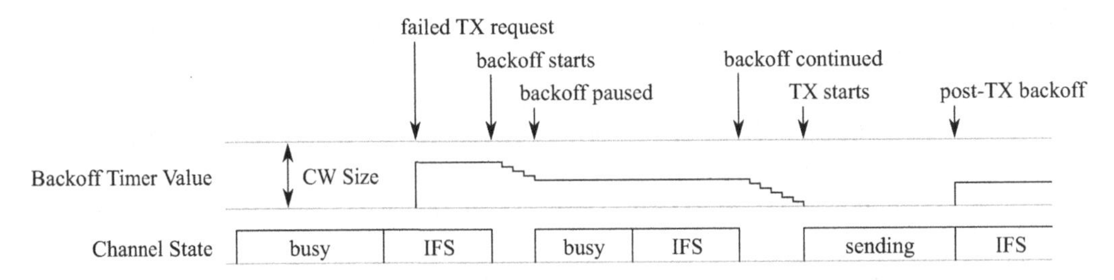

Figure 4.6 An illustration of the IEEE 802.11 backoff procedure: the MAC layer receives a request to transmit a frame, but the channel has not yet been idle long enough, causing it to back off. It chooses a random backoff timer value from its CW and counts down while the channel is free. When the backoff timer reaches zero, it transmits the frame, then enters a post-transmit backoff.

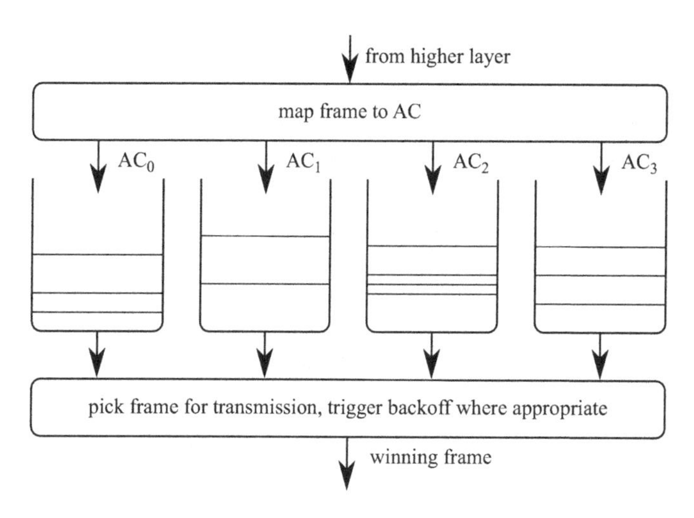

Figure 4.7 An illustration of EDCA queue management.

A 2005 extension to WLAN, called EDCA (short for *enhanced distributed channel access*) and standardized in IEEE 802.11e, further improves the channel-access procedure to provide QoS mechanisms to higher layers. Frames are classified into four distinct access categories (ACs), each corresponding to a different set of channel-access parameters, most importantly the minimum and maximum contention window size (so higher-priority frames are retried, on average, sooner) and the IFS length (so higher-priority frames can access the channel before others). For this it introduces four new IFS lengths, called AIFS (short for *arbitration interframe space*). As illustrated in Figure 4.7, frames in each AC are stored in separate queues, one per AC, each with its own backoff timer. This opens up the possibility of two or more queues' backoff timers reaching zero at the same time, which will be treated like a collision by the lower-priority queue. Queues are also associated with a transmission opportunity (TXOP) limit value, indicating the maximum consecutive time for which their transmissions may occupy the channel.

On the highest layer, a central concept of IEEE 802.11 is the basic service set (BSS), a set of devices that have synchronized parameters to communicate with each other. Each frame includes a BSS identifier, so that all stations that are not in the target BSS (but, by chance, use the same channel parameters) can discard the frame. Any station may only ever be a member of one BSS and be in infrastructure mode or ad-hoc mode.

In infrastructure mode, one or more members of the BSS are access points (APs), with which stations can *associate* to send data to (and receive data from) hosts on a wired network or other stations, possibly via other APs. In order to allow this data forwarding to work, a station may only be associated with one AP. With the exception of networks relying on the simplest security scheme, direct encryption of all frames using a well-known key, the second role that APs fulfill is that of IEEE 802.1X authentication and key management, which is performed during an involved handshaking procedure.

An AP can also coordinate transmissions in the BSS via what WLAN terms the point coordination function (PCF), that is, by polling data from stations (using a frame of second-shortest IFS, to which the station replies after a SIFS). The MAC layer ensures that such polling frames get priority access to the channel.

As an alternative, stations can also be in ad-hoc mode, forming an independent basic service set (IBSS). Here, each station can act as an authenticator for a new station. Data will be sent directly from station to station.

The use of IEEE 802.11 WLAN in the framework of an intelligent transportation system (ITS) was investigated early on. The first measurement campaigns date back to about 2005. Ott & Kutscher (2005) studied what they called the "drive-thru Internet". The main motivation was to investigate the impact of the speed of vehicles on the capability to connect to an AP to exchange data.

Still, without the further modifications in IEEE 802.11 WLAN would be of little use for ITS, as we will see in the following. Analogously to standard Ethernet that was adapted both on the physical layer and on higher layers to suit in-car communication (see Section 2.3 on page 32), WLAN was adapted as well. These modifications are the subject of amendment IEEE 802.11p, which we will now discuss.

4.2.2 IEEE 802.11p

Commonly, the beginnings of IEEE 802.11p are traced back to 1997 when the Intelligent Transportation Society of America (ITSA) petitioned the US Federal Communications Commission (FCC) for 75 MHz of dedicated bandwidth in the 5.9-GHz region (Uzcátegui & Acosta-Marum 2009). Existing technologies, such as those used for automated tolling, had too little bandwidth available to support modern ITS applications, and they had to share the spectrum with other applications.

In 1999, the FCC ruled to grant this request and reserve the requested frequency band for exclusive use by ITS applications, later also specifying its partitioning into seven 10-MHz channels (that is, channels 172, 174, ..., 184), with the option of combining two channels (channels 174/176 and 180/182 into 20-MHz channels) in FCC rulings §90.377 and §95.1511. Channel 178 was named the control channel (CCH), with the idea that this would be the well-known channel that all stations are tuned to by default. Access to channels 172 and 184 was reserved for governmental use for applications regarding the safety of life and property. All other channels remained free for public use, provided that the FCC rules for the use of these frequencies were followed. FCC rulings were later mirrored by other countries, which also reserved dedicated spectrum for ITS applications. For example, in 2008 the Electronic Communications Committee (ECC) and European Commission ruled that member states reserve three 10-MHz channels (channels 176, 178, and 180) for safety-related communications (decision 2008/671/EC), and the ECC recommended that two more channels (channels 172 and 174) be reserved for non-safety communications. Figure 4.8 gives an overview of the channel allocation.

With reserved spectrum allocated by the FCC in 1999, congress tasked the US Department of Transportation (US DOT) to develop a standard that would enable

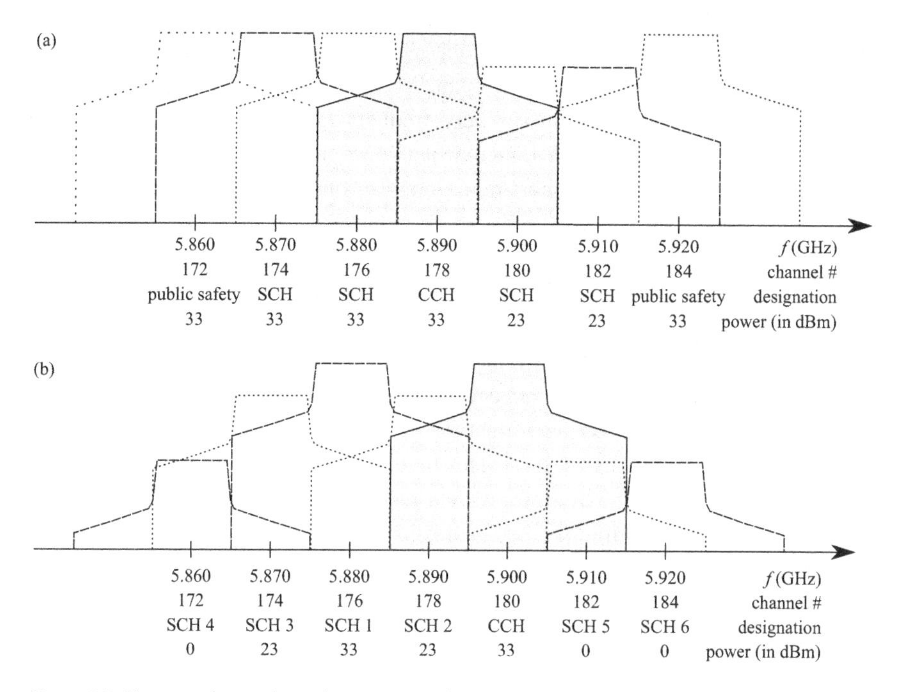

Figure 4.8 Frequencies and maximum transmit power as assigned in the USA (a) and in Europe (b), respectively, according to FCC ruling §90.377 and ETSI EN 302 663. The spectral masks are not to scale.

the efficient use of this dedicated frequency band, called the dedicated short-range communication (DSRC) band. Many of the existing technologies were viable candidates to operate in the DSRC band (and many technologies have in the past been claimed to be *the* new DSRC standard). The US DOT Federal Highway Administration (FHWA) agency thus chartered the American Society for Testing and Materials (ASTM) to investigate possible technologies. In 2002, the ASTM published a first standard specification (ASTM E2213) that recommended, on the basis of results from extensive research, testing, and feasibility studies, that ITS technology be based on a modified version of the IEEE 802.11a OFDM physical layer. While this standard substantially changed the physical layer, it took great care to allow the easy use of hardware designed for IEEE 802.11a. Since the ASTM feared that the new standard would be hard to implement and maintain as an independent add-on to both IEEE 802.11 and IEEE 802.11a, it asked IEEE 802.11 to form a study group for an eventual draft amendment. This study group was formed in 2004 and developed amendment IEEE 802.11p, which was published in 2010. Their standard amendment was later integrated into the 2012 release of IEEE 802.11.

Before we continue with the technical details of IEEE 802.11p, it will be helpful to reflect on why an unmodified deployment of earlier versions of IEEE 802.11 WLAN would be unsuited for ITS. A core problem of WLAN was that much of the MAC operation was geared towards long-term associations of stations with APs or other stations in relatively static BSSs. Synchronization, authentication, and association are very time-intensive – as anyone using WiFi to wirelessly connect to the Internet can attest to. Further, a station can, by definition, only be in one BSS at the same time, making these processes very frequent in even moderately mobile networks – and outright infeasible in vehicular networks. Lastly, all existing physical layers were geared towards stable, short-range connections under (comparatively) good channel conditions.

For these reasons, IEEE 802.11p extended IEEE 802.11 in two main ways. First, and least surprisingly, the physical layer was allowed to operate in the allocated 5.9-GHz ITS bands of the USA and Europe, specifying regulatory classes, power levels, receiver performance requirements, and transmit spectral masks for 5-MHz and 10-MHz channels. ITSs are expected to communicate on 10-MHz channels, using the OFDM physical-layer specification of IEEE 802.11a, but in a *half-clocked* variant introduced with IEEE 802.11j. This mode doubles all symbol times, as well as the CCA time, cutting the maximum raw data rates in half (3 Mbit/s to 27 Mbit/s). On the plus side, inter-symbol interference is drastically reduced, making transmissions much more robust with respect to, e.g., multi-path delay.

As a second extension, IEEE 802.11p drastically changed the MAC layer by introducing a completely new operation mode that stations operating in the ITS band switch to. This mode of operation is called OCB (short for *outside-the-context-of-a-BSS*) mode (sometimes also referred to as WAVE mode), allowing stations to operate without being part of a BSS at reduced functionality specifically geared towards vehicular networks. Consequently, a station performs no authentication or association procedures, since it never joins a BSS. Instead, stations are expected to transmit and receive frames using pre-agreed (that is, *well-known*) physical-layer parameters, or use higher-layer signaling to agree on such parameters, e.g., which channel to switch to for a particular service. MAC-layer confidentiality measures are not used, since one relies instead on upper layers to provide this functionality, if desired. The procedure sets the BSS identifier of all sent frames to a wildcard value, allowing all nearby stations to process the frame. Such frames may also be sent to groups of stations, by addressing them to special group destination MAC addresses (identified by a set most significant bit), realizing multicast support in addition to broadcast support on the MAC layer.

Smaller extensions of IEEE 802.11p include slightly longer AIFS values for stations in OCB mode, vendor-specific frames to exchange management information, a standardized way of measuring the received signal strength (RSS) on the channel, and time-advertisement frames to allow stations to provide means of establishing a common time base for higher-layer functions in the absence of a time-synchronization function (TSF) managed by the BSS.

Taken together, these extensions make IEEE 802.11p WLAN a communication technology that is well suited to the requirements of vehicular networks. This is evidenced by the results from numerous field operational tests (FOTs) that have been conducted in recent years, which we will discuss in Section 6.1.2. In the following, we will look at higher-layer protocols that make use of IEEE 802.11 to realize an intelligent transportation system (ITS).

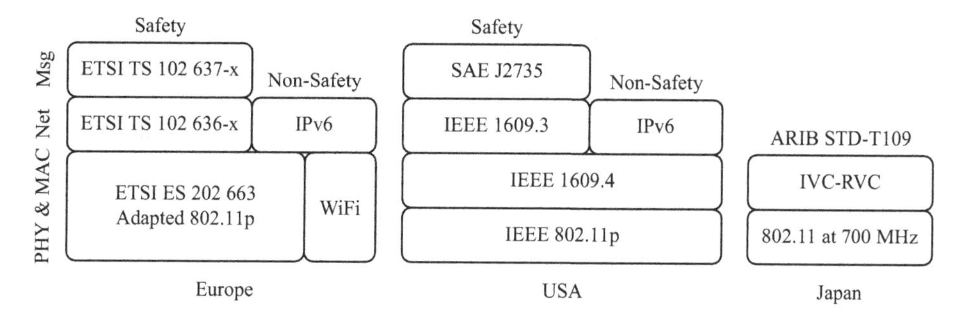

Figure 4.9 IVC protocol stacks as defined for Europe, the USA, and Japan.

4.2.3 Higher-layer protocols

Many countries' regulatory and standardization bodies are focusing on using WLAN as the basic communication technology for providing an ITS. The differences concern the number of available channels.

As mentioned previously, the FCC and ECC have allocated seven and up to five dedicated channels in the 5.9-GHz band for ITS, respectively, while in Japan only a single channel in the 700-MHz band has been allocated.

Depending on the number of available channels (single or multi-channel operation) and the number of mandated transceivers (single or multi-radio operation), three different systems can be thought of: single-radio single-channel, single-radio multi-channel, and multi-radio multi-channel. This basic choice informed three different standards that are being developed in Japan, the USA, and Europe. The Japanese family of standards assumes that only a single channel at 700 MHz is available. The US family of standards assumes that multiple channels are available, but only one radio might be installed in the car. The European family of standards, lastly, constitutes a classical multi-radio multi-channel system.

Figure 4.9 gives a very high-level overview of these families of standards. In the following, we will look at the two protocol stacks operating on multiple channels in more detail: First, the US family of standards, called IEEE WAVE; and second, the European family of standards, called ETSI ITS.

IEEE 1609: Wireless access in vehicular environments (WAVE)

IEEE 802.11p was developed to serve as the lower layers of a complete ITS stack for the USA, called WAVE (short for *wireless access in vehicular environments*). This stack was designed around the assumption that vehicles might be equipped with just a single radio, but that they should be able to communicate on all of the up to seven 10-MHz channels in the DSRC band, supporting communication not only on the CCH, but also on any of the other channels, called service channels (SCHs).

In order to still be able to reach all vehicles, even though they might be using services on different channels, functionality needed to be introduced that would instrument the IEEE 802.11 physical layer to switch all stations to the CCH during defined time intervals. WAVE adopts the following default principle (if not overruled by higher layers):

By default, a vehicle will spend the first 50 ms of every coordinated universal time (UTC) second tuned to the CCH, then it is free to tune to any channel for the next 50 ms, after which the cycle repeats. While switching between channels, the MAC layer has to take care to set appropriate EDCA parameters for each channel. This functionality is realized in WAVE by an additional MAC layer on top of IEEE 802.11 called the WAVE MAC. This allows WAVE not only to rely on, e.g., data received via the global positioning system (GPS), but also to use time-advertisement frames for fine-grained synchronization. To account for synchronization errors, the first 4 ms of each 50-ms interval serve as a guard interval and are left unused. This MAC layer was standardized as IEEE 1609.4, with a first version of the standard published in 2006. It was later amended in IEEE Std 1609.4-2010.

Aside from switching between channels, the WAVE MAC also takes care to route incoming packets from higher layers to appropriate EDCA queues – and to manage the EDCA queues according to the currently active channel. For this, the WAVE MAC has to either maintain separate EDCA queues for different channels or flush EDCA queues before each channel switch, so that packets never get sent on the wrong channel. Moreover, when packets are queued for a different channel than the one that is currently active, it has to make sure that this packet is not sent at the moment when the physical layer is tuned to this channel – otherwise, all vehicles with queued frames would try to access the channel at exactly the same time. As a simple remedy, the standard mandates that a backoff procedure is started for the first packet that would be sent at the time of the next channel switch. Since the contention window (CW) of many ACs is quite small, however, this can still lead to a large amount of synchronized collisions (Eckhoff *et al.* 2012), so the situation is much better remedied by making upper layers aware of the current channel, designing them to queue packets only while the right channel is active.

Services are identified by variable-length provider service identifiers (PSIDs), well-known numbers that identify the type of service that is provided. They are defined in IEEE 1609.12, which allocates, e.g., PSID `0x0c` to emergency warnings, `0x8002` to intersection safety and awareness, and `0xbfa0` through `0xbfdf` for private use.

To support the provision and use of services on multiple channels, IEEE 1609.3 defines WAVE service advertisements (WSAs), which can be broadcast on the CCH to notify other vehicles of services offered on SCHs. WSAs are composed of multiple blocks, any of which can be of type *service information*, *channel information*, or *routing advertisement*, as illustrated in Figure 4.10. The *service information* blocks contain fields like the PSID of the offered service, a priority, and a channel index. This channel index references a *channel information* block in the WSA, which contains fields like the channel number and data rate; it may also contain EDCA parameters for this service. Finally, a *routing advertisement* block can be used to indicate that this station offers connectivity to an IPv6 network. The dissemination of WSAs (by providers of a service) and their processing (by potential users of a service) is managed autonomously by a further component of the protocol stack, the WAVE management entity (WME) which is also defined in IEEE 1609.3. The standard also recommends that the WME serve as a coordination point that, e.g., measures channel load to allow smarter assignment of channels.

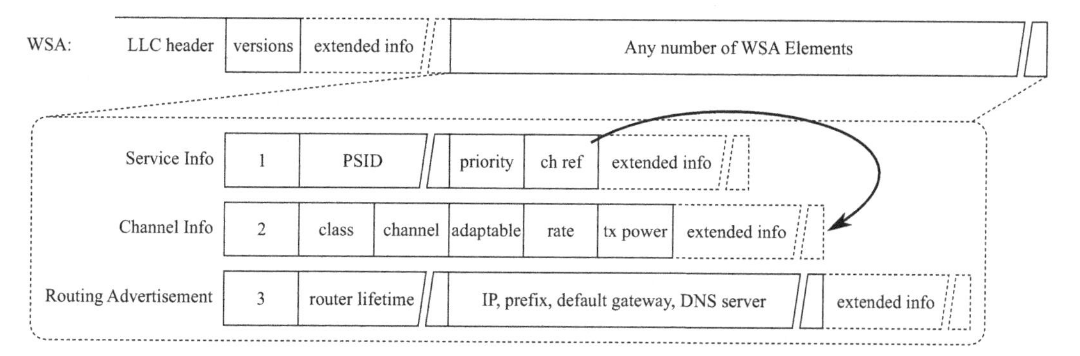

Figure 4.10 The WSA packet format. Optional *extended info* blocks transmit additional information like location (for the advertisement), EDCA parameters (for the channel information), or secondary DNS servers (for the routing advertisement).

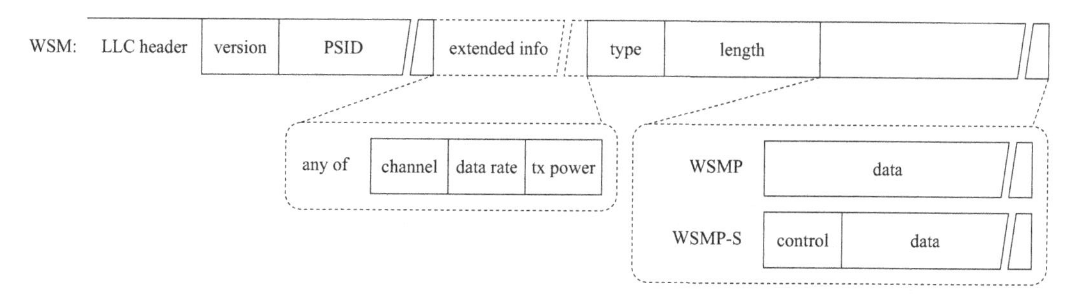

Figure 4.11 The WSM packet format.

A WAVE station supports the transmission of legacy IPv6 packets as well as the transmission of messages using a new lightweight WSMP (short for *WAVE short message protocol*), also defined in IEEE 1609.3. It supports the setting of individual transmission parameters for each WAVE short message (WSM), such as the expiry time, channel number, data rate, or transmit power. Among others, WSMs consist of the following core fields, illustrated in Figure 4.11: In addition to a PSID, a type field (indicating, e.g., whether the transmitted message is a plain WSM or one with enhanced information about local channel-switching decisions called WSM-S, short for *WAVE short message with safety supplement*), a length field, and the actual data, each WSM can also optionally include the used transmission parameters.

The last of the central services defined for the WAVE stack deal with security. They are defined in IEEE 1609.2. These services provide certificate management (in the CME, short for *certificate management entity*), manage information such as private keys required for sending secure data (in the PSSME, short for *provider service security management entity*), and provide the necessary service primitives to actually sign, encrypt, decrypt, or verify data (and, in particular, WSAs). Considerations relating to the privacy (in particular, location privacy) and trustworthiness of information are currently outside the scope of the standard.

Applications using this WAVE stack have been defined in further standards by the ISO, the SAE, and the IEEE 1609 working group. Two examples of applications defined by SAE J2735 are basic safety messages (BSMs) or signal phase and timing (SPAT) messages. An example application defined by IEEE 1609.11 is electronic payment.

ETSI ITS

In Europe, the European Telecommunications Standards Institute (ETSI) standardized a different higher-layer protocol stack for IVC. Like IEEE 1609 WAVE it is based on IEEE 802.11 WLAN and makes heavy use of the extensions introduced in IEEE 802.11p (although adapted for ECC regulations, therefore called ETSI ITS-G5 technology in the standards). The basic concept of the ETSI approach, however, is different and is illustrated in Figure 4.12.

Instead of trying to cope with a single transceiver system utilizing multiple channels, ETSI ITS mandates that a vehicle that wants to exchange safety information be equipped with one ETSI ITS-G5 (that is, one IEEE 802.11p) transceiver that stays tuned to the control channel (called the G5CC) at all times. Additional service channels (called G5SCs) may freely be used by additional IEEE 802.11p transceivers. Similarly, a vehicle may use additional services offered via regular WiFi or cellular technologies. ETSI ITS abstracts away from channel-access technologies, that is, the used MAC and physical layers of the various installed transceivers, introducing the concept of an *access layer*. Since the development of the ETSI ITS is heavily focusing on the use of IEEE 802.11p, we will be concerned solely with this technology in the following, and use ETSI ITS-G5 and ETSI ITS synonymously.

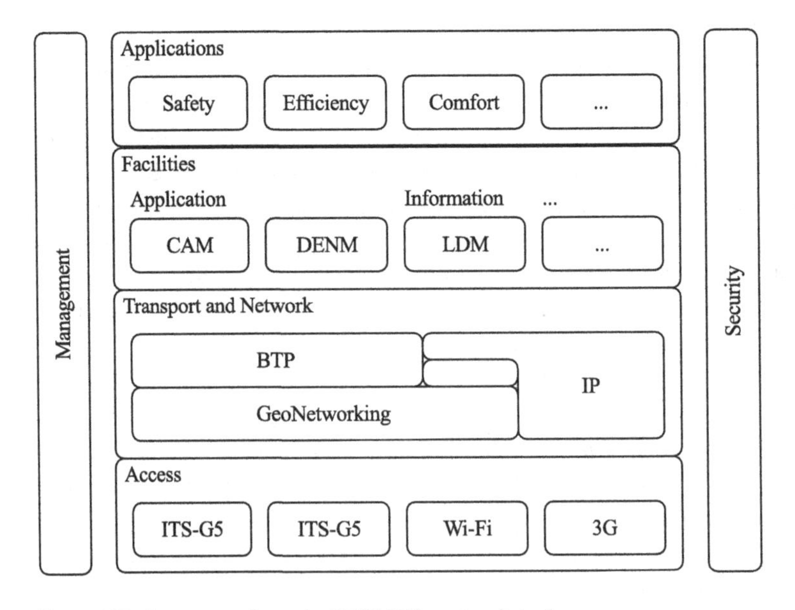

Figure 4.12 An excerpt from the ETSI ITS protocol stack.

Cross-layer services such as choosing an access technology and changing pseudonyms at ideal times (see Chapter 7) are provided by vertical *management* and *security* layers, respectively, which are available to all layers of the protocol stack.

Right below the *application* layer is the *facilities* layer (ETSI 2013a), which provides common support functionality for all levels of the ETSI ITS stack (subdivided into application support, information support, and communication support). Such example functionality concerns the sending and receiving of generic messages known as cooperative awareness messages (CAMs) and decentralized environmental notification messages (DENMs), or more specific messages like SPAT and road topology (TOPO) messages, which can be used by applications. Further, the facilities layer takes care of service-announcement-message (SAM) exchange, the equivalent of IEEE WAVE WSMs. Another central component is the local dynamic map (LDM), a database that stores information about the local neighborhood.

A *networking and transport layer* sits in between the access layer and the facilities layer to provide not only standard TCP/UDP over IP services, but also the ETSI ITS-G5 basic transport protocol (BTP) over a *GeoNetworking* service. GeoNetworking employs the access layer for multi-hop data dissemination to nodes that are identified solely by their geographic position. This caters to the most common use case of safety protocols: Instead of addressing messages to named nodes, they need to be delivered to nodes identified solely by their distance to the sender, the current road, or similar attributes. It should also be noted that, as an alternative to the BTP protocol, there exists a very lightweight protocol (ensuring neither reliable, de-duplicated delivery nor in-order delivery) that merely serves to deliver messages to appropriate upper layers on the basis of a 16-bit BTP address. GeoNetworking services can also be used by other applications, such as those using IPv6, by means of an adaption layer.

The GeoNetworking service relies heavily on neighbor tables, which are maintained via periodically transmitted beacons. Each vehicle's location table stores information such as neighbors' addresses and their type (such as vehicle or roadside station), along with the required geographic information like position, speed, and acceleration. The GeoNetworking layer uses this information to provide its services, which are explained in detail in Section 5.5.1 on page 197.

4.3 White spaces and cog radio

Spectrum is a scarce resource. Outside of the ISM bands (the most commonly known of which are at 2.45 GHz and 5.8 GHz), access to spectrum is tightly regulated. Ever since the advent of personal mobile communications, spectrum has become increasingly scarce – and increasingly expensive. To give just one example, *Auction 73*, the sale of spectrum in the 700-MHz range by the US FCC in 2008, raised a total of 19 billion US$ in winning bids for 62 MHz of spectrum.

Since the late 1990s, the common practice of coarse-grained fixed allocation of spectrum to users has repeatedly been found to be very wasteful. Just like other countries' regulatory bodies, in 2002, the FCC conducted an investigation into possible

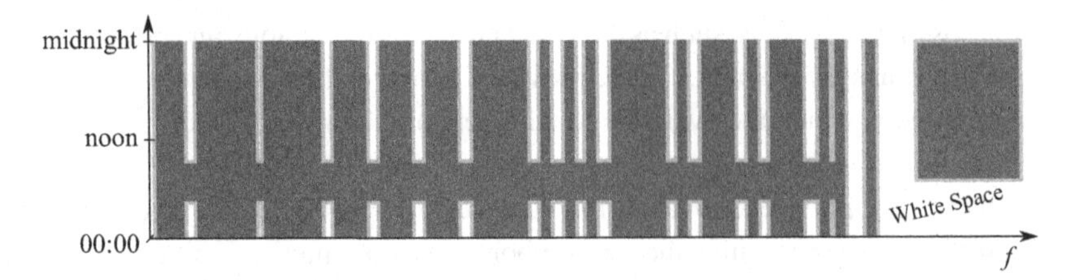

Figure 4.13 An illustration of radio spectrum *white space*: Spectrum, though fully allocated, is often unused. Guard bands between frequencies and temporally varying usage (black) leave large parts of the spectrum underutilized.

improvements in the way spectrum is managed. Among its chief findings was that many portions of spectrum are not in use for extended periods of time and/or in large geographic areas (FCC 2002). These unused parts of the spectrum, illustrated in Figure 4.13, are called *white space*. The FCC thus explicitly recommended that policies for spectrum sharing be introduced into regulations: spectrum should be licensed to *primary users*, but other users should be allowed to transmit if (and only if) the channel is unused.

Worldwide, regulatory bodies are currently rethinking how to manage the available spectrum (Nekovee *et al.* 2012). Instead of assigning spectrum statically, both in time and in space (current allocations are often country-wide and usually last for decades) and exclusively (e.g., to a MNO), regulatory bodies are now considering allowing spectrum to be shared or even sublet. This would mean that primary users could freely negotiate a price and extent of use with non-primary users.

4.3.1 Cognitive radio

This sharing of spectrum can be achieved by a device capable of dynamic spectrum access (DSA), that is, a device that can flexibly use different parts of the spectrum. Such a device is commonly referred to as a *cognitive radio*.

The term was first coined in 1999 for intelligent transmitters that can actively learn from, reason about, and plan according to their environment (Mitola & Maguire 1999). The name is inspired by the cognitive cycle, a well-established feedback model of cognition with a long tradition (Neisser 1976). Figure 4.14 illustrates a modernized version adapted to cognitive radios.

In order to plan ahead, a cognitive radio needs to know about its user's and its network's requirements as well as its own capabilities. Cognitive radio technology has thus been expected to implement learning and reasoning algorithms from the domain of machine learning (e.g., reinforcement learning or case-based reasoning). Further, in order to be fully adaptive to the radio environment, a fully reconfigurable transmitter, such as a software-defined radio (SDR), must be employed in a cognitive radio.

The term *cognitive radio* has also been much more loosely applied to refer to any radio that can adapt to its environment. Current designs for such radios are much less

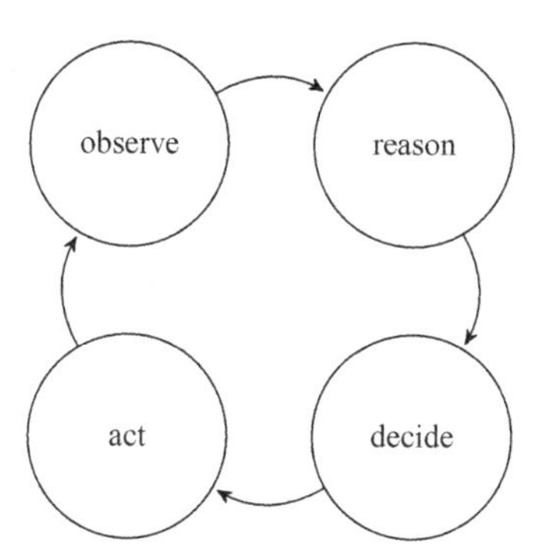

Figure 4.14 A simplified, adapted cognitive cycle, the basic concept of a cognitive radio

ambitious, operating on a narrow range of frequencies and according to very simple, fixed rules to exploit white space.

Still, how to accurately detect which frequencies constitute white space remains an active research field. The most direct approach is spectrum sensing, whereby each device autonomously monitors a range of frequencies to detect white space.

In the simplest case, a device performing spectrum sensing will rely on energy detection alone. While this is arguably the most direct approach, it relies on a very strong signal. In the presence of hidden nodes, however, signal levels can easily drop well below thermal noise, rendering transmissions sent by primary users undetectable. Therefore, more advanced techniques that can exploit known features of signals sent by primary users have been developed. Some examples are, in decreasing order of complexity, matched filters (that is, correlating the signal with a known template), cyclostationary feature detection (that is, exploiting periodicity in the signal or derived statistics), or autocorrelation detection (that is, exploiting the fact that OFDM signals carry a cyclic prefix). Further, considerable progress has been made in the domain of cooperative sensing, whereby multiple devices exchange spectrum-sensing information.

Spectrum sensing can further be supported, or obviated, with indirect approaches. Some mandate that easily detectable radio-frequency (RF) beacons (simple transmitters sending a sine wave) are operated at the location of primary users. As an alternative, these locations are directly made available to non-primary users in the form of a geolocation database. This database can also contain further information, like maximum power levels or time-dependent information, which is why this approach is safest for primary users.

4.3.2 TV white space (TVWS)

Flexible use of white space is slowly becoming commercially accepted. One example is TV white space (TVWS). Traditionally, spectrum planning for analog TV broadcast needed to make allowance for inefficient receivers that would be able to decode only high-powered signals at high signal-to-noise ratios (SNRs). This required broadcasters to operate a multi-frequency network: in order to avoid interference, different locations needed to be served on different frequencies, and large guard bands had to be introduced.

Today, with digital TV broadcast working with much more advanced receivers, such spectrum allocation would waste a large portion of the spectrum. With the digital switchover of TV broadcasting in countries around the world in the years from 2006 to 2024, the spectrum was (or will be) heavily consolidated, leaving large amounts of white space.

In the USA, unlicensed use of TVWS has been allowed by the FCC since 2008, initially under tight constraints like requiring all devices to continuously monitor for primary users. Since 2010 it has been allowed using a geolocation database alone, and since 2012 the first commercial systems have been deployed. Japan followed shortly after, and similar deliberations are currently being made by the ECC in Europe.

Aside from many newly introduced technologies and standards like the IEEE 802.22 wireless regional area network (WRAN) standard, existing standards are also being, or have been, amended with physical layers. One example is IEEE 802.15 Task Group 4m, which is tasked with specifying a physical-layer amendment to IEEE 802.15.4. IEEE 802.11 WLAN has also recently been extended with a physical layer that operates in TVWS. It is based on regular OFDM with channel widths of 6 MHz, 7 MHz, or 8 MHz (optionally allowing two adjacent channels to be grouped, and two non-adjacent groups to be grouped further). This physical layer was standardized in the IEEE 802.11af amendment in 2013. Stations in an IEEE 802.11af network are expected to request a list of available channels and channel schedules from other stations connected to a geolocation database, typically an access point. Which of multiple available channels to use and how the list of available channels and their schedules is obtained are both outside the scope of the standard.

4.3.3 Use of white space for IVC

Cognitive radio technology for exploiting TVWS also has promising applications in vehicular networks. The frequencies of the TVWS are much lower than the dedicated frequency bands for IEEE WAVE and ETSI ITS-G5 (in the MHz instead of GHz range), thus they profit from much better propagation conditions. This is especially true in urban environments, where buildings are the main factor limiting the range of IEEE 802.11p transmissions – allowing TVWS signals to be received at much greater distances. Using white space for IVC is also especially promising in countries where no or only small dedicated frequency bands could be set aside for IVC, such as Japan.

Figure 4.15 illustrates the principle of using white space for IVC: If vehicles, the non-primary users, are prepared to switch between frequencies when they detect active primary users, almost uninterrupted service can be achieved even on licensed spectrum. However, cognitive vehicular networks, as such networks have been termed (Di Felice *et al.* 2012), have to cope with unique challenges.

First and foremost, fast device motion adds another dimension to any scheduling problem. This not only requires much more frequent measurements, but also all but mandates cooperative sensing to predict spectrum occupancy on the route ahead.

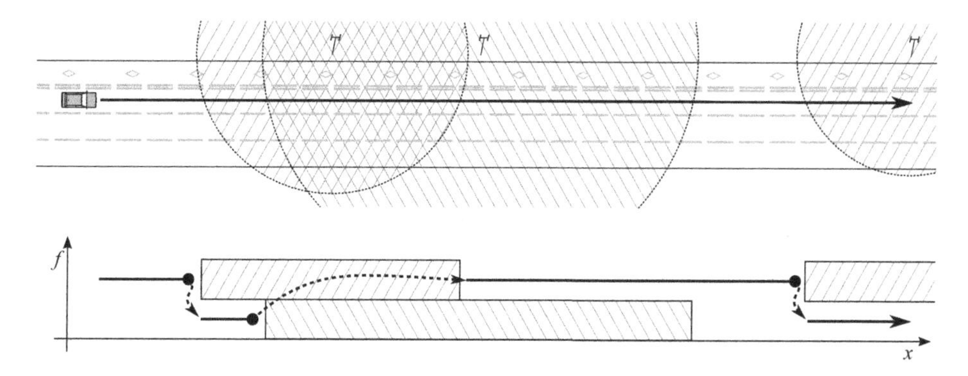

Figure 4.15 An illustration of dynamic use of white space in IVC: Licensed primary users utilize spectrum only in some areas and only at some times. The resulting white space can be exploited by non-primary users if they are prepared to switch between frequencies.

Second, efficient cooperative sensing is complicated by rapidly changing neighbor sets – and it is in direct conflict with privacy constraints (which we discuss in Chapter 7) by virtue of adding additional identifying information to exchanged data. Third, Doppler spread and short contact times further complicate spectrum sensing. Lastly, channel sensing requires quiet times in order to be able to detect primary users, which is in direct conflict with safety considerations.

On the positive side, vehicles' mobility also enables each individual device to perform spectrum sensing at multiple locations over time, reducing, e.g., the impact of shadowing effects. As an extreme case, cognitive radios could also give direct feedback to route planning to improve available spectrum, further broadening the solution space for planning decisions of cognitive radios.

Analytical evaluations of the feasibility of supplementing or replacing communication in the DSRC band with communication in TVWS by Chen *et al.* (2011b) gave promising results: On the basis of measurements conduced along interstate I-90 in Massachusetts, the authors reported that, for as much as 60 messages per second, channel access should be possible with mean delays of below 13 ms in 95% of cases (albeit with the notable exception of mean delays sometimes reaching 160 ms near Boston).

Early trials in metropolitan areas produced similarly promising results (Altintas *et al.* 2014). Here, vehicles have been successfully used as long-range message ferries in a presumed disaster scenario where public and mobile networks were no longer available.

Lim *et al.* (2014) conducted a simulation of the benefit of using TVWS in addition to a DSRC band for IVC emergency-message dissemination in a completely self-organizing system. For this, they assumed that each vehicle is equipped with two radios: one tuned to a DSRC channel, and one free to tune to any of up to 30 TVWS channels.

Their reasoning was that transmissions in TVWS channels are likely to reach much further than those in DSRC channels, so any of the available TVWS channels should preferably be chosen for at least the initial transmission of an emergency message. Yet, a TVWS channel might not always be available, and the chosen TVWS channel might be jammed for some vehicles. They thus proposed additional proactive re-transmissions

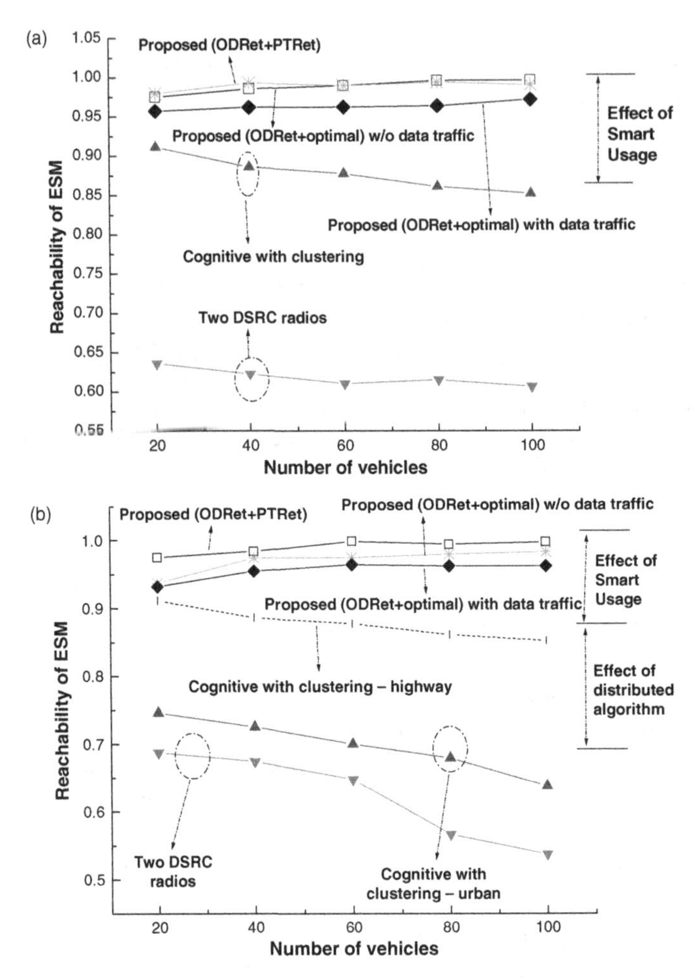

Figure 4.16 Benefit of using TV white space (TVWS) in addition to the DSRC band in two scenarios (a) highway and (b) Manhattan grid. Shown is the probability of receiving an emergency safety message ("ESM") when using two DSRC bands only ("two DSRC radios"), using one DSRC band and one TVWS band managed by cluster heads ("cognitive with clustering"), and using completely self-organizing TVWS approaches with DSRC for error recovery ("proposed"). © [2014] IEEE. Reprinted, with permission, from Lim *et al.* (2014) with correction.

via different TVWS channels, if available. Further, they proposed that one should also flood a short notification signal via a DSRC channel to enable vehicles to selectively request repeating the TVWS transmission via the DSRC channel.

They proposed using IEEE 802.11p for transmitting in the DSRC band and IEEE 802.11 on 5-MHz channels in TVWS. The rendezvous scheme of transmitter and receiver follows a straightforward scanning approach: Receivers continuously scan through all available channels, stopping the scan if they detect a reference tone. The transmitter can thus pick any of the available TVWS channels (ideally the best channel, e.g., the one available to most of the intended service area and having the lowest background signal level). The duration of the reference tone signal (here, repeated transmissions of the IEEE 802.11 preamble) need be only as long as it takes a receiver to scan through all channels to allow them to rendezvous. The authors estimate this to be at most 2.205 ms (from reference hardware and measurements). The transmitter can then transmit the emergency message.

Figure 4.16 illustrates the result of their simulations, conducted in both a highway scenario and a Manhattan-grid-type scenario. Here, Lim *et al.* (2014) compared their scheme with one using clustering to coordinate transmissions in a more straightforward fashion, demonstrating that an improvement of 17% could be attained in the highway scenario and 56% in the Manhattan-grid-type scenario. Even more interestingly, they also compared their proposed scheme with the straightforward use of two IEEE 802.11p radios operating in the DSRC band, demonstrating that a 64% improvement could be attained in the highway scenario and an 86% improvement in the Manhattan-grid-type scenario. This underlines the potential that the use of TVWS can offer for IVC communication.

5 Information dissemination

The objective of this chapter is to discuss data dissemination in vehicular networks in detail. We concentrate essentially on network-layer and application-layer protocols, which are often discussed and developed as a single protocol above the respective access technologies. The key objective is to achieve information exchange between any two vehicles, from one vehicle to all neighboring ones, from one vehicle to infrastructure components, from infrastructure to one or all neighboring vehicles, and dissemination from a vehicle to all those that are interested in the content.

We start by looking at ad-hoc routing protocols that were suggested in the early days of vehicular networks. Having identified their limitations, we explore alternatives, starting with geographic routing and geocast as communication primitives, and then exploring one of the most promising domains in the scope of vehicular networks: beaconing or one-hop broadcasting. In this framework, this chapter details the basic principles underlying such inter-vehicle communication (IVC) approaches, namely unicast, local broadcast, anycast, and multicast communication principles as used in vehicular networks. We of course investigate the current state of the standardization efforts, primarily focusing on the European Telecommunications Standards Institute (ETSI) standardization towards cooperative awareness messages (CAMs) and decentralized environmental notification messages (DENMs) as well as the ETSI GeoNetworking initiative. Furthermore, we explore options for exploiting available infrastructure such as roadside units (RSUs) or even parked vehicles to provide access to some backbone network or to help spread information among the vehicles. We conclude this chapter with a discussion of delay/disruption-tolerant network (DTN) approaches and the use of concepts known from Internet-based peer-to-peer networks.

This chapter is organized as follows.

- Ad-hoc routing (Section 5.1) – We start by exploring classical mobile ad-hoc network (MANET) routing algorithms in detail, because some basic knowledge of ad-hoc routing is needed in order to understand the more sophisticated vehicular ad-hoc network (VANET) routing options. In this section, we also discuss the applicability of MANET routing to VANETs as well as the specific challenges resulting from the underlying mobility pattern and delay constraints for vehicular safety applications.
- Geographic routing (Section 5.2) – In this section, we study a conceptual alternative to the address-based ad-hoc routing solutions: geographic routing. Instead of

relying on static identifiers such as IP addresses, geographic coordinates are used for greedy routing, which selects the next-hop forwarder that is geographically closest to the final destination. These geographic addresses can also be used to provide data management in form of a distributed hash table (DHT)-like structure. We will see that geographic routing faces a major problem, namely routing voids or dead ends. Virtual-coordinate-based routing solutions help to solve this problem by assigning virtual coordinates to all nodes.

- Beaconing (Section 5.3) – Beaconing or one-hop broadcast has become the premier technique for data dissemination in vehicular networks. In this section, we investigate simple static beaconing solutions as introduced in the scope of the SOTIS system and later standardized by the ETSI for CAMs.
- Adaptive beaconing (Section 5.4) – Following the introduction of beaconing, we explore in this section the most recent approaches for beacon-based information dissemination. The key challenge is distributed congestion control. We study an initial approach towards adaptive beaconing, the adaptive traffic beacon (ATB), which later inspired the development of the ETSI's decentralized congestion control (DCC). Besides the possibility of scheduling beacons in time, DCC also considers adapting the transmit power appropriately. Additionally, we investigate solutions trying to push the limits even further in order to be able to handle the high degree of dynamics in vehicular networks.
- Geocasting (Section 5.5) – In this section, we explore geocasting, a combination of broadcast and greedy forwarding. Focusing first on the ETSI's GeoNetworking approach, we investigate the potential of extending the CAM mechanism for targeted safety messages as implemented in the DENM protocol. Furthermore, we study the topology-assisted geo-opportunistic routing (TO-GO) concept, which allows one to select forwarding nodes according to the road network topology.
- Infrastructure support (Section 5.6) – In this section, we assess possibilities to support IVC by means of available infrastructure. This can be RSUs deployed along the street curb or at intersections complementing available traffic lights. These RSUs can help by providing access to a backbone network such as the Internet or can be used just to exchange information between the vehicles. If no RSUs are available, parked vehicles might take this role and continue, even if parked, to participate in the vehicular network.
- DTN and peer-to-peer networks (Section 5.7) – In this final section, we investigate the use of concepts known from the DTN and peer-to-peer networking domains. The store–carry–forward principle helps to interconnect disconnected groups or clusters of vehicles. After studying this DTN principle, we examine the use of BitTorrent-like information downloading in vehicular networks. For this approach, DTN and peer-to-peer concepts need to be combined to achieve optimized performance results. We conclude with a study of PeerTIS, a fully distributed traffic information system (TIS) making use of all the vehicles as an information store that collects and provides road traffic information to all participants of the system.

5.1 Ad-hoc routing

MANET routing has a long history in the domain of wireless networking (Abolhasan *et al.* 2004). Many of the concepts developed have even been taken to standardization – mainly in the context of Internet Engineering Task Force (IETF) standards.

In Section 3.3.3 on page 80, we briefly introduced the concept of ad-hoc routing as an early candidate for information dissemination in vehicular networks. Following the taxonomy provided there, we now explore the area of topology-based protocols in more detail.

Figure 5.1 revisits the general taxonomy discussed in Section 3.3.3 on page 80 but focuses on those routing solutions that are based on network topology information. Here, routing information updates (most commonly, about k-hop neighbors) are considered the basis for all routing decisions. As can be seen, this domain can be divided into three groups of solutions.

- *Proactive* routing refers to the classical Internet routing protocols. Information about the topology is collected before any data is available to be transmitted over the network. The advantage is that information on adequate paths is available instantly whenever data is ready for transmission. The downside is that the routing protocols have to continuously update the topology, which is especially hard in dynamic environments such as vehicular networks, even if no transmissions are scheduled.
- *Reactive* routing solutions prevent the continuous topology update if no data is to be transmitted. Thus, the network is trying to prevent maintenance if current topology information is simply not needed. Instead, topology management is started anew for each data packet to be transmitted.

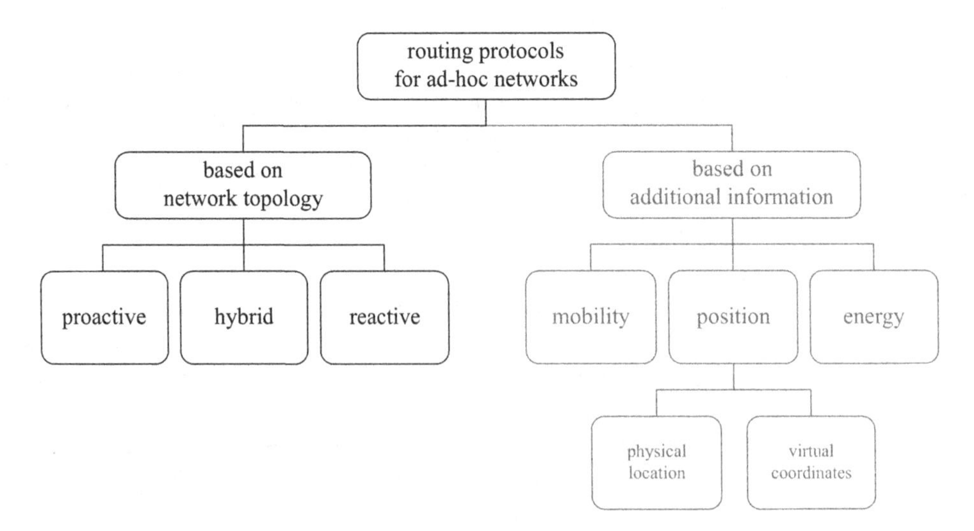

Figure 5.1 A taxonomy of routing protocols based on network topology in the taxonomy of ad-hoc routing protocols.

- Hybrid solutions try to combine the advantages of both worlds. The most direct approach is to follow a hierarchical concept, e.g., using reactive routing at the lower level in clusters of nodes but proactive routing among the different clusters.

In this section, we will explore selected examples in detail. For more information, we'd like to point the reader to some excellent textbooks on ad-hoc and sensor networks (Dressler 2007; Karl & Willig 2005).

5.1.1 Proactive routing protocols

The key idea of proactive routing protocols is to provide active routes to all possible destinations all the time. This is in line with the concept of Internet routing, which assumes routing tables to be established as soon as nodes join or leave the network, i.e., whenever the topology changes. In order to detect such changes, continuous topology maintenance must be active. This is usually achieved either by means of periodic exchanges of hello messages between neighbors in the network or by means of link monitoring in wired networks.

One of the first approaches to proactive routing in the scope of MANETs was the destination sequenced distance vector (DSDV) protocol (Perkins & Bhagwat 1994). DSDV is a typical table-driven protocol, which is based on the Bellman–Ford algorithm which is also used for the routing information protocol (RIP), one of the early routing protocols for local area networks in the Internet (Hedrick 1998). Improvements have been added to the standard algorithm according to the special needs of mobile ad-hoc networks.

The Bellman–Ford concept is to periodically send full routing-table updates to all neighboring nodes. The receiving node checks its local routing table to ascertain whether better routes to certain locations have been found. As a distance measure, usually the number of hops is used as the key metric. Therefore, this algorithm is also called distance-vector routing. If an update of the local routing table has been performed, the node broadcasts the updated routing table itself. This process continues until no further changes can be detected, i.e., the routing protocol converges to a stable state.

The key characteristic of all variants of the protocol is the convergence time. This can become problematic if the cost changes – one of the most critical problems is the so-called count-to-infinity problem, which is based on the characteristic that routes that are no longer available might be temporarily replaced by incorrect (stale) routes until the protocol converges again. We do not want to dig deep into the microscopic behavior of distance-vector routing. For more information, we refer the reader to standard computer networking textbooks (Tanenbaum & Wetherall 2011).

DSDV follows this scheme, but, in order to prevent too expensive updates following topology changes, incremental route updates have been introduced. This helps to reduce the amount of information transmitted during the convergence process but does not reduce the number of update messages necessary.

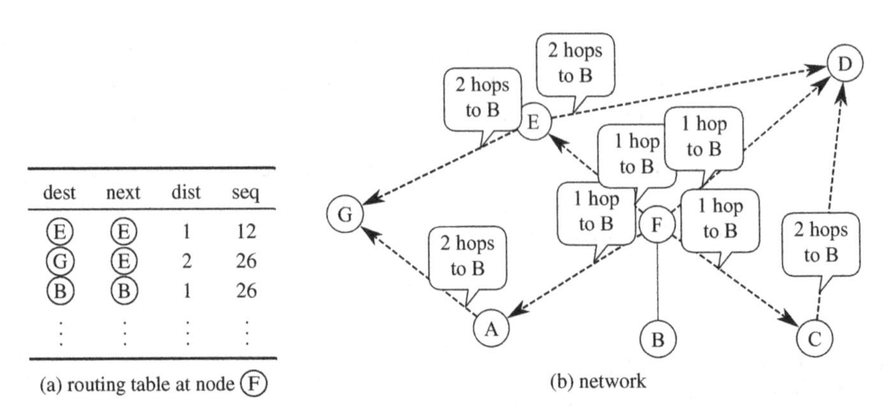

dest	next	dist	seq
Ⓔ	Ⓔ	1	12
Ⓖ	Ⓔ	2	26
Ⓑ	Ⓑ	1	26
⋮	⋮	⋮	⋮

(a) routing table at node Ⓕ (b) network

Figure 5.2 The principles of distance-vector routing as used by DSDV. The example focuses on the routing information to node Ⓑ, which is stepwise distributed through the entire network.

The propagation of an (incremental) update is depicted in Figure 5.2. Node Ⓕ detects a new neighbor (node Ⓑ) and adds this information to its local routing table. According to the rules, this knowledge is distributed to all neighbors immediately when the routing table changes. The update is then stepwise propagated through the network as each node receiving the information includes the new entry pointing to node Ⓑ in its routing table. If a copy of the information is received by any of the nodes, it checks whether it needs to update the routing information, i.e., if the newly propagated path information is better in terms of path length, then an update is made.

An abbreviated example of a routing table is also outlined in Figure 5.2. The table lists all known nodes in the network, i.e., full topology information. The table also includes information about the last sequence number. This is used to prevent routing loops. Each node receiving information about a certain destination first checks whether the information received is more current than the information stored in its local routing table. Furthermore, DSDV suggests that one should update unstable routes with some delay in order to prevent fluctuations.

Its applicability to routing in vehicular networks is obviously not straightforward. Most critical is the general problem of proactive routing protocols needing to keep track of frequent topology changes. Topology in vehicular networks, however, changes extremely fast. Such changes generate a huge amount of update messages that, in the extreme case, may prevent any other communication in the network. Let's discuss the alternative, namely reactive routing, in the next section.

5.1.2 Reactive routing protocols

Reactive routing protocols (they are also called on-demand routing) have been developed with the intention of obviating the need for the quite expensive periodic routing exchange, which may consume an essential amount of the available network resources. The main idea is to search for paths to specific nodes only if data packets have been scheduled for transmission. This leads to a substantial reduction of the control traffic in the network if one of the following assumptions holds:

- messages are sent infrequently or in short bursts of multiple packets, or
- messages are sent to a small subset of possible destination nodes only.

The general idea of reactive routing protocols is therefore to identify suitable routes to a destination node exactly at the time when it is needed. Thus, the routing protocol overhead is reduced to an absolute minimum while an additional delay is introduced for the setup of routing information.

No routing information exists at the time of packet forwarding (at least for the very first packet that is to be sent to this destination node), or, more precisely, initially, no information at all is available about the next hop. The general approach is to send the data packet (or a special route-discovery message) to all nodes in the network, i.e., flooding the network with this message. This flooding process allows all of the nodes involved (i.e., typically all nodes in the network) to learn a path to the source node. At some point, the flooded message reaches the destination, which, in turn, is able to reply to the received message by sending a message back to the source node. This reply can be sent on the shortest path that has just been learned and is also used for backward learning of the best route from source to destination.

Since the early days of MANETs, many different reactive routing protocols have been introduced. The best-known examples include dynamic source routing (DSR), ad-hoc on-demand distance vector (AODV), and dynamic MANET on demand (DYMO) (Basagni *et al.* 2004; Dressler 2007). We will investigate these protocols as case studies that are representative for the entire domain of on-demand routing. Especially AODV, including many extensions, and DYMO have also been investigated for their applicability in VANETs.

Dynamic source routing

One of the first reactive ad-hoc routing protocols was dynamic source routing (DSR) (Johnson & Maltz 1996; Johnson *et al.* 2001), which was finally standardized by the IETF (Johnson *et al.* 2007). In contrast to most typical routing protocols, DSR does not maintain routing tables. Instead, it relies on *source routing*, i.e., the idea of providing routing information directly in each transmitted data packet. Thus, the packet itself describes the path it wants to follow through the network in its header. The source node alone is responsible for establishing and maintaining a route through the network in order to create this header information. The advantage is that no topology maintenance is needed on a global scale, albeit at the cost of increased packet sizes (this can be substantial for longer network paths).

For the route-discovery process, DSR uses so-called route request (RREQ) messages. These RREQs are flooded through the entire network to discover a path from source to destination. All routing information is stored in the RREQ; that is, each node forwarding the RREQ adds its own address to the already-established path. All nodes are supposed to cache the RREQs for a while in order to detect loops and duplicates. The intention is that, at some point, the discovery packet will reach the destination node, which, in turn,

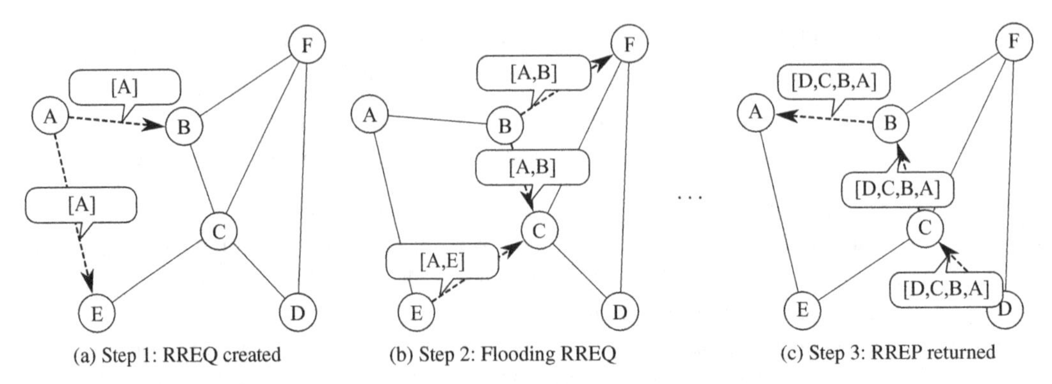

Figure 5.3 The route-discovery process of DSR.

will send an answer back. This route reply (RREP) message can be unicast using the routing information stored in the received RREQ but in reverse order.

The overall process is depicted in Figure 5.3. In this example, node Ⓐ starts the discovery process (step 1) sending an RREQ to all neighbors (in wireless networks, this is of course a single broadcast message). Each node receiving the message re-broadcasts the RREQ after adding its own address to the current path information stored in the header (step 2). If eventually the RREQ arrives at destination node Ⓓ, an RREP is sent back to node Ⓐ on the shortest path, informing the source about the discovered path (step 3).

Route maintenance is needed in order to discover and repair failing connections. If a link breaks, i.e., if a node was not able to forward a message using source routing, a route error (RERR) message is created and sent back to the source. There, route construction is re-initiated by creating and flooding another RREQ.

For optimizing the protocol performance, multiple concepts have been proposed. Route caches can be used to store information about nodes in the network and use this information to reply to a received RREQ on behalf of the destination node if a path is already known. This also helps to reduce the communication overhead caused by frequently flooded RREQs. In order to reduce the potential reply storm (many nodes might be aware of a path to the destination node), backoff techniques have to be implemented.

The negative side effect is that stale routes (that have to be updated by means of another RREQ) might be propagated. This effect becomes especially critical for fast topology changes. In high-mobility scenarios, the quality of route caches may degrade significantly. DSR has no built-in mechanism to locally repair broken links. Piggy-backing data messages on RREQs helps to overcome this issue to a certain extent. Yet, the available payload size is quite limited.

Ad-hoc on-demand distance vector

The best-known reactive routing protocol is ad-hoc on-demand distance vector (AODV) (Perkins & Royer 1999, Perkins *et al.* 2003). AODV combines table-driven routing according to the distance-vector concept with reactive route discovery. The protocol has been extended over the years to support quality-of-service (QoS) and multi-path routing. Here, we concentrate on the original version in order to get a basic understanding of reactive routing as used in the scope of MANETs.

The route-discovery procedure is similar to that of DSR. Routes are searched by means of flooding RREQ messages until the destination node has been discovered. It, in turn, replies with an RREP to establish the route. Yet, instead of storing all routing information in the discovery packets, all nodes maintain routing tables similar to the DSDV approach. For each route-discovery message (RREQ or RREP) received, the node checks the local routing table and updates the respective information pointing to the source or destination, respectively, if necessary.

Entries in the routing table are associated with a timeout to remove outdated routing information. However, entries are refreshed for each data packet forwarded; thus, timeouts are delayed until the path breaks or until data packets are no longer transmitted. Sequence numbers are used to ensure the freshness of routes and to handle stale routes.

The route setup process of AODV is outlined in Figure 5.4. In the example shown, node Ⓐ searches for a path to node Ⓓ. It therefore floods an RREQ looking for the destination node (step 1). During this discovery process, all intermediate nodes update their routing tables accordingly, i.e., the routing entry pointing to node Ⓐ is either updated or created. When node Ⓓ finally receives the RREQ, it replies with an RREP, which is unicast on the shortest path (step 2). This is possible because all nodes already have routing-table entries pointing to node Ⓐ. As the RREP is propagated back to the source, all involved nodes add or update routing-table entries pointing to the destination node Ⓓ. Once the source node has received the RREP, it may begin to forward data packets to the destination.

The resulting routing tables are summarized in Figure 5.4(c). As can be seen, all nodes in the network have an entry pointing to the source node Ⓐ, which flooded the RREQ through the network. Only those nodes on the shortest path, however, have an entry for node Ⓓ, because the RREP has been unicast to the source node.

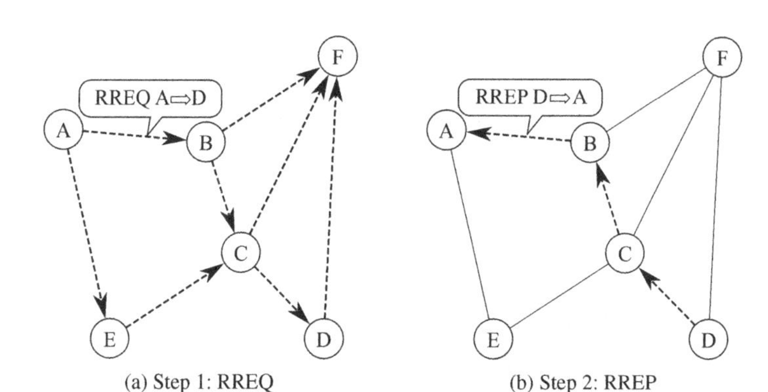

(a) Step 1: RREQ

(b) Step 2: RREP

node	dest	next	dist
Ⓑ	Ⓐ	Ⓐ	1
Ⓑ	Ⓓ	Ⓒ	2
Ⓕ	Ⓐ	Ⓑ	2
Ⓒ	Ⓐ	Ⓑ	2
Ⓒ	Ⓓ	Ⓓ	1
⋮	⋮	⋮	⋮
Ⓓ	Ⓐ	Ⓒ	3

(c) Resulting routing tables

Figure 5.4 The route-discovery process of AODV.

If an intermediate node receives an RREQ that it has already processed, it discards the RREQ and stops the flooding process. In order to reduce the overhead caused by flooding the RREQ, AODV uses an *expanding-ring search* strategy for the route setup. The RREQ will be sent with increasing time-to-live (TTL) values in order to affect only parts of the network. Furthermore, intermediate nodes are encouraged to answer an RREQ on behalf of the destination node if a path to the destination is already available in the local routing table.

If, during data communication, a message cannot be forwarded to the next hop node, i.e., a link failure has been detected, this information is advertised to the source node by means of an RERR message. After receiving the RERR, the source node can re-initiate the route discovery.

In typical ad-hoc networks, AODV performs well even in networks with limited mobility. Unsolicited RERR messages can improve the performance by announcing broken links even if no data packet needs to be forwarded (hello messages can be used to update neighborhood information – which is called beaconing in the standard but has no relation to beaconing protocols in vehicular networks). Since all intermediate nodes maintain a local routing table, they can make immediate use of this information both for data dissemination and for reducing the overhead caused by the flooding of RREQs. In the case of high node mobility, the overhead substantially increases because links tend to frequently break and the route discovery process needs to be re-initiated.

Dynamic MANET on demand

In order to overcome some of the limitations of AODV and to make it more robust with respect to mobility, the dynamic MANET on-demand (DYMO) protocol was developed, the first draft being released in 2005. After draft version 26 (Perkins *et al.* 2013), development continued as *AODV v2*. DYMO aims at a somewhat simpler design than AODV, helping to reduce the system requirements of participating nodes, and simplifying the protocol implementation.

Most importantly, DYMO introduces the concept of path accumulation as depicted in Figure 5.5. During the route-discovery process, RREQ and RREP messages are enhanced by including all intermediate nodes. This is similar to the source-routing principle used by DSR. This information can be exploited by each node processing the route-discovery messages for updating their local routing tables to point to all the nodes listed in the RREQ or RREP, respectively.

In Figure 5.5, node Ⓐ searches for a path to node Ⓓ. Each intermediate node adds its own address to the route-discovery message, so that finally the full path Ⓐ–Ⓑ–Ⓒ is listed in the RREQ. Similarly, the RREP is processed, showing eventually Ⓓ–Ⓒ–Ⓑ in the list. That means that each node in this simplistic networking example learns about all other nodes in the network just using a single RREQ and RREP exchange. In more complex ad-hoc scenarios, an essential part of the network can be discovered using just a few routing messages.

In the case of link failures, DYMO also uses RERR messages. Yet, these error messages are broadcast to all neighbors and then forwarded using multicast communication to all nodes that likely use the same routing information.

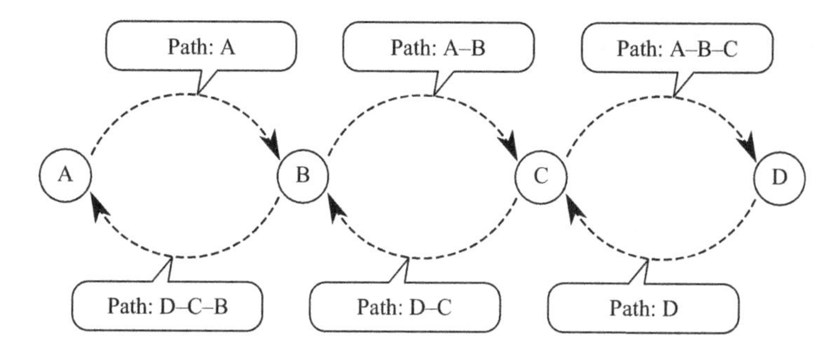

Figure 5.5 The route-discovery process of DYMO.

Compared with AODV, DYMO primarily benefits from the possibility of passively learning topology information of intermediate nodes. The network load due to protocol overhead can be significantly reduced in most scenarios.

5.1.3 Application in VANETs

The use of ad-hoc routing in VANETs has a long history. A very good survey on the initial approach to routing in vehicular networks was published by Chennikara-Varghese *et al.* (2006).

We start discussing the applicability of ad-hoc routing protocols with a case study for the use of DYMO in VANETs. The DYMO protocol provides all the means for communication between arbitrary vehicles but also for coupling of a MANET with the Internet. This makes an evaluation of communication connections between mobile nodes and static infrastructure especially attractive.

As we will see, it is difficult to say the least to apply ad-hoc routing to the area of vehicular networks. Two additional issues need to be addressed if one is to make MANET algorithms fit for VANETs.

- *Mobility patterns* are very different from the random-waypoint or similar models that were frequently considered during the development of MANET routing algorithms. Mobility in general is a critical issue for most of the established ad-hoc routing protocols.
- *Delay-constrained routing* needs to be addressed if we are considering ad-hoc routing for all possible vehicular networking applications. Not only do safety applications demand bounded delays, but also multimedia streaming applications imply certain QoS constraints.

Thus, we also explore some of the related problems and proposed solutions. Given the multitude of approaches and publications in this direction, our discussion cannot, of course, cover all the possible directions, but rather is meant to outline the concepts on a higher level.

Using DYMO to exchange road traffic information

As outlined by Sommer & Dressler (2007), a car taking part in a VANET scenario could already establish such connections in reach of one of the RSUs along major freeways or at the curbside in suburban environments. Apparently, this coupling of VANETs and the Internet is especially attractive for road users if it allows the utilization of virtually all existing resources of the Internet without relying on expensive dedicated channels provided by a cellular network.

We focus on selected results from a work in which the feasibility, performance, and limits of ad-hoc communication using DYMO were evaluated (Sommer *et al.* 2008). Special care was taken to provide realistic scenarios of both road traffic and network usage. This was accomplished by simulating a variety of such scenarios with the help of the Vehicles In Network Simulation (Veins) simulator, which couples two well-known simulation tools for network simulation and road traffic microsimulation, respectively (Sommer *et al.* 2011b).

The key objective of this study was to understand the behavior of the routing protocol DYMO under varying environmental conditions (vehicle density) and with different application-layer logics (centralized vs. fully distributed).

Evaluation scenario

In order to understand the protocol behavior of DYMO in vehicular environments, two scenarios were investigated, as illustrated in Figure 5.6. The common background is to implement a TIS used to inform other cars about unplanned stops by means of IVC. In particular, vehicles that have stopped, i.e., those with a speed of zero, start disseminating this information to other cars in order to help them to avoid this road segment. When

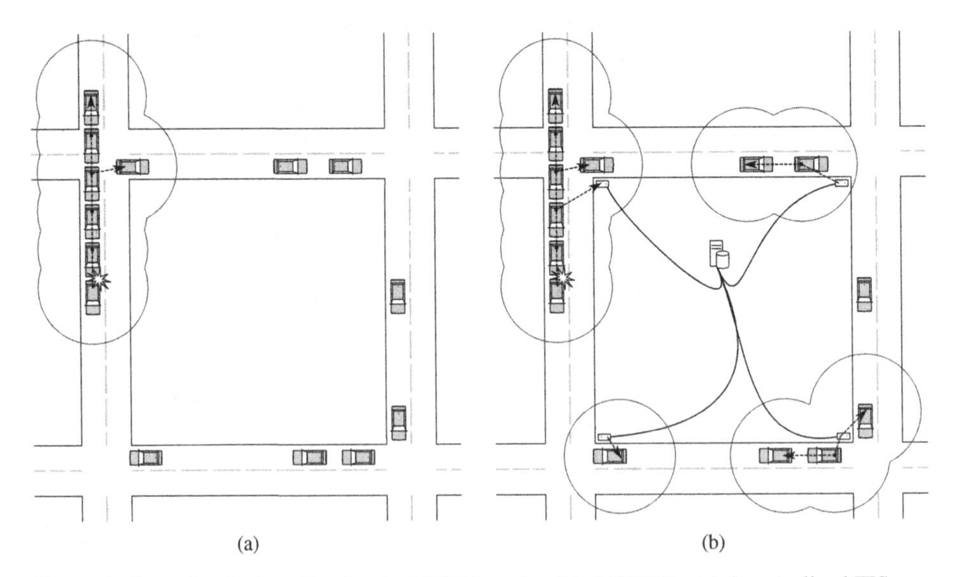

Figure 5.6 Scenarios for investigating the DYMO protocol in VANETs: (a) decentralized TIS using UDP to exchange information and (b) centralized TIS using TCP to connect to a TIC.

the originating vehicle resumes its journey, it notifies other vehicles that the road can be used again.

The scenarios have been configured as follows.

- *Decentralized TIS* (Figure 5.6(a)): Here, traffic information is disseminated by means of flooding incident warnings through the VANET over 5 or 25 hops (configured by means of a TTL), depending on the vehicle density selected. The vehicle receiving the message further connects to the originator to check whether the message is still relevant. If this can be confirmed, it marks the road segment as blocked in its local database and recalculates the route on the road network, avoiding this segment.
- *Centralized TIS* (Figure 5.6(b)): In this scenario, a TCP connection to a central TIC is established. The connection is provided by RSUs that have been placed at intersections. Vehicles maintain a TCP connection to the central server, which is used to publish and revoke incident information. Two different update intervals are used, depending on the vehicle density. The TCP connection is used to retrieve a complete list of traffic information related to the vehicle's route.

For the road network, a simple Manhattan grid has been used. Two setups have been investigated. First, a low-density scenario, in which a small 5×5 grid is established to let vehicles drive from the top left corner to the bottom right one. A second, higher-density, scenario covers a larger area of 16×16 roads.

Traffic obstructions are introduced by stopping the lead vehicle for 60 s or 240 s, depending on the scenario. Since each road offers a single lane per driving direction, nodes cannot overtake each other and, hence, need to find a way around blocked roads by means of IVC, or get stuck in traffic.

Table 5.1 lists the values used to parametrize the vehicles of the road traffic microsimulation, modeling dense downtown traffic with inattentive drivers. For all communications, the complete IP network stack is simulated and wireless modules are configured to closely resemble IEEE 802.11b network cards transmitting at 11 Mbit/s with RTS/CTS disabled. For ad-hoc routing among the nodes, the DYMO implementation by Sommer & Dressler (2007) has been used.

Simulation results

We concentrate on a subset of the results from this study; further information can be found in Sommer *et al.* (2008). In particular, we look at a metric that is not related to the

Table 5.1 Parameters used for road traffic microsimulation

Parameter	Value
Maximum vehicle speed	14 m/s
Maximum vehicle acceleration	2.6 m/s^2
Maximum desired deceleration	4.5 m/s^2
Assumed vehicle length	5 m
Driver imperfection σ ("dawdling")	0.5

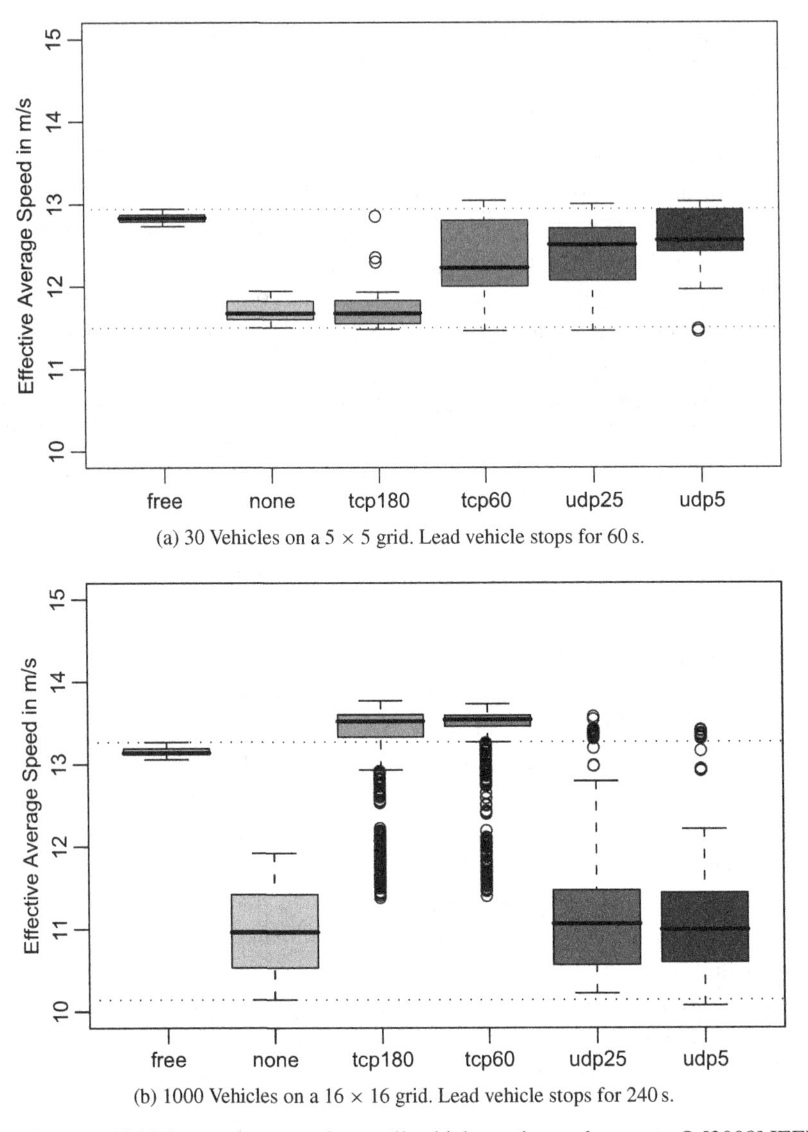

(a) 30 Vehicles on a 5 × 5 grid. Lead vehicle stops for 60 s.

(b) 1000 Vehicles on a 16 × 16 grid. Lead vehicle stops for 240 s.

Figure 5.7 Vehicle speed averaged over all vehicles and complete route.© [2008] IEEE. Reprinted, with permission, from Sommer *et al.* (2008).

wireless communication among the vehicles: the effective average speed. Essentially, it reflects the resulting travel time for the individual cars and, therefore, the effectiveness of the implemented TIS.

The results are summarized in Figure 5.7. Both graphs show the traveling speed in the form of a boxplot for six configurations. The baseline cases are *free* and *none*, representing the case of there being no traffic obstructions and the traffic behavior with no TIS, respectively. The UDP configurations differ in terms of the TTL used, whereas the two TCP scenarios differ in terms of the timeout used.

The first observation is that in free-flowing traffic the vehicles' speed is certainly higher than in the case where cars block the traffic for a short period. Also, the speed distribution among simulated vehicles in both scenarios is almost homogeneous, as could be expected.

Most importantly, however, depending on the scale of the simulation, different IVC scenarios performed differently at helping vehicles avoid this artificially generated incident. The conclusion, without yet looking at the details, is that DYMO is seemingly able to help disseminate TIS data between the vehicles, but the performance depends strongly on the scenario the vehicles are in. It has been shown that this can be generalized to the extent that only a few configurations can be found in which ad-hoc routing protocols help at all in disseminating messages between vehicles.

Let's have a brief look at the results in detail. In the small-scale setup (see Figure 5.7(a)), a polling interval of 180 s for TCP communications proved to be too long to significantly influence road traffic performance, but a polling interval of 60 s already led to a noticeable improvement. The performance was even better for UDP communication scenarios, where almost a quarter of the vehicles did not suffer from increased travel times due to the simulated incident if the TTL was reduced to five hops.

In the large-scale scenario (see Figure 5.7(b)), however, the results were almost reversed. UDP communications could only insignificantly improve road traffic performance. Instead, both TCP configurations performed very well. When a small polling interval was used, almost all vehicles reached their goal even faster than they could in the case of unobstructed traffic without IVC, thanks to a large number of vehicles taking alternate routes, which reduced traffic densities and helped avoid micro-jams that automatically happen before vehicles turn at intersections.

Mobility pattern

The key problem for all MANET routing algorithms is mobility in general. Depending on the specific mobility pattern, ad-hoc routing protocols quickly tend to become overreactive, i.e., to spend most of the available communication resources on topology maintenance and discovering routing paths. This is independent of the specific functionality of the routing protocol. Both reactive and proactive routing protocols eventually end up in this situation.

Depending on the specific concept that is fundamental to a routing protocol, certain mobility patterns or degrees of mobility (i.e., the speed at which nodes move about the scene) can be tolerated. For ad-hoc networks in general, this has been explored in detail (Camp *et al.* 2002, Bai & Helmy 2004). There is a common agreement that high mobility always leads to routing problems even for mid-sized networks.

A fundamental change in the routing paradigms can, however, completely change the situation. In one of the groundbreaking papers in the MANET community, Grossglauser & Tse (2002) identified even positive aspects of high node mobility: Mobility helps to increase the capacity of ad-hoc wireless networks. General capacity bounds have been identified earlier by Gupta & Kumar (2000). These bounds can be shifted if one considers node mobility as a helper function.

This concept was later exploited in the development of the last-encounter routing (LER) protocol (Grossglauser & Vetterli 2006). LER explicitly relies on (a high degree of) node mobility to collect contact information between nodes. This spatio-temporal information can be very helpful to route data packets to specific destination nodes. In particular, LER greedily forwards messages to nodes that met the destination (or a node that has met the destination) most recently. By fully exploiting this contact information and the *store–carry–forward* communication principle, LER outperforms any other MANET routing protocol, given that nodes move frequently and uniformly in a certain geographic area.

This protocol has never been successfully applied to VANETs, because the mobility patterns are completely different. Regarding mobility in vehicular networks, the mobility of vehicles is constrained not only by the road network topology but also by traffic rules, by other vehicles, and even by pedestrians. Thus, the characteristics of random-waypoint models do not apply in this case. Back in 2008, we investigated a mobility pattern being used in simulation-based performance evaluation of data dissemination in vehicular networks (Sommer & Dressler 2008). Indeed, the mobility of vehicles needs to be investigated in a microscopic way that takes into account all the interactions between each vehicle and the road network, other vehicles, and the respective traffic rules. We will investigate this in more detail in Chapter 6.

Investigations in this direction led to the development of mobility-pattern-aware routing concepts. Examples include mobility-pattern routing (Hung *et al.* 2008) and ideas for routing in sparse vehicular networks (Wisitpongphan *et al.* 2007a). Especially the second example addressed one of the most general problems of ad-hoc routing in vehicular networks: The network frequently becomes disconnected. The authors investigated the average re-healing time of the network, i.e., the time taken for nodes to get re-connected to the network, as a fundamental criterion. By characterizing the probability distributions of inter-arrival time and inter-vehicle spacing between vehicles on freeways on the basis of empirical data, they developed a first routing protocol that specifically exploits the mobility characteristics of vehicular networks. Yet, only specific scenarios can be handled using this protocol. To date, no generalized ad-hoc routing protocol that can operate in all scenarios of vehicular networks has been developed.

Delay-bounded routing

The second issue to talk about is delay-bounded routing. QoS-aware routing has been a major research issue in the field of MANETs for a long time. It is, however, not obvious how specific situations or scenarios have to be treated in order to guarantee certain upper bounds for the communication delay.

In MANETs, the best-known solutions focus on extensions to the already-discussed AODV routing protocol, which, even given its major weaknesses and limitations, has become a *de facto* standard in the mobile ad-hoc networking domain. It is not within the scope of this textbook to explore these limitations in detail. Instead, we focus on two of these extensions.

Ad-hoc QoS on-demand routing (AQOR) uses the end-to-end delay as the routing metric (Xue & Ganz 2003). It is based on a resource-reservation routing and signaling technique. Besides the end-to-end delay, it also supports the reservation of a demanded data rate. AQOR has been developed to support the increasing demand to use MANETs also for multimedia applications such as voice and video, which need certain QoS support in order to provide this service at an adequate level. The idea behind AQOR is to estimate the available bandwidth and end-to-end delay in an unsynchronized wireless environment. Given this estimate, adequate paths can be selected.

Adaptive dispersity QoS routing (ADQR) goes one step further and even supports multiple disjoint paths (Hwang & Varshney 2003). The idea is to use a route-discovery algorithm that is able to find multiple disjoint paths with longer-lived connections. Using this information, network resource reservations can be combined with data dispersion. A route-maintenance algorithm is used to proactively monitor network topology changes, thus providing all the relevant information for resource-management protocols. ADQR also provides a fast re-routing algorithm to significantly reduce data-flow and QoS disruptions.

The applicability of both concepts to VANETs is limited due to, again, the much higher and very specific mobility pattern. The case of assuming a vehicular network with at least partial infrastructure support, i.e., the use of RSUs or access points (APs) along the streets has been investigated by Skordylis & Trigoni (2008). In this work, two approaches have been developed: delay-bounded greedy forwarding (D-Greedy) and delay-bounded min-cost forwarding (D-MinCost). The general objective is to deliver messages originating from a vehicle to an access point with bounded delay while minimizing the number of wireless transmissions. D-MinCost requires knowledge of global traffic conditions (this can be provided by means of statistics), whereas D-Greedy requires no such knowledge.

D-Greedy constantly calculates the available delay budget for each message and the distance to the closest AP. It assumes that the remaining delay budget can be uniformly distributed among all the road edges that compose the shortest path to the AP. On the basis of this assumption, each edge is assigned a delay budget that is proportional to its length. Now, the algorithm assumes further that each vehicle periodically sends a beacon containing information about its position, speed, heading, etc. On the basis of the calculated delay budgets and the received information, the vehicle can switch between two modes: greedy forwarding to other vehicles that will arrive earlier at the AP or data muling, i.e., itself carrying the message towards the AP. The algorithm, of course, cannot incorporate road traffic information other than the speed information which is currently being measured and received by means of beacons. D-MinCost integrates such traffic statistics into the calculation to improve the accuracy of the estimated delay budgets for each edge.

QoS constraints have been reconsidered in the context of vehicular networks most recently when it comes to video streaming applications (Wisitpongphan & Bai 2013, Park *et al.* 2006, Asefi *et al.* 2010). With new emerging applications such as live video streaming to help following cars "see" through a big truck in order to improve the safety of overtaking maneuvers, the delay and jitter of a wireless transmission become key parameters. Yet, ad-hoc routing with integrated QoS features has never been realized in MANETs or VANETs on a large scale.

5.2 Geographic routing

Another approach to routing in MANETs is geographic routing. This concept is also a very promising candidate for information dissemination in vehicular networks (Mauve *et al.* 2001). Figure 5.8 revisits the general taxonomy discussed in Section 3.3.3 on page 80 but focuses on routing that is based on specific resources – in our case, nodes' positions.

Geographic routing is conceptually quite simple. It is based on *greedy forwarding* of messages towards a certain position (physical or virtual). Greedy forwarding means selecting a next-hop node to make as much progress towards the destination as possible. Using purely geographic coordinates, this means selecting a node that is closest to the destination in terms of a geographic distance.

An alternative is to rely on virtual coordinates, which can be seen as replacement of geographic coordinates but having better characteristics for greedy routing. Both geographic and virtual coordinates can also be used as keys in a DHT. Combining routing and information management in the form of a DHT provides even more capabilities for many applications to store and retrieve information from within the mobile network. DHTs have been developed in the scope of peer-to-peer networks to store information (in replicated form) in a fully distributed manner in the Internet. The same concept has now been integrated in ad-hoc and vehicular networks.

In this chapter, we introduce geographic routing as it has been discussed in the mobile ad-hoc networking domain. In the framework of this discussion, we will see that greedy routing is not as simple as it appears to be. Thus, in a second part, we study greedy forwarding in the context of virtual coordinates. This field is still a very active research area aiming at finding solutions having the nice properties of virtual coordinates for greedy forwarding but keeping the helpful geographic coordination

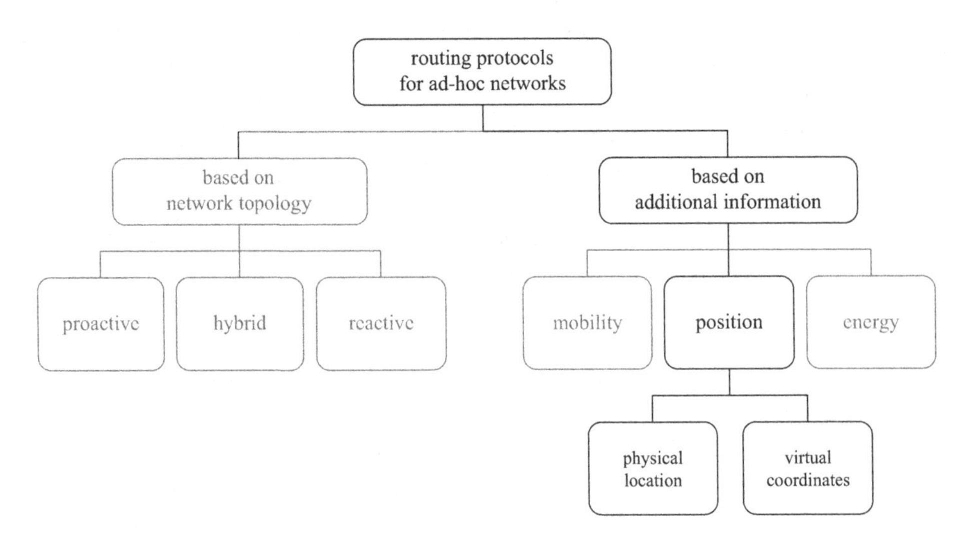

Figure 5.8 Geo-routing in the general taxonomy of ad-hoc routing.

for identifying destination locations. We conclude the chapter with an overview of geocasting concepts that have been introduced in the domain of vehicular networks.

5.2.1 Geographic routing

We have seen in the context of ad-hoc routing that routing tables are used to obtain information regarding to which hop a packet should be forwarded in order to make progress towards the final destination. These routing tables need to be constructed *explicitly*. The alternative is to *infer* this information from the physical placement of the communicating nodes, i.e., their geographic positions. We do not discuss here how the geographic positions can be determined; we assume the use of such technologies as the Global Positioning System (GPS) for this purpose. The granularity and accuracy of the GPS coordinates is sufficient for all the techniques discussed in this chapter.

The obvious question, which has not been addressed by most of the geographic routing protocols, is that of how to identify the location of the destination node or area. Location services can be used to map nodes to node positions.

The first approach that helped make geographic routing useful for practical applications was greedy perimeter stateless routing (GPSR) (Karp & Kung 2000). More information on this topic can be found in Kim *et al.* (2005). In the following, we study geographic routing on the basis of this example, which already addresses many of the conceptual problems of geographic routing in general. We briefly list more advanced solutions when discussing the use of virtual coordinates in Section 5.2.2 on page 157.

Greedy perimeter stateless routing

Greedy perimeter stateless routing (GPSR) is one of the best-known approaches to geographic routing (Karp & Kung 2000). As has already been mentioned, the idea is to achieve as much progress as possible towards the destination in each forwarding hop. This concept has been named greedy routing. The alternatives are to select the nearest node with any progress in order to minimize the transmission power in sensor networks or to use a directional routing concept selecting the next-hop node that is angularly closest to the destination.

Figure 5.9 outlines the greedy-forwarding concept of GPSR. In our example, node Ⓐ is about to transmit a message to destination node Ⓖ. Always selecting the neighbor as a next-hop node that is closest to the destination, node Ⓐ first forwarded to node Ⓒ and then to node Ⓓ. As can be seen, three nodes (Ⓑ, Ⓒ, and Ⓔ) are within communication range of node Ⓓ, which is about to select an appropriate forwarder. Node Ⓔ is obviously the one that is closest to the destination node Ⓖ. Therefore, it will be selected, and then will continue to directly transmit to the destination node Ⓖ.

This simple strategy, however, may lead to the situation that packets are sent into a dead end (or a routing void). Such a situation is illustrated in Figure 5.10. Here, node Ⓐ greedily forwards the message to node Ⓑ and then to node Ⓓ. At this position, no more progress towards the destination is possible. The basic idea employed to get out of such a dead end is to use the "right-hand rule" – like in a labyrinth (only with paths and

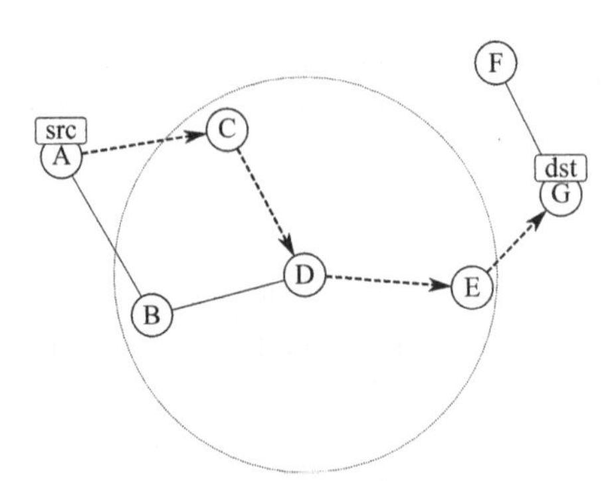

Figure 5.9 The concept of greedy routing as used by GPSR.

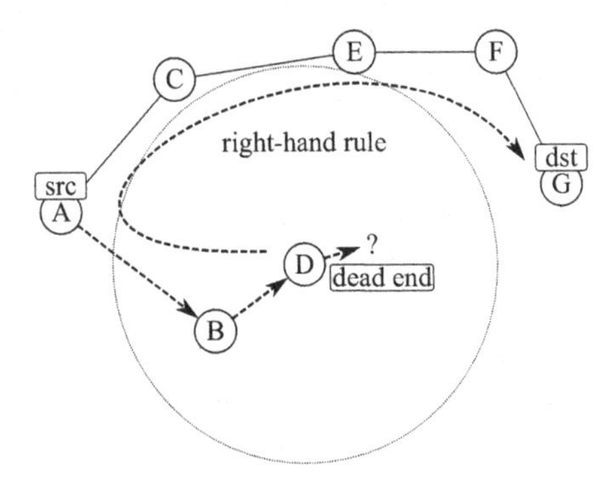

Figure 5.10 The recovery technique employed if greedy routing fails (routing void or dead end).

walls replaced by connectivity and non-connectivity). Following some imaginary wall may lead in the depicted example out of the dead end and into a situation where greedy routing can be continued to the destination node Ⓖ. However, this procedure does not work on some inner walls. In such a case, a routing loop is created.

GPSR supports a mechanism called *face routing*. If no more progress is possible using greedy routing towards the destination, the protocol switches to face routing. Here, a face describes the largest possible region on a plane that is not cut by any edge of the graph. Face routing uses the right-hand rule to send the packet around the face. The position at which the face was entered and the destination position are used together to determine when the face can be left again, i.e., switching back to greedy routing.

Figure 5.11 outlines this principle. The greedy-routing strategy lets node Ⓐ forward the message to node Ⓒ and further to node Ⓓ. This represents a routing void or dead end. Using the face information, the inner face spanned by nodes Ⓒ, Ⓓ, and Ⓔ can be passed using the right-hand rule. Starting at node Ⓔ, the face is left, and greedy routing ensures delivery at the destination node Ⓖ.

The establishment of the face information is based on a planar graph. Topology-control techniques can be used to determine the graph. This, however, makes the resulting protocol rather sensitive to node mobility. If node mobility is to be expected, a

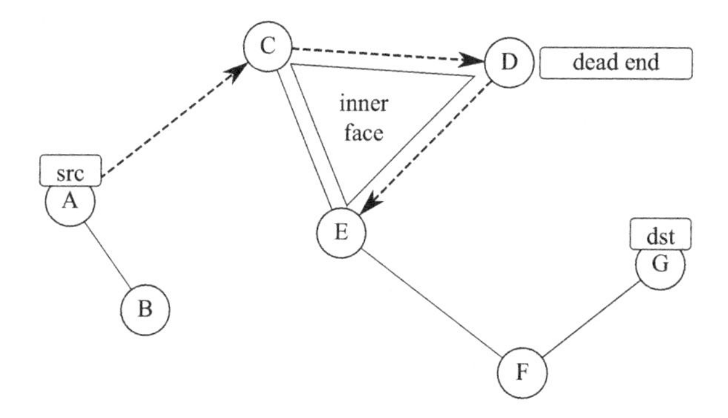

Figure 5.11 Face routing as employed by GPSR to overcome routing voids.

compromise between fast recovery from dead ends and the overhead induced by continuous topology control has to be made.

Geographic hash tables

The idea of geographic hash tables (GHTs) is to combine the advantages of DHTs with geographic routing. Thus, the objective is to provide not another routing concept but instead more means for identifying locations, i.e., specific nodes, to store data at and to retrieve data from. Hence, it is a data-centric approach using geographic coordinates as identifiers for keys and nodes.

The GHT concept is to hash keys identifying data items into geographic locations, so that only those data items which are geographically nearest to the hash of its key are stored on the sensor node (Ratnasamy *et al.* 2002, Shenker *et al.* 2003, Ratnasamy *et al.* 2003). If needed, replication can be used to improve the reliability of the system in the case of node failures or to compensate for node mobility. Similarly to typical Internet-based DHTs, GHT essentially provides an overlay using any (geographic) underlay routing protocol for data dissemination between the involved nodes. Following the discussion by Ratnasamy *et al.* (2003), we assume GPSR to provide these underlay geographic routing capabilities.

In particular, GHT uses GPSR's face (or perimeter) mode to identify *home nodes* for each data item. Again, a hash algorithm provides means of hashing a data item's key to a geographic position. The home node is meant to be located closest to this position. For replication purposes, the face that is spanned around the destination's coordinates and that by definition includes the home node, is used to identify potential locations for replicas. GPSR is able to provide exactly this information.

Ratnasamy *et al.* (2003) proposed a perimeter-refresh protocol to accomplish replication of key-value pairs and their consistent placement at appropriate home nodes when the network topology changes. This protocol stores a copy of each data item at each node on the home perimeter. This can easily be realized by sending the data item around the face that contains the home node, as illustrated in Figure 5.12. Node failures and updated data can be detected using refresh packets, which are generated by the home node and, again, sent around the defined face. In Figure 5.12, node Ⓐ generated a data item that

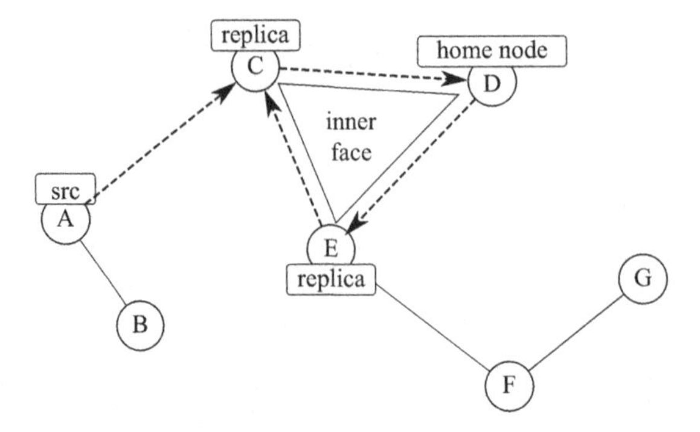

Figure 5.12 Data storage and replication as provided by GHT using the perimeter refresh protocol.

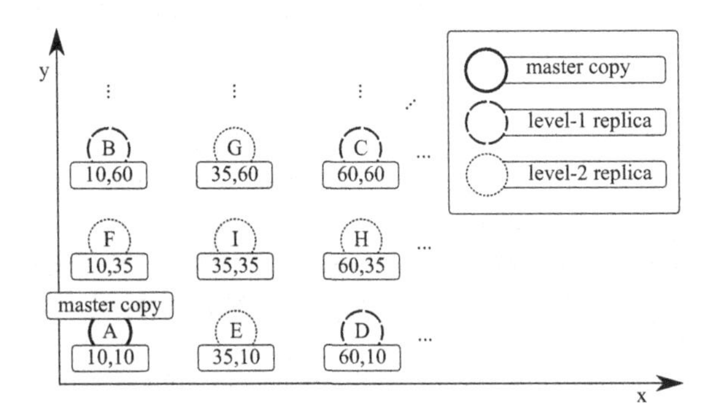

Figure 5.13 Data storage and replication as provided by GHT using structured replication in a geographic area.

is sent using GPSR's greedy routing to the home node Ⓓ, a node closest to the real geographic coordinates of the data item's hash key. Node Ⓓ, in turn, sends the item around the face including nodes Ⓔ and Ⓒ, both of them storing a replica of the data item.

A second scheme supported by GHT is called structured replication. In contrast to the perimeter-refresh protocol, structured replication supports multi-level replicas even in (geographically) more distant locations. Furthermore, this can be used for load balancing within the network.

In particular, structured replication uses a hierarchical decomposition of the key space. The concept is shown in Figure 5.13. The original key still hashes to a specific location that is associated with the closest node, which maintains the master copy. For a hierarchy depth d, $4^d - 1$ images of this master are computed. New items can be stored at the closest image node. Queries are routed to all image nodes, starting at the master and continuing to all replicas. Figure 5.13 shows a $d = 2$ decomposition and a subset of the respective level-1 and level-2 replicas.

Some additional notes are necessary in order to understand the possibilities of using GHTs in practice. In addition to the location information of a node, which is assumed to be available by means of GPS or similar localization techniques, the geographic dimension of the deployment area must be known in advance. This is needed to calibrate the hash algorithm appropriately. This is most likely the most difficult part when it comes to application in vehicular networks. Assuming an entire city (or even a larger area), it might lead to the situation that a home or master node is exactly on the opposite side of the area, thus resulting in a huge overhead for the transmission of data items to their master replica.

The second issue is related to radio communication in realistic environments. First of all, nodes might be located at the maximum transmission range, resulting in spontaneous connectivity (De Couto *et al.* 2003, Kim *et al.* 2005). Also, nodes (in vehicular environments) will not likely be uniformly distributed, which has been a key assumption during the development of GHTs. If this assumption does not hold, significant performance degradations are to be expected. The use of virtual coordinates instead of geographic ones is considered to overcome these problems. We study the performance of GHTs in comparison with such virtual-coordinate-based techniques in the scope of the following section.

5.2.2 Virtual-coordinate-based routing

As we have seen, it is tempting to make use of geographic coordinates for greedy routing, yet the problem of routing voids often causes major problems. It has therefore been investigated whether this issue can be countered without losing the greedy-routing properties, which, in the end, make the routing approach fully distributed or self-organizing (Dressler 2007).

This research line led to the development of so-called virtual-coordinate-based routing techniques. The common idea is to assign coordinates to nodes in such a way that greedy routing can be used (in some cases limited to searching in large DHTs). In the following, we study this concept and also investigate a selected example, the virtual coordinate protocol (VCP), in more detail.

From geographic to virtual coordinates

One of the first concepts towards this objective was geographic routing without location information (GRWLI) (Rao *et al.* 2003). Instead of relying on geographic coordinates, this protocol creates synthetic coordinates by means of an iterative relaxation algorithm that embeds nodes in a Cartesian space. The approach therefore supports situations in which GPS coordinates are not available and other self-localization techniques (Eckert *et al.* 2011) would consume too much time or resources. The construction of the coordinates starts by identifying the perimeter nodes and projecting them onto an imaginary circle. Two nodes are assigned as beacon nodes. All other nodes then determine their location with respect to the perimeter using a heuristic that is based on the measured hop count from the beacon nodes. A relaxation algorithm is used to specify the resulting virtual locations of the nodes in the network. Thanks to the projection on the perimeter, well-known DHT algorithms can be used in combination with GRWLI.

Coordinate transformation

Protocols such as GPSR and GRWLI scale extremely well. Unfortunately, greedy forwarding cannot guarantee the reachability of all destinations because of possible dead ends (Mauve *et al.* 2001). It has, therefore, been investigated how geographic coordinates can be transformed in a way that fully eliminates these dead ends.

Transformation of geographic coordinates can be considered as a first step towards the use of virtual coordinates (Flury *et al.* 2009). Furthermore, hierarchical approaches using a mixture of geographic coordinates and a virtual overlay have been considered (Tan *et al.* 2009). Later, this hierarchical approach was extended to completely switch to virtual coordinates on the overlay using additional coordinate transformation techniques. Comprehensive surveys of such virtual-coordinate-based solutions have been provided by Tsai *et al.* (2009) and Watteyne *et al.* (2007).

Virtual ring routing

One of the first approaches to using solely virtual coordinates was virtual ring routing (VRR) (Caesar *et al.* 2006). This protocol combines overlay routing as used in DHT-based peer-to-peer networks with an underlay routing concept using virtual node addressing. This leads to small routing tables and ensures guaranteed delivery – albeit not necessarily on the shortest path. We will study some of the performance characteristics of VRR in our discussion of another virtual-coordinate-based routing system, VCP, in the next section.

VRR uses unique keys to identify nodes. Each such key is a location-independent integer value, which can later be used to span a numeric space for a DHT. Inspired by the DHT protocol Chord (Stoica *et al.* 2001, 2003), VRR organizes all nodes in a virtual ring. For routing, each node maintains a finger table that stores routing information to a set of r virtual neighbors. In addition, each node also keeps a physical neighbor set identifying all nodes that are within direct communication range. Both tables are proactively maintained.

The concept is illustrated in Figure 5.14, which shows the physical network as well as the virtual overlay. Please note that node addresses are virtual identifiers that are ordered in increasing order in the overlay. Additionally, the finger table for node Ⓐ is shown in the virtual overlay.

If node Ⓐ wants to transmit a message to node Ⓕ, it first selects the node in its finger table that is closest to node Ⓕ (in our example, this is node Ⓔ) according to a greedy forwarding within the overlay network. It then forwards the message to node Ⓔ using the underlay routing information. Node Ⓔ continues this procedure. It will likely have destination node Ⓕ in its finger table and forwards the message to its final destination. As illustrated in this simple example, adjacent nodes in the virtual overlay might be rather far from each other in the physical underlay network. VRR is not able to maintain or to identify shortest-path information.

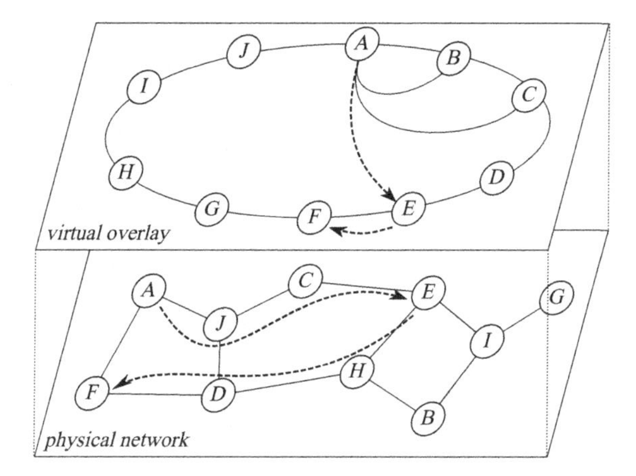

Figure 5.14 Nodes with their representations in the virtual overlay and physical underlay as used by VRR. Finger tables are used on the overlay for efficient routing.

Further improvements

Several approaches addressing one or more of the remaining shortcomings have been proposed. For example, GSpring improved the performance of greedy forwarding by assigning the coordinate system in a specific way (Leong *et al.* 2007). It makes use of a system of springs and repulsion forces to balance the coordinates in the system. In a similar way, hop-ID routing assigns multidimensional coordinates according to the distance of some landmark nodes (Zhao *et al.* 2007). Even though the system supports randomly chosen landmarks, the careful selection of these nodes improves the resulting routing performance substantially. Dead ends are handled by means of detouring via the landmark nodes. Finally, the virtual-coordinate-assignment protocol ABVCap_Uni needs to be mentioned (Lin *et al.* 2008). This protocol specifically supports unidirectional links by forming virtual rings containing the unidirectional links and treating these as extended nodes. Routing is then performed using both real nodes and the extended nodes. Finally, VCP that carefully constructs virtual node identifiers such that dead ends are prevented by design needs to be mentioned. We study this protocol in detail in the following section.

Virtual cord protocol

The idea behind the virtual coordinate protocol (VCP) is to combine data lookup with routing techniques in an efficient way (Awad *et al.* 2008, 2009a, 2011). VCP accomplishes this by placing all nodes on a virtual cord, which is also used to associate data items with nodes. The routing mechanism relies on two concepts: First, the virtual cord can be used as a path to each destination in the network; additionally, locally available neighborhood information is exploited for greedy routing towards the destination.

Constructing the virtual cord

The VCP protocol initially constructs and later maintains a virtual cord that connects all nodes in the network and serves two purposes:

- the cord provides a guaranteed path between any two nodes in the network; and
- the cord helps to assign IDs to all nodes that are ordered from a start node S to an end node E, both spanning the numbering range $[S, E]$.

Periodic hello messages are used to establish neighborhood information. This, together with the cord position, is the only information available for routing in the network. Before we discuss the routing procedure, let's have a quick look at the establishment of the cord.

One node must be pre-programmed as an initial node, i.e., it is assigned the start position S. All other nodes are dynamically incorporated into the cord on the basis of the hello-message exchange. These hello messages are exchanged by means of local broadcast messages. Thus, each node is able to receive hellos and to further decide about possible reactions.

The join process is triggered if a node is not yet part of the cord but has received one or multiple hellos from cord members. The hellos contain information about the neighbors' IDs and thus, their position in the cord. Furthermore, successor and predecessor information is included.

VCP suggests a pre-defined delay T_{ps} before joining a VCP cord. This delay is needed in order to obtain hellos from as many active neighbors as possible.

Joining the cord essentially means obtaining a virtual cord position P and updating the cord structure by informing both the successor and the predecessor in the cord about the changes.

The join process works as illustrated in Figure 5.15 (Awad *et al.* 2011). The first node in the cord is pre-programmed with the start position S. Each other node joining the network has to receive at least one hello message from a node that has already joined the cord in order to get a relative position in this cord.

If a node can communicate only with one end of the cord, i.e., either node S or node E, the new node takes over this position. The old node gets a new position between the end value and its successor or predecessor, depending on its old position. In the special case that only a single node S exists in the cord, the former start node gets the other end of the cord E assigned as a new position.

If a node can communicate with two adjacent nodes in the cord, the new node is integrated into the cord by taking a position in between the two adjacent nodes, i.e., the new node is assigned a number in between the two cord neighbors and becomes the successor to the old node with the lower position value and the predecessor of the node that has the higher position value.

Finally, if the new node can communicate with only one node in the network, which is neither at S nor at E, then this node is asked to create a virtual node. As illustrated in Figure 5.16, the virtual node gets a position between the position of the real node and that of its successor or predecessor. The new joining node can now continue the normal join process and take a position in between the real position and the virtual position of the node in the cord.

An acknowledgement procedure is used for reliable position exchange among the nodes involved.

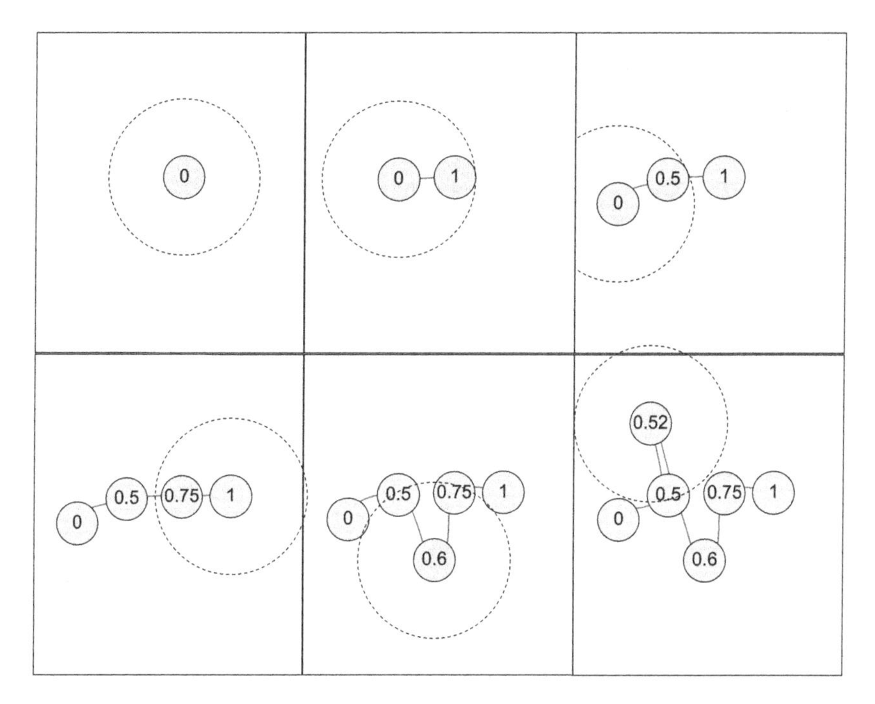

Figure 5.15 The basic join operation in VCP illustrated for six nodes. © [2011] IEEE. Reprinted, with permission, from Awad *et al.* (2011).

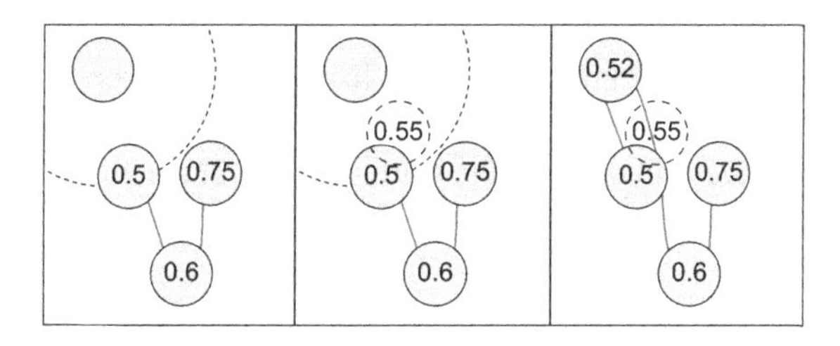

Figure 5.16 The creation of a virtual node. For a new node that finds only one cord member, this node has to create a virtual node at its location to support the normal join procedure. © [2011] IEEE. Reprinted, with permission, from Awad *et al.* (2011).

Let's have a look at the join process depicted in Figures 5.15 and 5.16 in more detail. The large dashed circle indicates the joining node and its radio transmission range. In our example, we use the numerical range [0, 1] for the cord addresses, i.e., $S = 0$ and $E = 1$. After setting up a new cord by pre-programming a node with the start address $S = 0$ on the following four steps (see Figure 5.15), nodes are placed in the cord either at an end or in the middle of the cord. In the last step, when the sixth node joins the network, only one neighbor, node 0.5, is within communication range. Thus, this node is asked to create a virtual node as shown in Figure 5.16. The main idea is simple: The cord must connect all the nodes in the network in an ordered way. Thus, if no two nodes

Table 5.2 Initial parameters for the join process

Parameter, default value	Description
Start $S = 0.0$	Lowest position on the cord
End $E = 1.0$	Highest position on the cord
Position $P =$ N/A	Current position in the cord
HelloPeriod $T_{\text{h}} = 1$ s	Time interval between `hello` messages
SetPosDelay $T_{\text{ps}} = 1$ s	Time interval before re-requesting a new position
SetVPosDelay $T_{\text{vps}} = 1$ s	Time interval before requesting a virtual position
BlockDelay $T_{\text{b}} = 1$ s	Blocking period to prevent assigning the same position to more than one node

adjacent in the cord are within direct communication range of the new node, a virtual node is created to balance the cord again.

The final result of the join process is a virtual cord that interconnects all the nodes in the network, i.e., it guarantees at least one possible path between any two nodes. As we have already seen in our simple example, the cord will quickly become rather complex, especially if multiple virtual nodes have been created. In fact, the cord does not need to fulfill any specific requirements, i.e., it does not need to be efficient in any sense. We will explore how greedy routing can be used in the resulting virtual structure in the following.

Please note that the cord spans the numeric range $[S, E]$ which can be exploited by a hash function for data storage or service discovery.

Several initial variables are initialized in the startup phase. These are listed together with default values in Table 5.2.

Greedy forwarding

VCP uses greedy routing similar to what we studied for geographic routing before. In particular, two sets of information are used for data forwarding: first, information about the cord, i.e., the local position in the cord as well as information about successor and predecessor; and second, information about other nodes in communication range (including their virtual positions), which is actively maintained using the periodic hello messages.

Greedy forwarding in VCP works as follows. A node is about to send a message to a destination node D, which can either be a specific node address in the virtual-coordinate system or a hash value of a data item. In each step, the current node N selects as a next-hop node the one node N_i out of its neighbor nodes that has a position in the cord closest to the destination D.

Forwarding is terminated if no more progress is possible, i.e., the local coordinate P is closest to D. On the basis of the established cord, VCP routing will always lead to a path to the destination – it is not possible to run into a dead end.

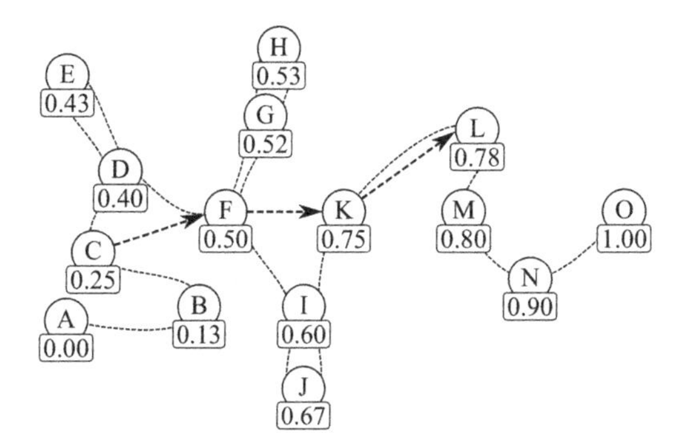

Figure 5.17 An example routing path using greedy routing along the virtual cord and exploiting local neighborhood information.

Figure 5.17 illustrates VCP's greedy routing for a simple network containing 15 nodes. In our example, we assume that VCP has already established the entire cord, so that all nodes have all the relevant routing information readily available.

In our example, node 0.25 has a message to transmit to node 0.78. Thus, it will forward the message towards the destination node via node 0.50, which has the closest position to the target value among all physical neighbors. Afterwards, node 0.50 will send it to node 0.75, and finally node 0.75 will forward it to node 0.78. As can be seen, both nodes 0.25 and 0.50 greedily shortcut the cord using information about their physical neighbors.

The resulting path is, in our example, also the shortest path. This is not necessarily the case, so we have to assess the typical deviation from the shortest path to understand the performance of the protocol.

Failure-handling mechanisms are in place to handle situations like node failures and node mobility. Furthermore, data-replication strategies have been integrated to further improve the reliability of the overall system. We do not study these in detail but refer the reader to the original publication of VCP instead (Awad *et al.* 2011).

Performance aspects

In order to acquire a better understanding of the advantages of virtual-coordinate-based routing, we study the performance of VCP compared with VRR and GPSR/GHT. All of the results cited have been taken from Awad *et al.* (2009b) and Awad *et al.* (2011). We investigate just a simple scenario where nodes are deployed in a grid structure but for different numbers of nodes.

Stretch ratio

As a first metric, we study the effectiveness of the routing concept of VCP. For this, we investigate the stretch ratio, i.e., the difference of paths used for VCP routing compared with the shortest path. Obviously, the optimal value would be 1, characterizing exactly a shortest path. In the selected grid scenario, the average path length is $\sqrt{n} - 1$ and the maximum path length is $2\sqrt{n}-2$, where n is the number of nodes. For different network sizes, we measured the path length from all nodes to the upper-left node.

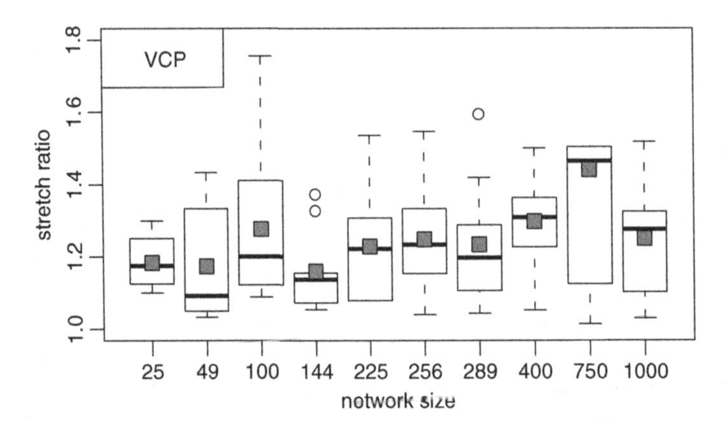

Figure 5.18 Stretch ratios for different network sizes. © [2011] IEEE. Reprinted, with permission, from Awad *et al.* (2011).

Figure 5.18 plots the stretch ratio for network sizes of 25 to 1000 nodes. The graph is shown in the form of a boxplot showing the median as well as the first and third quartiles, whereas the filled squares depict the average value. As can be seen, the stretch ratio slightly increases with the network size; however, this increase is reasonable and the median stays below 25%. In comparison, the stretch ratio of VRR increases to over 40% already for network sizes larger than 200 nodes (data not shown). This is because successors and predecessors, which are included in the routing table, can be far away from each other and, therefore, messages may be routed over many unnecessary nodes.

Success rate

Let's now focus on the protocol's behavior in the presence of frequent node failures. We consider as a node failure any event that prevents communication to a particular node at a given time, e.g., complete node failures or interrupted communications due to node mobility. Node failures are modeled as a uniformly distributed on/off process. For this experiment, the proportion of nodes toggling their state investigated ranged from 0% to 100%. This configuration follows the setup described by Ratnasamy *et al.* (2003). We compare the performance of VCP with that of VRR.

Figures 5.19(a) and (b) depict the observed success rates for VCP and VRR, respectively. As can be seen, the ratio of successful transmissions degrades with the number of failing nodes.

VCP is able to maintain a success rate of about 70%–80%. In contrast, the success rate degrades much faster for VRR (down to 50%). A look at the network load reveals much higher values for VRR, which explains the reduced success rate of VRR compared with VCP. This is mainly due to the mugh higher overhead for maintaining the finger tables in the case of node failures.

Data-replication performance

Finally, we investigate the performance of the data replication. As analyzed for example in the context of GHTs, a clear improvement can be expected in the case of frequent

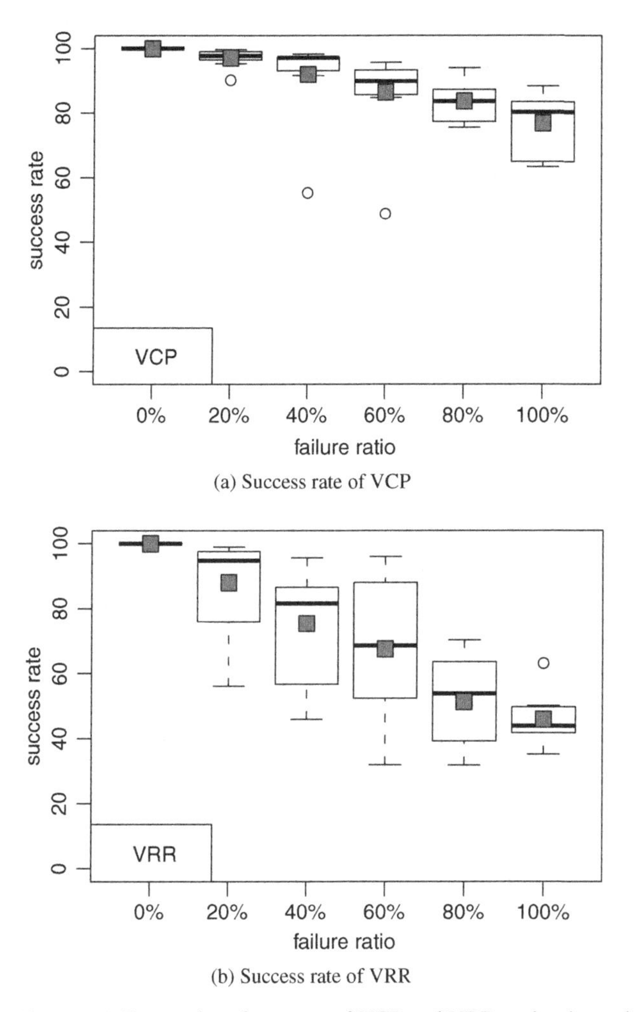

(a) Success rate of VCP

(b) Success rate of VRR

Figure 5.19 Protocol performance of VCP and VRR under dynamic network conditions. Depicted is the success rate for increasing node-failure ratios. © [2011] IEEE. Reprinted, with permission, from Awad *et al.* (2011).

node failures. The design of the experiment follows again the recommendations by Ratnasamy *et al.* (2002). Continuous queries are performed for certain items while again simulating increasing failure rates of all the nodes.

Compared with the performance results reported for GHTs (Ratnasamy *et al.* 2003), VCP performs very well – the results are compared numerically in Table 5.3. As can be seen, GHT yields a success rate of only about 83%, whereas VCP is able to successfully support more than 92% of the queries.

All the results presented demonstrate the capabilities of both geographic and virtual-coordinate-based routing techniques. The more dynamic the network becomes, the greater the extent to which virtual-coordinate-based systems perform better than geographic-position-based systems. It is, however, important to understand that only

Table 5.3 Success ratios and the necessary refresh messages for VCP and GHT

Failure ratio	GHT success rate	GHT refresh messages	VCP success rate	VCP refresh messages
0	100 %	1.6	100 %	0.00
20	99.7 %	1.5	99.2 %	0.02
40	98.6 %	1.6	95.2 %	0.03
60	97.3 %	1.8	95.3 %	0.05
80	94.2 %	1.8	95.2 %	0.08
100	83.3 %	1.6	92.1 %	0.08

a certain degree of mobility or dynamics can be supported, necessitating additions for deployment in vehicular networks.

Towards inter-domain routing

On looking at the vehicular networking domain, we frequently encounter scenarios in which the relative mobility of vehicles is rather limited for a group of cars. This is, for example, the case for clusters of vehicles driving in the same direction, such as on a freeway. Another example is provided by parked vehicles that represent rather perfect conditions for a distributed information storage that is used and updated by passing vehicles. For all these applications, additional concepts such as capabilities for routing between different groups or clusters of vehicles are needed. We briefly study such techniques in the following.

Inter-domain routing in MANETs was first studied by Chau *et al.* (2008). The motivation was to understand how multiple ad-hoc networks could be connected to exchange information. Conceptually, the Internet border gateway protocol (BGP) routing concept (Rekhter & Li 1995) was extended to make it useful even in the mobile networking case. This idea was later improved to reduce the routing protocol overhead (Train *et al.* 2011).

A completely orthogonal approach has been investigated by Zhou *et al.* (2009). The authors proposed the geographic inter-domain routing (GIDR) protocol, which was developed especially for geographic routing concepts. GIDR achieves scalability in large networks by using geo-routing-based packet forwarding together with clustering techniques. Essentially, the entire network is split into clusters. Each cluster is coordinated by a cluster head, which advertises its connectivity, its members, and the domain information to the rest of the network. Geographical routing is then used to greedily forward messages between the clusters. If greedy forwarding fails, the packet is "directionally" forwarded to the most promising node along the advertised direction.

Inter-domain routing for virtual-coordinate-based concepts was explored first by Dressler *et al.* (2009). The idea is to assign cluster or domain IDs to each (virtual-coordinate-based) network. Internally, protocols such as VCP or VRR are used to forward messages. The inherent ability of the network to also store arbitrary information in the DHT is exploited to manage and maintain inter-domain information (Dressler *et al.* 2009, 2010). Essentially, each node that comes into contact with another domain (e.g., by means of VCP's hello messages) stores this information in the local DHT. Nodes that are about to transmit messages to a neighboring domain look up possible gateway nodes and rely on greedy forwarding to send the message to this gateway first and then further towards the destination node.

That approach, however, is still limited in terms of finding efficient routes over multiple transit networks. This work was later extended by defining a framework for optimized inter-domain routing (Dressler & Gerla 2013). In particular, ant colony optimization (ACO) has been used as a routing heuristic for optimizing routes between multiple network domains.

5.3 Beaconing

We briefly studied *beaconing*, i.e., one-hop broadcasting, in Section 3.3.4 on page 85. In this section, we investigate two selected *static beaconing* concepts, SOTIS and ETSI CAM, in more detail. Static beaconing has become the term used in the vehicular networking community for periodic, *fixed-interval* beaconing. That means, a broadcast is created using a fixed frequency. This terminology is sometimes a bit complicated: Please note that we sometimes refer to a frequency (e.g., 10 Hz) and sometimes to an interval (e.g., 100 ms). This terminology is also very well established in the literature – both versions are absolutely correct, but it is important to carefully check which option is being used in any particular case.

5.3.1 Self-organized traffic information system

The field of beaconing was pioneered by the self-organizing traffic information system (SOTIS) (Wischhof *et al.* 2003, 2005) approach. In this system information about current traffic conditions is disseminated by means of repeated application-layer broadcasts of cars' knowledge. This is accomplished by each car keeping a local knowledge base that integrates both locally generated sensor data and data received from other participating cars. Cars will then periodically send a beacon, each containing information from the local knowledge base. We will study this system in detail in the following.

Concept

The idea of SOTIS is to realize a fully distributed and self-organizing system to exchange traffic information among the participating nodes. This is a fundamental change from the historical view of a TIS: In the past, we always observed a direct interaction between sensors and a (central) TIC, as well as a direct communication interface between this TIC and the customers. With SOTIS it was suggested, for the first time, that one should install all the necessary components directly in the car.

The SOTIS concept is outlined in Figure 5.20. Each car maintains its local knowledge base. As mentioned already, the knowledge base contains both locally generated entries and those received from other vehicles. By means of the periodic exchange of

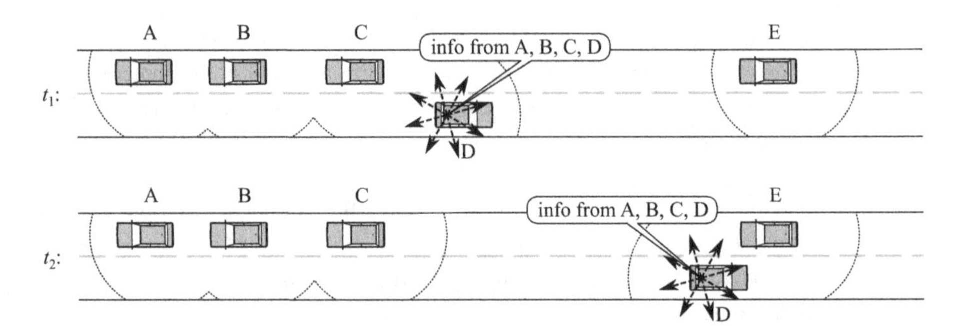

Figure 5.20 Broadcast-based information exchange in SOTIS.

knowledge bases, SOTIS enables all the participating vehicles to update their knowledge about traffic-related information. Assuming an infinite time for the exchange, one can assume that eventually each car receives all parts of all available knowledge bases.

In reality, there are constraints that make this impossible. First, the time for exchange is finite and contacts between vehicles are discrete over time, and depend on the mobility of the vehicles. Second, the communication channel is limited in its capacity and throughput. Thus, it is impossible to exchange the entire knowledge base in a single interaction. We have to check which changes are needed in order to make SOTIS applicable to real-world scenarios.

Assumptions and operation principles

Several assumptions have to be taken into consideration for the SOTIS system.

- All decisions taken by the driver are primarily influenced by information about his or her very local neighborhood. Thus, SOTIS must be able to provide reliable traffic information and meta data for all events within a rather small range, e.g., within 50 km of the current position of the vehicle.
- Each SOTIS-equipped vehicle must also be equipped with a GPS and a navigation unit, which provides map data and computational resources for simple algorithms. Of course, the vehicle also needs a radio communication device.
- The dissemination of traffic information should work already at penetration rates of about 2%. This implies that ad-hoc routing in a multi-hop environment is not an option.
- There is a fundamental relation between the relevance and the necessary precision of traffic information and the distance of the events from the current position of the vehicle. The information granularity can therefore be reduced with increasing distance (graceful degradation).

With the given assumptions, SOTIS internally operates essentially on a digital map. This map can be annotated with road traffic information. An example is depicted in Figure 5.21. Besides the event description, each tag is also annotated with time information, for example describing the initial time at which the event was recorded, the

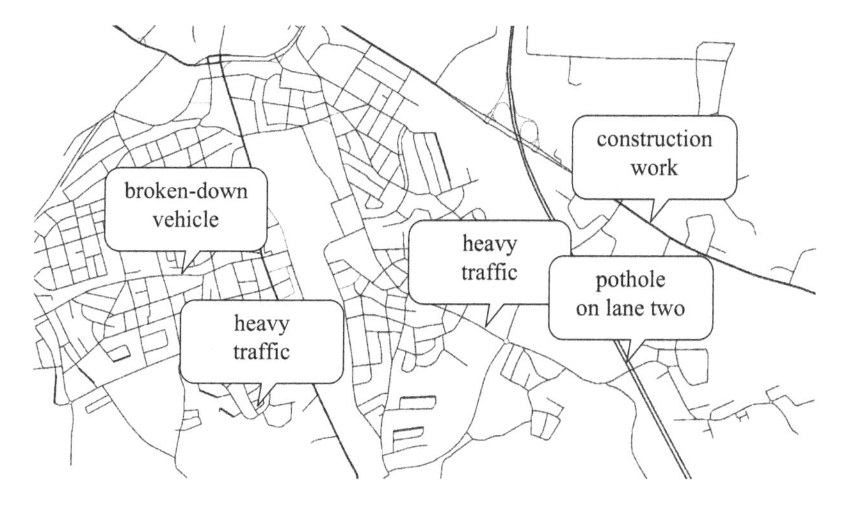

Figure 5.21 Internal data management in the form of annotations on a digital map.

last update, and the estimated duration. Annotations that are no longer valid, have been timed out, or are simply too old can be removed from the knowledge base.

The communication, i.e., the exchange of complete knowledge bases or parts of those, was assumed to be handled using WiFi. SOTIS distinguishes between two types of reports.

- *Periodic reports* – The basis for periodic reports is a broadcast message. All known information (or a selected subset thereof) is sent in a single beacon message to all neighboring SOTIS stations. The beacon interval is configurable.
- *Emergency reports* – In addition to periodic reports, emergency messages can be sent if necessary. The main difference is that no congestion-control techniques have to be used in this case.

The default beacon interval is envisioned to be on the order of seconds (Wischhoff *et al.* gave 5 s as an example). Emergency reports can be sent immediately. For this, it is assumed that the MAC-layer protocol guarantees immediate access to the channel, e.g., by always reserving some capacity for emergency reports.

Segment-oriented data abstraction and dissemination

The basis for information dissemination is segment-oriented data abstraction and dissemination (SODAD). The authors considered SODAD the basis for a wide range of comfort and traffic-efficiency applications (Wischhof *et al.* 2005). Events to be disseminated are typically associated with a certain geographic location. Therefore, in order to conserve network resources, each car stores and transmits information pertaining not to specific points, but to small segments of roads.

Each car determines the length of a segment (and, thus, the granularity of information) depending on its distance from the segment. Figure 5.22 shows an example for the segment-oriented operation of SODAD. The size of each segment is automatically

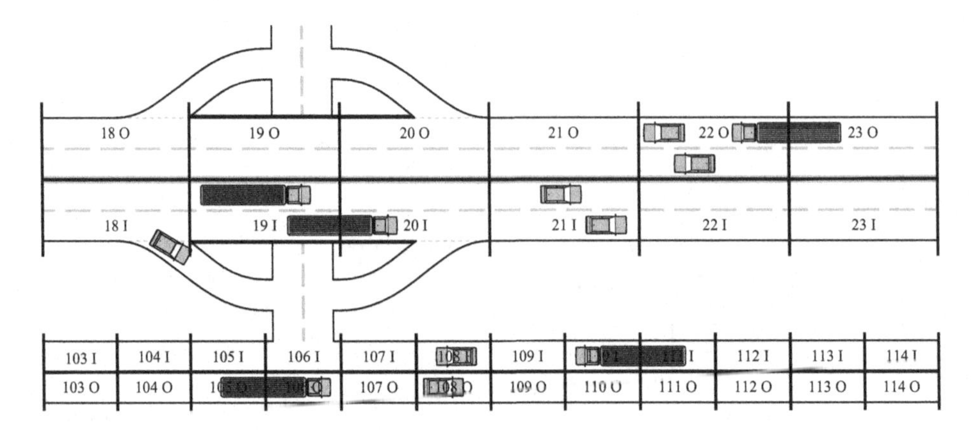

Figure 5.22 The representation of abstract map data in the form of single segments.

chosen by the vehicle according to whether it is traveling on a freeway or on a country road. Together with segment-based addressing, this road identifier allows space-efficient encoding. The optimal segment size is a function of the desired granularity of the associated information. The larger the segments become, the coarser-grained the information which can be stored but the higher the possible aggregation gain to compress the information. Because cars are envisioned to be equipped with compatible digital maps, they are able to assign a unique identifier to each road.

Besides the encoding of information, the dissemination of per-segment information is the second objective of SODAD. The protocol uses two principles.

- *Local broadcast* – This is the simple periodic beaconing we have already referred to.
- *Application layer store-and-forward* – The application logic can become quite complex when it comes to the selection of knowledge-base items to be included in a beacon; SODAD allows one to distinguish between more important messages and others.

SODAD is fully based on WiFi (see Section 4.2.1 on page 119).

Autonomous map creation

For cases in which (compatible) maps cannot be assumed to be available in each participating car, the authors later proposed an extension (Wischhof *et al.* 2006) to SOTIS, which was evaluated (Eckhoff *et al.* 2011a) and adapted for use in the sim^{TD} (Stübing *et al.* 2010) project. This extension enables cars to generate road maps on the fly, using GPS information, as well as to exchange information about them in a highly efficient manner via IVC.

The process is based on a reference grid of virtual lines spaced D_{grid} apart, as illustrated in Figure 5.23. Each time a vehicle crosses a grid line (① in the figure), it generates a *geo-reference point* at the location of intersection. In order to avoid creating spurious points caused by positioning errors, no point is created when crossing a line

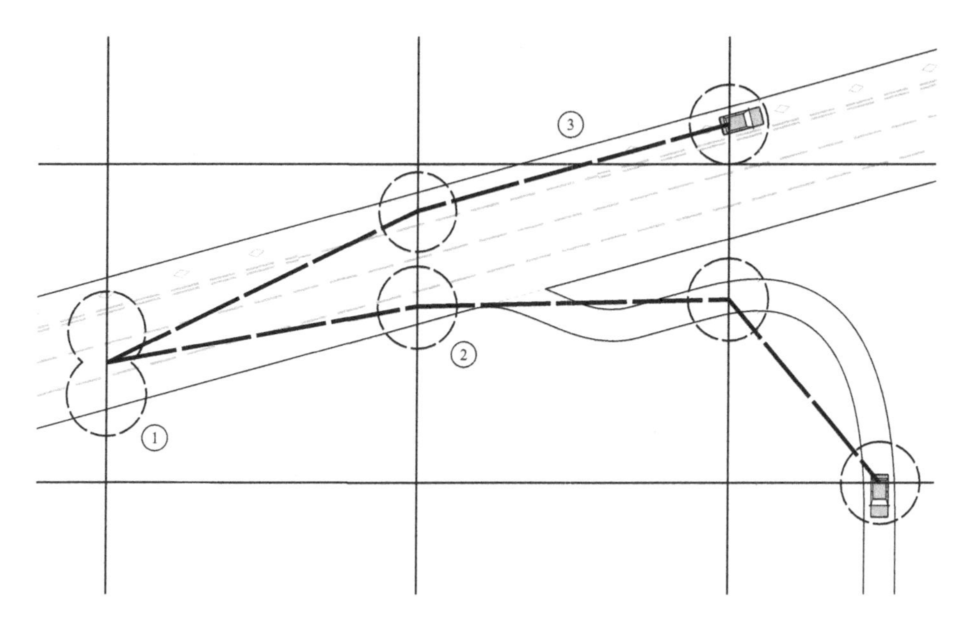

Figure 5.23 On-the-fly generation of road maps for SOTIS.

at an angle smaller than a pre-configured angle α (③ in the figure). In a second step, sequences of points are assembled into one *geo-reference list* each; from these lists an approximate digital map can be derived, which can then be exchanged between cars.

Creating and storing duplicate points is prevented by the application of a simple distance metric: If a new point is closer than a pre-configured positional inaccuracy D_{acc} to an existing one (and the associated list runs in a compatible direction), the new point is assumed to be referring to the same point and the existing one is used instead. While this process cannot guarantee matching of map data if positioning errors are unexpectedly high or lanes are unexpectedly wide (② in the figure), it comes reasonably close: In a simulation experiment (Wischhof *et al.* 2006) the authors claimed to have achieved a mean deviation of generated map data of approximately half the root-mean square error (RMSE) in position estimation.

An IVC communication system based on this (extended) SOTIS approach is thus able to operate without deployment of infrastructure on the road and without digital maps on the vehicles (much less the deployment of identical ones). Information can be shared using a highly efficient data encoding that is based on road segments, and no information is lost during sporadic network disconnections.

Broadcast protocol

During the development of SOTIS, it turned out that the fixed beaconing might cause critical problems when it came to synchronization of the MAC protocol of multiple vehicles. As mentioned before, the SOTIS system is based on WiFi, which was the only feasible option at the time of development. The authors therefore suggested, for the first time, that one should consider adaptation of both the transmission rate and the transmission power – we will study such schemes in Section 5.4 on page 174. What was realized in the end is a technique for delaying events and to increase the variance of the time between two subsequent beacons.

The basic idea is to distinguish between two kinds of events.

- *Provocation* – This is an event that reduces the time until the next broadcast packet is transmitted.
- *Mollification* – This event increases the time until the next broadcast packet is transmitted.

Now, for each message that has been received, the knowledge base is checked, if a provoking or a mollifying event has occurred. This is done by comparison of the received data and its time stamp within the knowledge base for each individual road segment. The decision regarding whether an information value is significantly newer or different than the previously available information is based on the time stamps.

As an additional parameter, the transmission distance to another car is used. If it is larger than a certain threshold, provocation is used since it can be assumed that the other car will quickly pass out of communication range. SOTIS thus prefers large hops in terms of communication distance.

5.3.2 Cooperative awareness messages

With the development of intelligent transportation systems (ITSs) and in particular the IEEE 802.11p protocol, new applications have been investigated and finally submitted for standardization. The ETSI is one of the big players in this field, and is about to standardize some basic services for these applications.

Cooperative awareness is one of the most important applications in the ETSI protocol suite. It means that all road users and roadside infrastructure are informed about each other's position, dynamics, and attributes (ETSI 2013b). In the ETSI vision, this includes all vehicles like cars, trucks, motorcycles, bicycles, and even pedestrians. Cooperative awareness is achieved by the regular exchange of information among these road users. For the exchange, a message format and a protocol have been defined in the scope of CAM.

In this section, we concentrate on the initial standardization of cooperative awareness messages (CAMs) (ETSI 2013b). In contrast to SOTIS, which uses WiFi, all the ETSI protocols are based on ETSI ITS, i.e., use IEEE 802.11p at the MAC layer. For more information on how the channel access is organized, please refer to Section 4.2 on page 118.

CAM protocol

The initial protocol for exchanging CAMs was rather simple. CAMs are assumed to be generated periodically with a frequency that is controlled by the cooperative awareness basic service. The generation frequency is assumed to be in the range of 1–10 Hz. In some documents, frequencies of up to 40 Hz have been considered. Still, the frequency is fixed and does not depend on the current channel characteristics. We therefore talk about a static beaconing protocol.

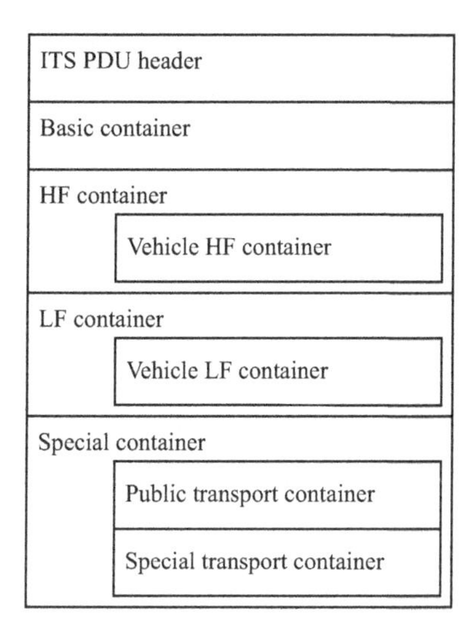

Figure 5.24 The general structure of a CAM.

It quickly turned out that the fixed beaconing interval leads to problems when considering different and, most importantly, dynamic scenarios. Thus, the concept was further developed during the standardization process. This led to the development of DCC, which we discuss in detail in Section 5.4.2 on page 185.

Messages are broadcast on the control channel (CCH) to all neighboring vehicles. The message is scheduled for transmission by IEEE 802.11p. If a new CAM has been generated while the old one was still queued for transmission, the old message is discarded and the new one is scheduled instead. This situation may occur if the channel is very busy so that the MAC protocol was not yet able to win contention for the channel. Upon reception of a CAM, the receiver makes the content available to the local ITS application, which, in turn, can inform the driver about potentially critical situations, or may even trigger automated reactions in emergency situations.

CAM message format

A CAM is composed of a common ITS message header and multiple containers. The header contains information about the protocol version used, the message type, and the identification of the sender of the message. According to the most recent version of the standard, at least a basic container and a high-frequency (HF) container must be present in a CAM. In addition, a low-frequency (LF) container and other special containers may be added. All the containers mentioned contain different types of information about the vehicle originating the CAM.

The general structure of a CAM is depicted in Figure 5.24. The different containers have been defined as follows:

- Basic container: vehicle type and current geographic position
- Vehicle HF container: all fast-changing parameters such as heading and speed of the vehicle

- Vehicle LF container: static parameters (such as the dimensions of the vehicle) or slowly changing parameters (such as status of the exterior lights)
- Special containers: containers have been defined for public transport, special transport, dangerous goods, road work, rescue operations, emergency vehicles, and safety cars

5.4 Adaptive beaconing

Even though the design of SOTIS was a milestone development for beaconing information dissemination, three issues limit its usefulness for direct deployment in vehicular networks.

- Incorporation of infrastructure: Information dissemination becomes noticeably slower when market penetration of the system is still low, as the authors remark. An approach that could incorporate available infrastructure in addition to being able to operate in a completely infrastructure-less fashion would help bridge this gap.
- Medium sharing: The MAC-layer access protocol that SOTIS is deployed on is required to be able to guarantee immediate access to the channel, so that the medium can be shared between both comfort-oriented and safety-relevant messages. It is noted that this can easily be achieved by always reserving some capacity for emergency reports, yet this static allocation means that a substantial portion of the channel capacity is wasted. An approach that could incorporate safety-relevant messages to be transmitted over the same protocol, as well as coexist with other and future protocols, would allow a more efficient use of the channel capacity, while making sure that low-delay transmission of safety messages can be achieved.
- Adaptivity to scenarios: The authors of SOTIS note that it might be favorable to determine the beacon interval in an adaptive fashion, rather than using a fixed schedule for the dissemination of information. An approach that could take into account indications of past, present, and future channel use, as well as the usefulness of information to the network as a whole, would help achieve the fastest possible dissemination rates without ever overloading the channel.

These observations stimulated a large body of research activities aiming to develop new protocol variants that can be summarized as *adaptive beaconing*. The type of adaptation varies for each of these approaches. The objective, however, is the same. It is about achieving diversity in time and space.

In the following, we study the ATB protocol, which was one of the first approaches towards adaptive beaconing. The concepts developed in this protocol have been used to develop a new cooperative awareness protocol, which relies on CAM but changes the access scheme. The resulting DCC technique supports adaptation not only of the beacon interval but also of the transmit power. We also study a rather new protocol named dynamic beaconing (DynB), which revisits the problems addressed mainly by asking the question of whether strict congestion control may lead to issues related to the communication delay of safety-critical applications.

5.4.1 Adaptive traffic beacon

The adaptive traffic beacon (ATB) protocol was designed to ensure that one can maintain an uncongested channel, i.e., to prevent packet loss due to collisions, and to reduce the end-to-end delay of the information transfer (Sommer *et al.* 2010e, 2011c). Similarly to SOTIS and ETSI CAM, ATB uses periodic beacons to exchange information among neighboring cars. The novel aspect is the attempt to continuously adapt the beaconing interval so as not to overload the wireless channel. The objective of ATB is to carefully use only the remaining capacity of the wireless channel, without influencing other protocols, i.e., to be most conservative in its behavior.

The benefits that adapting beacon intervals can bring to beaconing schemes are obvious. The feasibility, but also the quality, of transmissions depends mainly on the vehicle density and penetration rate (Wisitpongphan *et al.* 2007a). Increasing the performance of static beaconing schemes to a level comparable to that of adaptive beaconing schemes causes packet collisions to rise to intolerably high values. In fact, the channel load can be up to several orders of magnitude higher (Sommer *et al.* 2011c).

Concept and system architecture

In Figure 5.25 we outline ATB's system architecture. Vehicles continuously exchange beacon messages containing TIS data. The interval between two messages is adapted according to two metrics: the perceived *channel quality* and the importance of the messages to be sent, i.e., the *message utility*. As a simple rule, ATB tries to send beacons as frequently as possible to ensure fast and reliable delivery, but always checks the channel quality to prevent collisions and interference with other protocols using the same wireless channel.

The quality of the wireless channel is estimated in three different time horizons.

Observing recent overload: By counting the number of collisions, the recent overload situations can be detected.

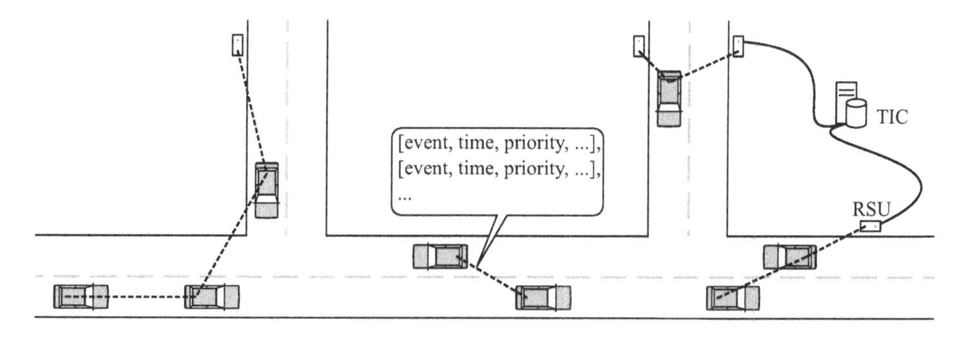

Figure 5.25 ATB's system architecture.

Measuring the current quality: The signal-to-noise ratio (SNR) provides a rough estimate of the current channel conditions.

Estimating the future channel load: The vehicle density is an indicator of the number of transmissions to be expected in the next time interval.

For the message utility, ATB further takes a message priority into account, which is based on the importance of the message and the estimated benefit to other vehicles. In short, the message utility is a function of the distance to an event and its criticality. The locally maintained knowledge bases are sorted w.r.t. the message utility. Thus, ATB inherently supports delay-sensitive data (e.g., information about approaching emergency vehicles) by means of message priorities.

In addition to supporting fully decentralized information exchange among participating vehicles, ATB can also make use of available infrastructure. As illustrated in Figure 5.25, support may be completely absent or range from stationary support units (SSUs), i.e., disconnected units participating only in the wireless network, up to a network of RSUs and servers.

In the following, we briefly introduce the different metrics ATB uses to assess channel quality and message utility. Each metric is derived by considering one particular measure of either channel quality or message utility and calculating its value relative to a fixed maximum value. We further illustrate how the metrics work together to adapt the beacon interval and present how nodes manage their local knowledge bases.

Beacon interval adaptation

Internally, ATB calculates the so-called interval parameter I, which is later used to adapt the beacon interval. ATB uses the *channel quality* C and the *message utility* P to calculate this interval parameter. Like all metrics of ATB, smaller values of C and P represent a better channel and a higher utility, respectively. The interval parameter I (in the range [0, 1]) is calculated according to

$$I = \left(1 - w_I\right) \times P^2 + \left(w_I \times C^2\right) \tag{5.1}$$

including C and P in squared form to make the relationship sensitive enough to environmental conditions. The relative impact of the two parameters is configured using an interval weighting factor w_I that can also be used to calibrate ATB for different MAC protocol variants.

Figure 5.26 shows the behavior of I for different values of C and P for $w_I = 0.75$ (from empirical data, this value has been shown to make ATB work properly in a wide range of scenarios). As can be seen, the interval parameter becomes 1 only for the lowest message utility and the worst channel quality. In all other cases, I quickly falls to values below 0.5.

From the interval parameter, the beacon interval ΔI is then derived according to

$$\Delta I = I_{\min} + (I_{\max} - I_{\min}) \times I \tag{5.2}$$

where $I_{\min}$ and $I_{\max}$ represent the minimum and the maximum beacon interval, respectively.

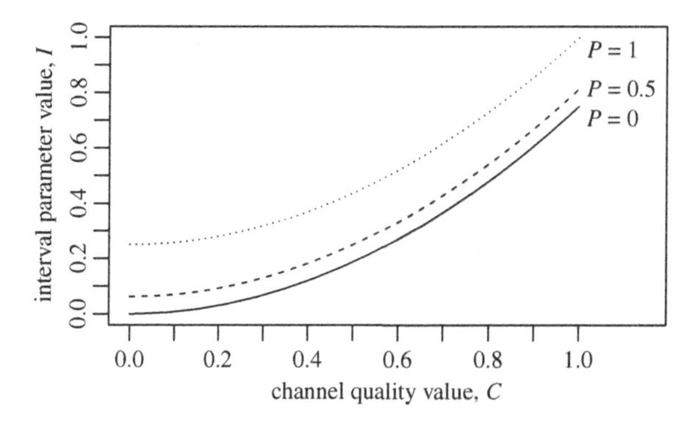

Figure 5.26 The ATB interval parameter I for different values of the channel quality C and message utility P, plotted for an interval weighting of $w_I = 0.75$.

Channel quality C

The *channel quality* C is a metric designed to indicate the availability of channel resources for ATB transmissions. ATB chooses three complementary parameters from the tremendous number of possible measures. Even though additional and more advanced metrics might offer even better performance, these parameters already serve ATB very well in estimating the channel quality on the three already-mentioned time scales.

From the number of collisions or packet errors observed in the last time interval, the load on the channel in the recent past can be estimated. The key objective of ATB is to ensure congestion-aware communication with no collisions or only a negligible amount of collisions, i.e., not to interfere with other protocols using the same wireless channel or with other cars using ATB. This metric is modeled according to

$$K = 1 - \frac{1}{1 + \#\text{collisions}} \tag{5.3}$$

An estimate for the current transmission quality is the SNR as perceived for the last transmission. In measurements, it has been shown that the error rate of WiFi communication quickly increases if the SNR drops below 25 dB (Rivera-Lara *et al.* 2008). Therefore, this metric is modeled according to

$$S = \max \left\{ 0;\ \left(\frac{\text{SNR}}{\text{SNR}_{\text{max}}} \right)^2 \right\} \tag{5.4}$$

with $\text{SNR}_{\text{max}} = 50\,\text{dB}$, so that the metric already decreases to $S = 0.25$ for an SNR equal to 25 dB.

ATB finally needs to predict the probability of other transmissions during the next time interval. To this end, it uses the density of vehicles (i.e., the number of neighbors) to estimate the congestion probability (the more neighbors, the higher the probability of simultaneous transmissions). This metric is modeled according to

$$N = \min \left\{ \left(\frac{\#\text{neighbors}}{\#\text{neighbors}_{\text{max}}} \right)^2;\ 1 \right\} \tag{5.5}$$

Finally, the channel quality C can be calculated according to

$$C = \frac{N + w_C \times \dfrac{S+K}{2}}{1 + w_C} \tag{5.6}$$

where the factor $w_C \geq 1$ is used to weight the measured parameters K and S higher than the estimated congestion probability N. This makes the ATB algorithm extremely sensitive to collisions and poor channel conditions.

Message utility P

The *message utility P* is an indicator of the importance or priority of a message. The higher the message utility, the higher the demand to broadcast messages early and frequently. In short, the message utility allows nodes to schedule the next transmission in such a way that nodes having high-priority messages will be able to transmit first.

Essentially, this metric is derived from two measures. First, we can measure the distance from a vehicle to an event D_e, which is the most direct indication of the message utility. The metric takes the current speed v of the vehicle into account to measure proximity in the form of an estimated travel time. D_e, which is calculated according to

$$D_e = \min\left\{\left(\frac{\text{distance to event}/v}{I_{\max}}\right)^2;\ 1\right\} \tag{5.7}$$

is calibrated according to the maximum beacon interval $I_{\max}$. This distance estimation can be further enhanced using map and location information (Lee *et al.* 2010b).

Second, ATB accounts for the message age A, thus allowing newer information to spread faster. The older the information is, the less frequently it should be distributed (bounded by the maximum beacon interval $I_{\max}$). The message age is calculated according to

$$A = \min\left\{\left(\frac{\text{message age}}{I_{\max}}\right)^2;\ 1\right\} \tag{5.8}$$

D_e and A are of equal value for determining the compound utility metric P.

In order to incorporate infrastructure elements such as RSUs and to make full use of their ability to send messages to other vehicles approaching the same intersection or to those following on a freeway, ATB also allows one to use the distance to the next RSU D_r as an additional measure. D_r describes the proximity to the next RSU as given in

$$D_r = \max\left\{0;\ 1 - \sqrt{\frac{\text{distance to RSU}/v}{I_{\max}}}\right\} \tag{5.9}$$

Internally, ATB will have to observe how well the information has already been disseminated. This measure, which is used only if the last beacon was received from an RSU, ensures that messages are quickly forwarded to the local RSU if it lacks information carried by the vehicle. Taking into account how much of the information to be sent was not received via an RSU, this factor is calculated according to

$$B = \frac{1}{1 + \#\text{unknown entries}} \tag{5.10}$$

The message utility P can then be calculated according to

$$P = B \times \frac{A + D_{\mathrm{e}} + D_{\mathrm{r}}}{3} \tag{5.11}$$

The value is always calculated for the TIS data with the highest utility in the local knowledge base.

A detailed sensitivity analysis of all the parameters discussed here can be found in (Sommer *et al.* 2010b).

Flexible use of infrastructure elements

ATB has been designed keeping in mind the possible use of all available infrastructure elements such as RSUs, and even centralized TICs are inherently supported by ATB. In principle, ATB-enabled vehicles and ATB-enabled RSUs operate in a similar fashion.

All RSUs participate in the beaconing process and adapt the beacon interval according to the same rules. In general, ATB supports both RSUs fully connected by a backbone network and SSUs operating as standalone systems, e.g., with an attached solar cell for autonomous operation as described by Lochert *et al.* (2008).

If RSUs are connected by a backbone network, this connection is used to inform other RSUs about received traffic information. In turn, the other RSUs update their local knowledge base accordingly, using the same procedure as when receiving a regular beacon by means of wireless communication. It is assumed that all RSUs know not only their own geographic position but also the positions of the neighboring RSUs.

The main difference between RSUs and vehicles is the calculation of the beacon interval. RSUs are not able to estimate their travel time to an event such as a point of traffic congestion. Thus, the message utility for an RSU P_{RSU} is derived ignoring all the distance-based measures and is finally calculated according to

$$P_{\mathrm{RSU}} = B \times A \tag{5.12}$$

The resulting interval parameter I_{RSU} is calculated as before.

$$I_{\mathrm{RSU}} = (1 - w_I) \times P_{\mathrm{RSU}}^2 + (w_I \times C) \tag{5.13}$$

Last but not least, the RSUs can be connected to one or more central TICs. A TIC disseminates received TIS data differently than the vehicles and RSUs. Using the available topology information of the connected RSUs, only relevant (i.e., geographically related) information is transmitted to each RSU.

TIS data management

Following the conceptual design of SOTIS, ATB maintains at each participating vehicle or RSU local knowledge bases that contain all received traffic information. ATB also follows the design rules for self-organizing systems. In order to maintain the scalability of the TIS, the transmission of irrelevant information needs to be suppressed. In

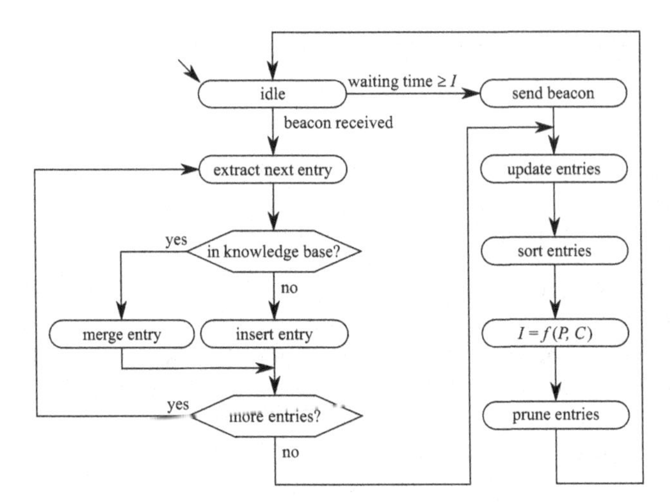

Figure 5.27 The core functionality of the ATB knowledge-base handling.

addition, even assuming on-board units (OBUs) with sufficient memory and processing capabilities, aggregation of the various received entries is used to maintain coarsely grained information for all data but fine-grained information for local events only.

The operation of ATB, however, is independent of the scheme used for the selection of knowledge-base entries and their aggregation, so any of the numerous approaches in the relevant literature can be used in an implementation. We will not go into the details of aggregation techniques. An elaborate treatment of these topics, along with a probabilistic aggregation scheme for message-store maintenance, can be found in the literature (Lochert *et al.* 2007; Schwartz *et al.* 2014; Lee *et al.* 2010b).

The initial proposal of ATB, illustrated in Figure 5.27, takes a more direct approach and stores only the most recent information for each route segment, i.e., new information elements either update records for an existing route segment (either in part or as a whole) or they are appended to the knowledge base. The knowledge base is updated with every received beacon, each of which may contain multiple information elements. Entries are prioritized to be transmitted in a beacon according to their age $\delta t_{\text{entry}} = t - t_{\text{entry}}$, the proximity to the event $t_{\text{c}} =$ distance to event$/v$, and the proximity to the next RSU $t_{\text{r}} =$ distance to RSU$/v$. From these measures, the priority of each entry can be calculated according to

$$p_{\text{entry}} = \delta t_{\text{entry}} - t_{\text{c}} + t_{\text{r}} \tag{5.14}$$

Using the calculated priorities, beacon messages can be generated by selecting as many entries as can be accommodated in a single MAC frame. Initially, ATB has been using WiFi employing a maximum transmission unit (MTU) of 1500 B. Entries are selected from the top of the list, i.e., those with highest priority.

This behavior will obviously help to spread the most important messages quickly, but it might also lead to the situation that some entries that just do not fit into a single broadcast packet will not be transmitted at all. Therefore, more elegant selection strategies are needed, as discussed for example by Schwartz *et al.* (2014).

A single entry in a beacon comprises at least the following elements: event type, time, position, priority, and RSU identifier. Thus, after receiving a beacon, each of its entries can simply be compared with the local knowledge base. If the event is not yet known, the entry is appended. Otherwise it is updated appropriately. Each update results in the recalculation of the priorities of all entries and the calculation of the next beacon interval.

Behavior on a microscopic level

All ATB instances work according to their local rules, thus, the overall behavior of ATB on the system level is, according to the rules of a self-organizing system, dictated by the local behavior of individual nodes. We first study how ATB behaves on the microscopic level by means of a simple example.

Let's use the simple scenario shown in Figure 5.28 (Sommer *et al.* 2011c). At the beginning of the observed time interval, car A senses a new event, which it will distribute to all nearby cars by means of ATB's beaconing mechanism. Suppose that, some time later, car F, which is just out of range of car A, will sense a new event, too, which it will proceed to distribute. In our example, cars B to E are all within communication range of car A.

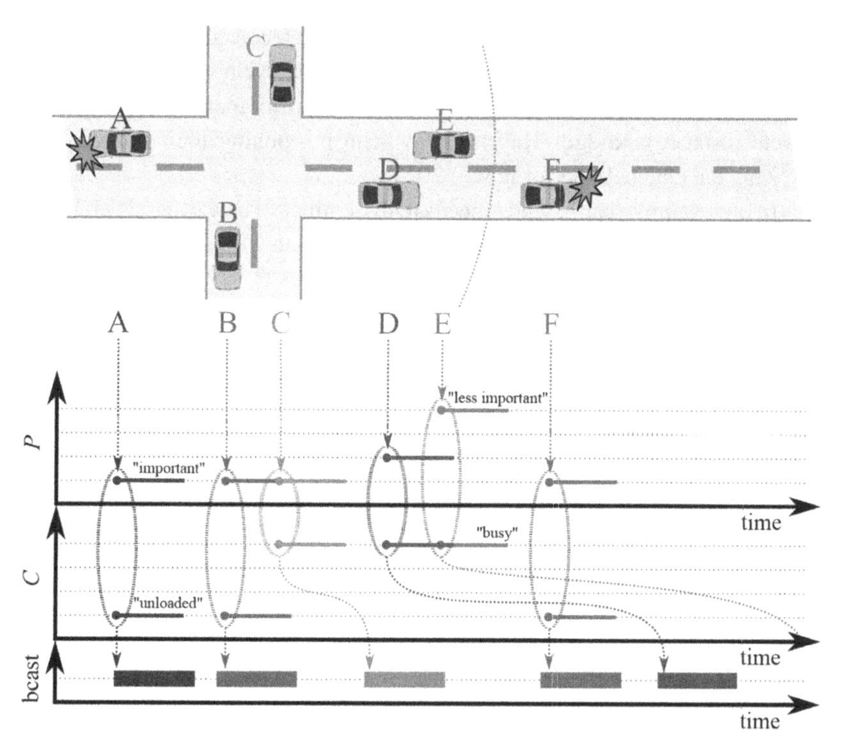

Figure 5.28 The behavior of ATB on the microscopic level using the simple scenario (top); and, in the timeline, we plot the metrics P and C (bottom). © [2011] IEEE. Reprinted, with permission, from Sommer *et al.* (2011c).

Figure 5.28 also shows how this scenario would play out in the style of a timing diagram. Please note that small values for C and P denote a free channel and a high message utility, respectively. From top to bottom, we plot the following four metrics of ATB: first, the time when an entry is inserted into a vehicle's knowledge base, i.e., the time when a new event was created (cars A and F) or when a broadcast was received (cars B, C, D, and E). Second, we plot the values of the message utility metric P and the channel quality metric C, which are used by ATB to calculate the beacon interval. Lastly, we indicate the time when an event is broadcast on the channel.

As can be observed, when car A registers the new event, it will be marked with the highest message utility, i.e., $P \approx 0$. In our example, the channel is free at that point in time, i.e., $C \approx 0$. Car A therefore chooses its minimum beacon interval for its next transmission, broadcasting the event almost immediately.

Without loss of generality, we assume that car B is the first vehicle to process the reception of this broadcast (taking the transmission delay, the operating system, and other system aspects into account), i.e., it becomes ready to re-broadcast first, followed shortly by car C. Car B therefore transmits its beacon instantly since both the message utility and the channel quality suggest a small beacon interval. Car C, however, observes the beacons of cars A and B in a short time interval and thus deduces a higher value for the channel quality metric C. Following the presented algorithm, it reacts by increasing its beacon interval, slightly postponing its broadcasting of the event.

Now, the beacons of cars A to C have also been received by cars D and E, which have not yet re-broadcast the message. Both cars now derive not only a further reduced channel quality metric C, but also a lower message utility metric P, owing to the increased event distance and age. This leads to them postponing their broadcast even further, leaving the channel idle for now.

In our example, car F, which needs to disseminate a new event, is presented with ideal channel conditions. Thus, it will estimate the channel quality as $C \approx 0$. Taken together with the message utility metric P, which assumes $P \approx 0$ because of the event being newly created, this leads to car F choosing its minimum value for the beacon interval. It can hence instantly broadcast a beacon containing the new event.

This example thus illustrates how the chosen metrics work in harmony to keep nodes' beacon frequency high whenever necessary, but the number of collisions on the channel low, helping ATB adapt to highly dynamic network conditions – both proactively and reactively. Of course, each message may include more than one event.

Behavior on a macroscopic level

The impact of ATB on a macroscopic level was investigated by Sommer *et al.* (2010e). For this experiment, vehicles were simulated using the Veins simulator (see Section 6.2.3 on page 262). An artificial traffic obstruction was used in order to let the vehicles exchange information about blocked roads via ATB so that the receivers could dynamically re-route around these obstructions. Similarly, instances of resolved traffic congestion triggered a recalculation of the route to check whether there was now a shorter route to the destination. The objective was to assess how ATB performs when compared with static beaconing.

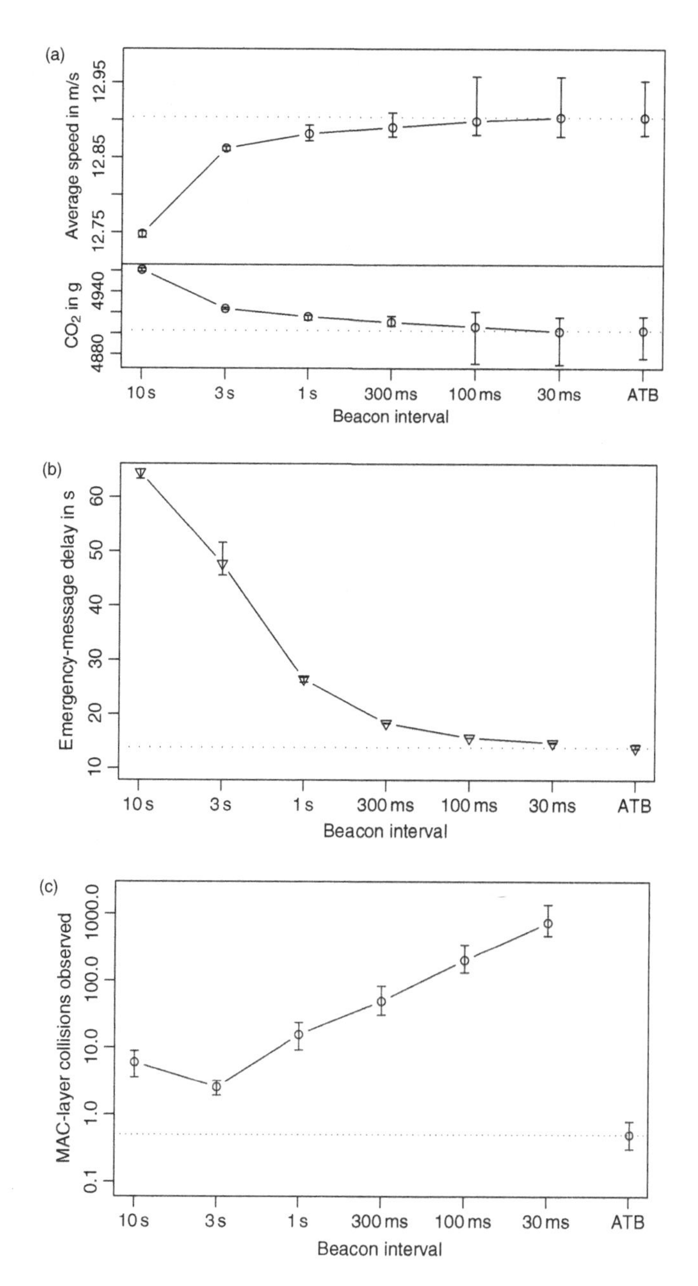

Figure 5.29 The performance of ATB compared with static beaconing for various beacon intervals: (a) traffic performance, (b) end-to-end delay, and (c) MAC-layer collisions. © [2011] IEEE. Reprinted, with permission, from Sommer *et al.* (2011c).

Figure 5.29 depicts the results of this experiment. The performance of ATB was compared with that of static beaconing for various beacon intervals in the range of 30–10 000 ms. ATB's limits had been configured accordingly to $I_{\text{min}} = 30$ ms and $I_{\text{max}} = 10$ s.

Let's first examine the impact of TIS operation on vehicles' speeds and CO_2 emissions. The results are plotted in Figure 5.29(a), which shows the mean values as well as the 10% and 90% quantiles for both metrics. Only when beacon intervals of 1 s are used do the results become comparable to those of ATB. On the other hand, ATB seems not to improve the situation compared with static beaconing with rates higher than or equal to 10 Hz.

We therefore have to have a look at the resulting delay as a second measure. The delays depicted in Figure 5.29(b) represent the typical store–carry–forward behavior in VANETs, which is greatly influenced by the mobility of vehicles. It also needs to be mentioned that the absolute measures need to be carefully evaluated because lost messages do not contribute in the graph shown. On comparing the end-to-end delays of generated traffic information for the same setups, we see that the performance of ATB is even a little better than that of static beaconing using an interval of 30 ms.

This result seems to be counter-intuitive. We therefore also examined the load and congestion of the wireless channel by measuring the number of collisions per packet received. Figure 5.29(c) depicts the results on a log-scale graph: Periodic beaconing always leads to a significant number of collisions, especially for beacon intervals smaller than 1 s. In contrast, ATB carefully manages the channel to operate below the congestion threshold. Hardly any collisions can be observed for ATB. This also explains the delay measurements. Since ATB always keeps the channels well below the congestion threshold, messages can easily be forwarded mainly according to their message utility.

From these results, we see that static beaconing with a period clearly smaller than 1 s allows a similar range and quality of the TIS information exchange. However, as can be seen from Figure 5.29(c), the number of collisions caused by the static beaconing exponentially increases for smaller periods. ATB is clearly able to perform well in all the investigated scenarios. Thus, we can conclude that ATB succeeds at managing access to the radio channel – which, according to the quality metrics used, also holds if other devices or applications start sharing the same wireless channel.

The impact of ATB was also compared with that of distributed vehicular broadcast (DV-CAST) (see Section 5.7.1 on page 219), thus comparing adaptive beaconing with a combined flooding and DTN approach (Sommer *et al.* 2011c). ATB clearly outperforms DV-CAST in terms of the speed at which new events are spread through a sparse network, and it is able to cover a much larger area in dense network scenarios.

Fairness

Still, many questions remain. The ATB protocol is assumed to behave in a fair way for all participating nodes. Yet, this is not guaranteed for the selection of messages to be exchanged. Fair and adaptive data dissemination (FairAD) focuses on exactly this issue in order to support fair distribution of the available channel capacity to all nodes (Schwartz *et al.* 2012b, 2014).

FairAD aims to achieve a *fair* distribution of data utility throughout the network while controlling the network load. It consists of two main components:

- a distributed fair data selection mechanism based on fair data dissemination (FairDD) (Schwartz *et al.* 2012a), and
- an adaptive periodic protocol based on ATB (Sommer *et al.* 2011c) to control the rate at which messages are broadcast into the network.

The message selection is based on a *utility function*. For a given application, the utility of a data message refers to the benefit that a vehicle can gain by receiving that message. A message utility is calculated on the basis of the current level of "interest" that a vehicle has in the message content depending on the vehicle's current context. For instance, if a message contains information about the vehicle's final destination, the application may consider giving a high utility to this message. However, from the perspective of another vehicle moving towards a different destination, the same information might be considered almost irrelevant.

This contextual knowledge can be classified into the following categories:

- the *mobility context*, which ranges from the complete route of a vehicle to the vehicle direction, speed, mobility history, etc.; and
- the *data context*, which includes the priority of the data message, age, geographic region, etc.

To achieve utility fairness in the neighborhood, a distributed data selection mechanism that considers the individual interests of vehicles has been proposed. FairAD relies on the Nash bargaining solution from game theory (Nash Jr. 1950). This solution achieves a compromise between fairness and efficiency. Fairness refers to the symmetry of utility distribution among vehicles and efficiency refers to the total utility distributed.

5.4.2 Decentralized congestion control

Since the findings showed the need for adaptive beaconing, this was picked up in the developments within the ETSI ITS standardization efforts. The objective was to provide congestion control on a large scale by making the access to the wireless channel sensitive to the current channel conditions. At the time of writing, this process has not been concluded yet, so we can report only about the current state.

ETSI DCC

Both concepts, adaptive beacon intervals and adaptive choice of transmit power, have found their way into current standardization efforts during the development of decentralized congestion control (DCC). ETSI CAMs are specified as repeated transmissions of beacons containing obvious information like vehicle ID, position, speed, but also less obvious data, such as the status of turn signal indicators (ETSI 2013b). In order not to overload the channel, the ETSI ITS DCC access control mechanism (ETSI 2011) takes care to limit the transmission rate, power, modulation, etc.

DCC access control works below the network layer (at the access layer), continuously monitoring a node's transmit statistics and its surrounding channel conditions as follows. Transmit statistics keep track of the number of packets sent (including repetitions as well as RTS/CTS and ACK frames). Channel conditions are monitored by measuring the channel busy fraction b_t over the course of fixed time intervals (e.g., 1 s) and keeping a short history of these measurements.

How to determine b_t is implementation-defined. The following reference procedure is outlined in ETSI (2011). During the course of a measuring interval (e.g., 1 s), the channel is repeatedly probed to check whether it is busy. Each probe consists of averaging the current power level on the channel over a very short duration (e.g., 10 μs) and comparing the average power level with the default clear channel assessment (CCA) threshold (the ETSI value is −85 dBm). The channel busy fraction b is calculated as the percentage of probes which reported a busy channel.

ETSI ITS DCC supports the adaptation in various dimensions.

- *Transmit rate control (TRC)* is very similar to the approach introduced by ATB. The idea is to reduce the beacon interval when the channel becomes congested.
- *Transmit power control (TPC)* refers to controlling the transmit power instead of the beacon interval. This allows one to increase the spatial diversity and thus, to increase the capacity of the wireless channel within a smaller communication range.
- *Transmit data rate control (TDC)* controls the data rate of the wireless link. Lower rates support higher reliability at the cost of reduced capacity.
- *DCC sensitivity control (DSC)* adapts the CCA mechanism to resolve local channel congestion (note that this modified CCA threshold does not apply to the measuring of b_t; here, always the same value is used).
- *Transmit access control (TAC)* introduces a scheme for transmit queue management to support different message priorities. TAC actually mirrors the queues of the enhanced distributed channel access (EDCA) queues of the lower layer.

In the following, we mainly concentrate on ETSI TRC and TPC, which essentially control the beaconing process.

Initial approach to TRC

ETSI ITS has standardized the DCC TRC mechanism (ETSI 2011) to adapt I based on a state machine, a simplified version of which is depicted in Figure 5.30. State transitions are driven by the observed channel busy time b_t. Given a sampling interval T_m, this means that b_t is the fraction of time the channel has been sensed busy between $t - T_m$ and t.

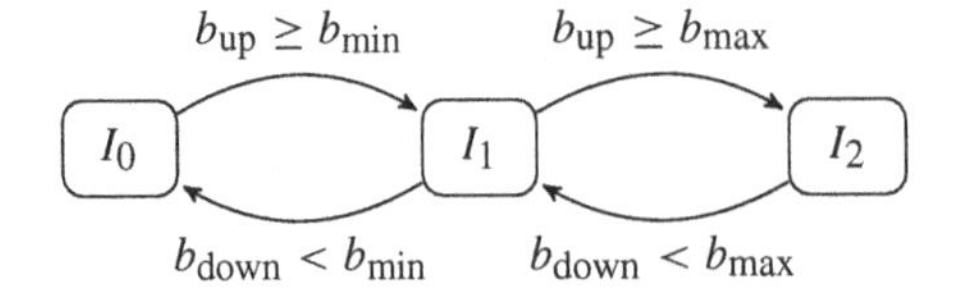

Figure 5.30 A simplified version of the state machine of the TRC algorithm.

Table 5.4 Sample parameters for simplified TRC

Parameter	Value
I_0, I_1, I_2	0.04 s, 0.5 s, 1 s
$b_{\text{min}}, b_{\text{max}}$	0.15, 0.40
$T_{\text{M}}, T_{\text{DCC}}$	1 s, 1 s
$T_{\text{up}}, T_{\text{down}}$	1 s, 5 s

T_{DCC} is the inter-decision interval (i.e., state transitions occur after every T_{DCC}), while T_{up} and T_{down} are filtering (time) windows applied to take the decisions on whether to increase or decrease the interval, respectively. The times T_{DCC}, T_{up}, and T_{down} are integer multiples of T_{m}.

The decision variables of the algorithm are

$$b_{\text{up}} = \min\left\{b_{t-T_{\text{up}}}, \ldots, b_t\right\} \tag{5.15}$$

and

$$b_{\text{down}} = \max\left\{b_{t-T_{\text{down}}}, \ldots, b_t\right\}. \tag{5.16}$$

Both are compared against threshold values b_{min} and b_{max}. Table 5.4 reports the ETSI ITS default values for the parameters. Not considering more involved DCC operations at the moment, each of the three states can correspond to a different interval in $\{I_0, I_1, I_2\}$, as shown in Figure 5.30.

In a static scenario, this scheme was shown to successfully manage channel access, albeit at the cost of synchronized oscillations in channel load and a pronounced under-utilization of channel capacity (Werner *et al.* 2012). However, in highly dynamic scenarios, the algorithm can also lead to the opposite: a pronounced over-load of the channel and, thus, packet loss. This has been addressed during the development of a more sophisticated scheme.

Fully featured DCC

The currently enforced limits of DCC access control are determined by which state the channel is currently in. For this, DCC access control maintains a state machine that assigns one of three basic states to the channel: *active*, *relaxed*, or *restricted*. Each state is associated with a set of parameters both for the MAC and for the PHY, controlling the power, modulation, carrier sense threshold, and periodicity of the beacons. For the active state, any of a number of different sub-states (four for the service channels, one for the control channel) is selected if channel conditions change.

Figure 5.31 shows the state machine. While for the CCH the states *relaxed* and *restrictive* do not change settings for each access category separately, transmit power settings in the *active* state affect access categories differently.

State transitions are still triggered according to the channel busy time b_t, i.e., depending on the observed busy time of the medium. Switching between states is

Table 5.5 Reference parameters corresponding to states of the ETSI ITS DCC state machine for the CCH

State switching threshold ($b_{\text{min/max,1s/5s}}$)	Relaxed 15 %	Active	Restrictive 40 %
TPC: Maximum power	33 dBm	20 dBm[a]	−10 dBm
TRC: Minimum interval	0.04 s	Unchanged	1 s
TDC: Maximum rate	3 Mbit/s	Unchanged	12 Mbit/s
DSC: CCA threshold	−95 dBm	Unchanged	−65 dBm

[a] Applies to AC_BE only; unchanged for AC_VI; 25 for AC_VO; 15 for AC_BK.

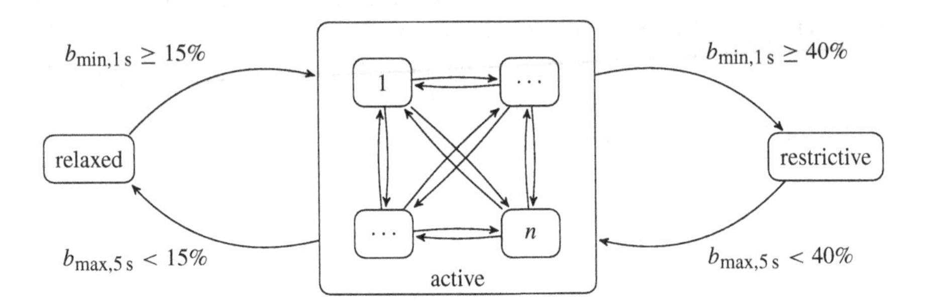

Figure 5.31 The ETSI ITS DCC state machine.

done by checking $b_{\text{min,1 s}}$, the minimal value of b in the past second, and $b_{\text{max,5 s}}$, the maximal value of b in the past five seconds, and then changing states as outlined in Figure 5.31.

The current channel state informs several subsystems of DCC access control. These are the aforementioned control mechanisms transmit rate control (TRC), transmit power control (TPC), DCC sensitivity control (DSC), transmit data rate control (TDC), and transmit access control (TAC). Reference parameters for these limits are given in ETSI (2011) and are listed in Table 5.5. Taken together, these mechanisms allow one to flexibly adapt a node's compound channel use to available resources.

Protocol performance

Several works investigating the protocol performance of ETSI ITS DCC have been published in the literature (Werner *et al.* 2012, Eckhoff *et al.* 2013b, Vesco *et al.* 2013).

Adaptivity to the channel load

In the following, we look at the experimental results reported by Eckhoff *et al.* (2013b) to illustrate the ability of DCC to control the load on the wireless channel.

In a first step, the authors studied the channel load in order to better understand how the MAC-layer mechanisms affect the channel conditions. Figure 5.32 shows the measured channel busy ratio b_t for a three-lane motorway junction scenario with a medium vehicle density and different penetration rates.

Observed channel busy times for the ETSI ITS DCC are controlled well, so that the channel load remains well below saturation (see Figure 5.32). On average, the channel

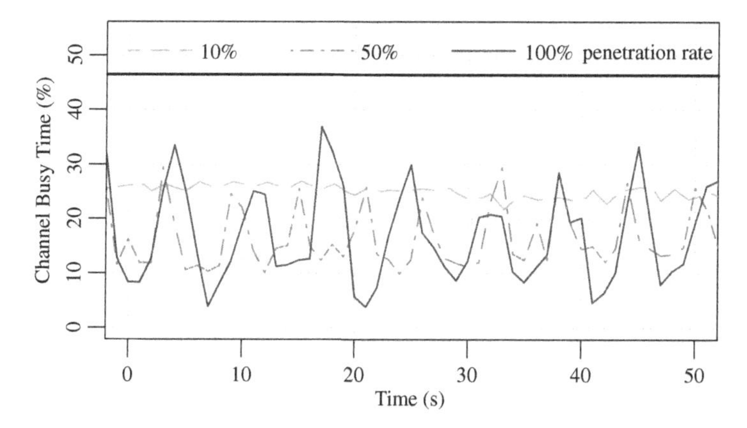

Figure 5.32 The average channel load measured in the form of the channel busy ratio b_t. © [2013] IEEE. Reprinted, with permission, from Eckhoff *et al.* (2013b).

busy time is on the order of 10%–25%, irrespective of the vehicle density and the penetration rate.

However, the system shows some oscillations. While at a low penetration rate the curve is almost a straight line at about 25%, the channel load increasingly oscillated with higher penetration rates. Although the full channel capacity is available, DCC does not always efficiently utilize the available bandwidth at higher penetration rates. The average channel load observed at high penetration rates was even lower than when only 10% of all vehicles were equipped with on-board units (OBUs).

The results can be discussed in two directions. First, ETSI ITS DCC clearly helps by adapting the beaconing system such that the channel does not become congested. It carefully uses the available resources and follows a quite conservative approach. Second, the observed oscillations suggest that there are periods in which the beacon interval is clearly too long. This results in additional delays that might be prohibitive for safety-critical warning messages. We will return to this question in Section 5.4.3 on page 191.

Potential oscillations

One of the most interesting results is that, given a high enough node density and penetration rate, DCC frequently oscillates between its states. This behavior is fully scenario-independent and can be observed on a local scale as well as from a global perspective.

Figure 5.33 shows this behavior. The state machine continuously switches its states from *relaxed* via *active* to *restrictive* and back, with the switching intervals approaching the minimum delay necessary for a state transition (see Figure 5.33(a)). These transitions instantly affect the channel load, causing the system to go into a loop as long as the observed channel load repeatedly exceeds the channel-load threshold.

These oscillations can also be observed on a global scale (see Figure 5.33(b)). Within a cluster of connected vehicles, nodes tend to synchronize their DCC state transitions, causing the globally observed channel load to periodically increase and decrease.

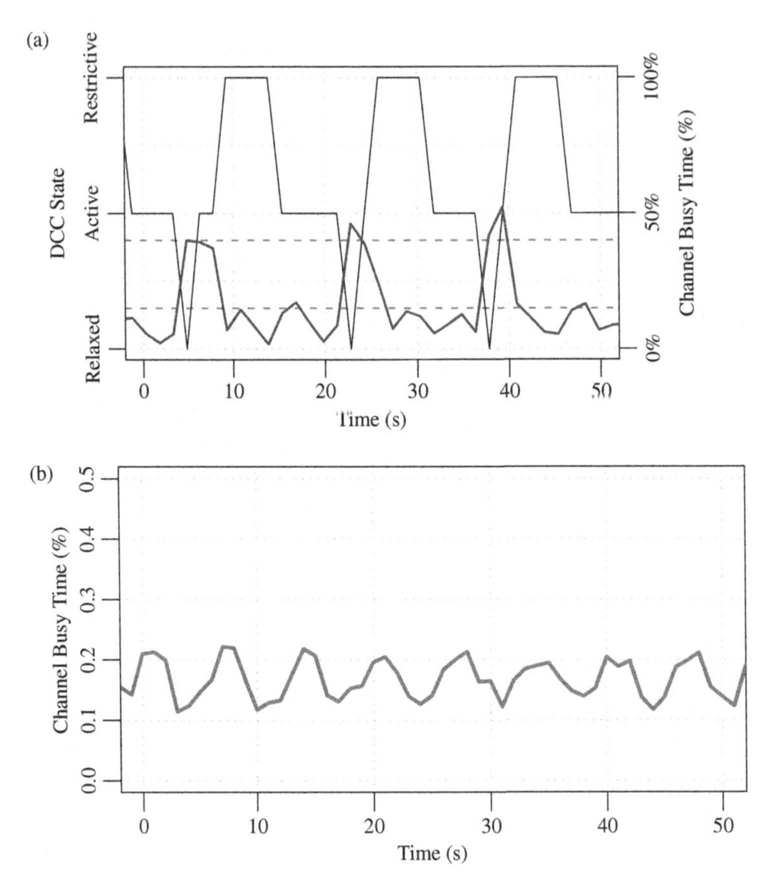

Figure 5.33 Local and global oscillation caused by ETSI ITS DCC's state machine: (a) DCC state (thin black line) and channel load (thicker gray line) for a random vehicle; and (b) the average channel load of all vehicles. © [2013] IEEE. Reprinted, with permission, from Eckhoff *et al.* (2013b).

The reason for this lies within the parametrization of the different DCC states. The *restrictive* state has a large influence on the medium access of a node, reducing the number of possible packets to one per second and also changing the transmit power to a value of −10 dBm. Vehicles in *restrictive* state hardly try to access the channel any more, possibly rendering them temporarily invisible to other vehicles.

Joint transmit rate and power control

Even given the various approaches to congestion control for beaconing-based systems, the extent to which they can help to bring vehicular safety communication from theory to practice remains unknown. The most critical concern is to ensure scalability while not sacrificing the functionality of safety-critical applications.

Tielert *et al.* (2013) have taken one step back and studied the underlying fundamental optimization problem, i.e., how the transmit power and transmit rate of periodic beacon messages can be adapted in order to prevent channel oversaturation on the one hand and to optimize reception performance for safety-critical applications on the other hand.

The main metric used in this work was the average (or percentile) inter-reception time which essentially describes the spatio-temporal behavior of the awareness messaging system. In a parameter study for rate and power combinations, it was observed that, for each distance and channel-load target, there is a specific transmit power optimizing the overall protocol performance. Furthermore, for all of the control strategies studied, significant compromises have to be made in terms of reception performance.

Tielert *et al.* came up with a new proposal for a combined power- and rate-control strategy, which is less complex than the standard ETSI ITS DCC system. It allows one to efficiently approximate the optimal parameters. The system first selects an appropriate rate and then uses the transmit power control as a secondary mechanism. The results are very promising, leading to close-to-optimal results in terms of the average inter-reception time. Yet, no guarantees for safety-critical warning messages can be derived.

The average or percentile does not reveal effects that can be observed for individual vehicles. Thus, it is even possible that important deadlines for safety-critical applications are missed. Furthermore, knowledge about the receiver position must be available for the algorithm (or a definition of the targeted distance). For receivers at different distances or complex shadowing situations in urban environments, there is no such optimum. Therefore, as concluded also by Tielert *et al.* (2013), transmit power control is not of much help and transmit-rate control becomes the main control parameter.

5.4.3 Dynamic beaconing

All of the adaptive beaconing approaches considered so far have assumed optimal (i.e., unobstructed) channel conditions. However, theoretical modeling approaches supported by measurement campaigns have clearly shown that signal shadowing and fading could substantially impact the wireless communication between neighboring vehicles (Sommer *et al.* 2011a; Mangel *et al.* 2011b; Boban *et al.* 2014). This is not only due to shadowing by buildings and other vehicles but also arises as a result of the all-but-homogeneous radiation patterns of antennas in realistic settings. In the following, we study this impact. We show that fixed-period beaconing as well as moderately reactive adaptive approaches as currently suggested by standardization bodies such as the ETSI cannot cope with the increased network dynamics caused by shadowing.

Motivation

Shadowing effects not only lead to new *challenges*, such as how to ensure low transmission latency for safety-critical applications and wide-range data dissemination for efficiency applications, but also provide *opportunities* due to an inherently reduced channel load. As a result, a new dynamic beaconing approach, dynamic beaconing (DynB) (Sommer *et al.* 2013), has been developed. It performs much better than the mechanisms adopted by the ETSI, namely the TRC. DynB paves the way for a new generation of dynamic beaconing solutions that react more aggressively to dynamics in the network – caused, for example, by time-varying signal shadowing effects.

The DCC TRC mechanism provides weak performance due to its poor adaptation properties and the coarse design of the controlling algorithm. During the development of

DynB, all the sampling and windowing parameters were removed, because the beacons themselves offer a natural and very convenient sampling process. Moreover, elementary control theory shows that, in a sampled system, using sampling processes other than the fundamental one can lead to instabilities, which is exactly what is observed with ETSI ITS DCC TRC (see Section 5.4.2 on page 185).

Operation principles

Similarly to DCC TRC, DynB uses the fraction of the busy time b_t in the last time interval $[t - I, t]$ to calibrate the beacon interval. In addition, the number of one-hop neighbors N is used as a measure for the expected channel load. These two measures are used to calculate the next beacon interval I. In particular, DynB forces the beacon interval I to be as close as possible to a desired value I_{des} as long as the channel load does not exceed a desired value for the busy ratio b_{des}.

The beacon interval is calculated according to

$$I = I_{\text{des}}(1 + rN) \tag{5.17}$$

The rationale is the following: I should increase as the network becomes denser (more neighbors), and it must do so only when the channel load is above a target value.

The channel occupancy is captured in the form of an overload ratio r given by

$$r = b_t/b_{\text{des}} - 1 \tag{5.18}$$

which is clipped in $[0, 1]$. It measures by how much the actual channel load b_t exceeds a desired load b_{des}.

The algorithm is fully distributed and each node adapts its beaconing interval to the local conditions.

Now we have to discuss the feasibility of measuring N and b_t, the two metrics used by DynB to adapt the beacon interval in practice. Measuring N is straightforward since each node, according to all standardization bodies, will be identifiable by a unique ID. Even if this ID might be changed (see Chapter 7 for a detailed discussion), it will be unique for the short beacon interval. N can then be measured directly as the number of unique nodes whose beacons have been received in the time interval $I_{\max}$. The busy ratio in exactly the same time interval $I_{\max}$ is more complicated to measure. Actually, there are two options for determining b_t. First, the time during which all successfully received beacons occupied the channel can be summed. This, however, will be just a lower bound, since collisions might have occurred. The second option, therefore, is to include those periods during which the power on the channel was going up but the signal was simply not decodable.

Calibrating the desired channel load

The desired channel load b_{des} depends on the payload length, the data rate used, and the modulation scheme. Let's check this for a typical example. We start by calculating an upper bound t_{busy} for transmitting a payload of $l = 512$ bit at 18 Mbit/s according to the operation of the PLME-TXTIME.*confirm* primitive described by the IEEE (2007):

$$t_{\text{busy}} = T_{\text{preamble}} + T_{\text{signal}} + T_{\text{sym}} \left\lceil \frac{16 + l + 6}{N_{\text{DBPS}}} \right\rceil. \tag{5.19}$$

With default values of the preamble duration $T_{\text{preamble}} = 32\,\mu\text{s}$, the SIGNAL symbol duration $T_{\text{signal}} = 8\,\mu\text{s}$, the duration of a symbol $T_{\text{sym}} = 8\,\mu\text{s}$, and the number of data bits per OFDM symbol $N_{\text{DBPS}} = 72$, we obtain $t_{\text{busy}} = 104\,\mu\text{s}$.

For a packet transmitted on the CCH with an application-layer priority that maps to AC_VO, default parameters according to IEEE (2010c) dictate a transmission opportunity (TXOP) limit of one frame, resulting in a minimum idle time of one arbitration interframe space (AIFS), $t_{\text{aifs}} = T_{\text{sifs}} + \text{AIFSN} \times T_{\text{slot}}$. With default values of $\text{AIFSN} = 2$, $T_{\text{sifs}} = 32\,\mu\text{s}$, and $T_{\text{slot}} = 13\,\mu\text{s}$, we obtain $t_{\text{aifs}} = 58\,\mu\text{s}$.

For a true channel idle time t_{idle}, b_t can be calculated as

$$b_t = \frac{t_{\text{busy}}}{t_{\text{busy}} + t_{\text{aifs}} + t_{\text{idle}}} \tag{5.20}$$

For $t_{\text{idle}} = 0$ (i.e., a continuous stream of data without respecting the contention protocol) we obtain a theoretical maximum busy ratio $b_t \approx 0.64$. Taking into account contentions, as a first approximation, one can add the average initial backoff counter to $t_{\text{idle}} = 1.5 \times 13\,\mu\text{s} = 20.5\,\mu\text{s}$ ($CW_{\text{min}} = 3$) and obtain $b_t \approx 0.57$.

The impact of collisions on safety-critical applications is catastrophic, thus we want to keep the channel load to a level that guarantees a marginal collision rate, let's say $p_{\text{coll}} \leq 0.05$. The computation of collision rates in IEEE 802.11 networks is complex, and to the best of our knowledge there are no simple models available to do it. However, disregarding the backoff freezing on successive attempts, we can approximate it with the probability that two or more stations have a beacon to transmit while the channel is busy multiplied by the probability that at least two stations chose the same backoff within the contention window. Easy combinatorics leads to a desired channel busy ratio of $b_{\text{des}} = 0.25$ for values of N compatible with vehicular networks.

Performance comparison

In order to compare the performance of DynB with that of DCC TRC, two scenarios have been simulated. The first is a freeway with two lanes in each direction, where trucks are allowed to drive only in the rightmost lane. In this scenario, two different vehicle densities have been investigated: regular traffic (low density) and a jam scenario (high density). The second is a suburban scenario using real-world geodata from OpenStreetMap for the city of Ingolstadt, Germany. The geodata provided information not only about the road network but also about the shape of buildings needed to simulate radio signal shadowing correctly. In order to study realistic vehicle-caused radio shadowing, a typical mix of different vehicles (90% cars and 10% trucks) was used.

For all the experiments, we present results for four types of shadowing models:

- Free space (F)
- Vehicle shadowing (V)

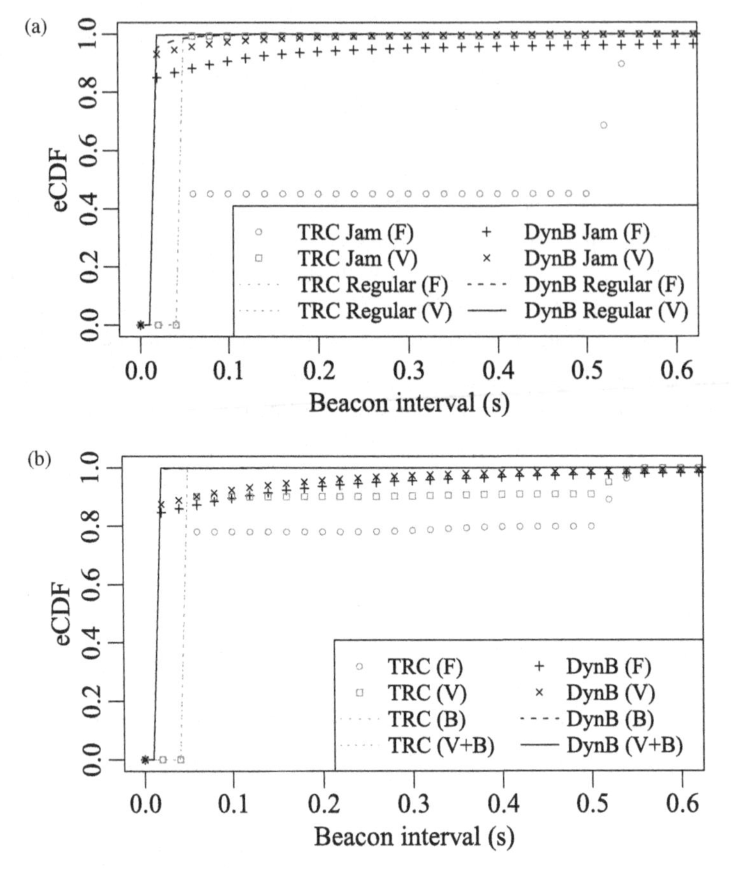

Figure 5.34 Adaptation of the beacon interval for DynB and TRC: (a) freeway scenario and (b) suburban scenario. © [2013] IEEE. Reprinted, with permission, from Sommer *et al.* (2013).

- Building shadowing (B)
- Shadowing by both vehicle and buildings (V+B)

Benefits of dynamic beaconing

The overall behavior of both algorithms, DynB and TRC, is directly reflected in their dynamic and adaptive selection of the beacon interval. Figure 5.34 plots this metric as an empirical cumulative density function (eCDF) of chosen beacon intervals – by virtue of the nature of the algorithms this distribution is highly irregular.

The comparatively low density of vehicles in all regular freeway scenarios (plotted as lines in Figure 5.34(a)) allows TRC to always send at its lowest configured interval ($I_{\min} = 40$ ms). Similarly, even DynB can almost always pick its shortest configured interval ($I_{\min} = 10$ ms). In the case of DynB, however, it is evident that this is possible only because other vehicles shield receivers from interference by neighbors: Simulation runs that ignore shadowing by vehicles can be seen to force DynB to pick larger beacon intervals, albeit infrequently (5% of recorded observations).

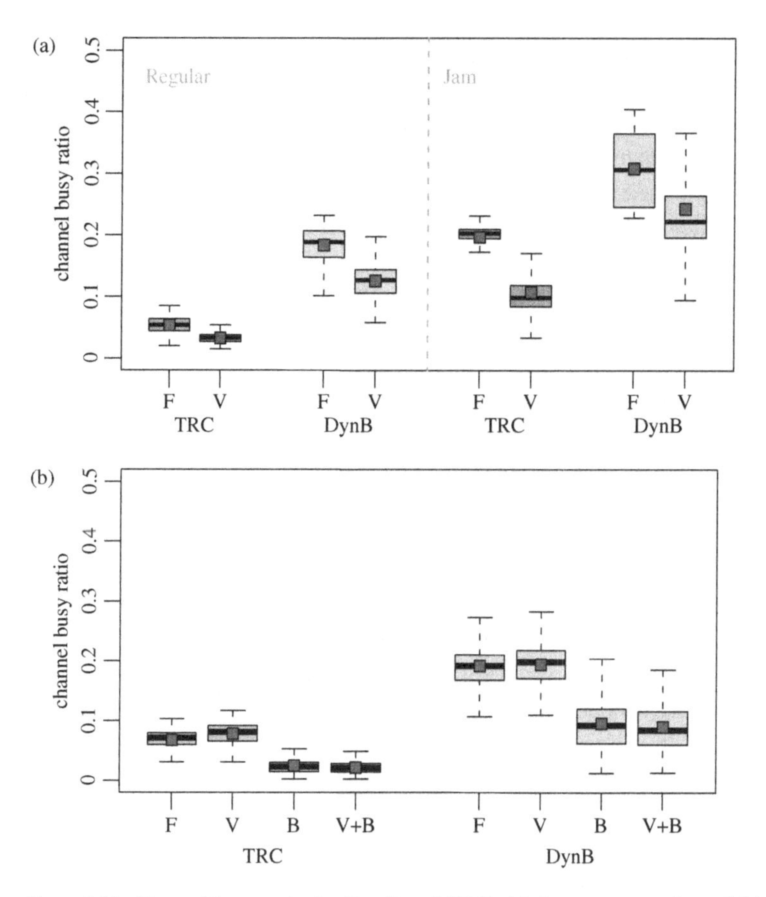

Figure 5.35 Channel busy ratio for DynB and TRC: (a) freeway scenario and (b) suburban scenario. © [2013] IEEE. Reprinted, with permission, from Sommer *et al.* (2013).

The benefit of these shadowing effects is even more evident in jammed freeway scenarios (plotted as dots in Figure 5.34(a)). Here, simulations ignoring shadowing by vehicles would consistently suggest that much higher beacon intervals need to be picked. TRC, in particular, would send more than half of all beacons at intervals of 0.5 s. The benefit of these shadowing effects is also very pronounced in suburban scenarios (see Figure 5.34(b)). Again, simulations ignoring shadowing by vehicles and/or shadowing by buildings would suggest that up to 20% of beacons need to be delayed – in the case of TRC by up to 0.5 s.

The channel load that results from these choices is illustrated in Figure 5.35. The aggressive channel use of DynB is shown to lead to a channel busy ratio much closer to the value of $b_{des} = 0.25$, a value that will keep the number of collisions at an acceptable level. TRC, however, is not able to make use of the full capacity of the wireless channel. It, following also the rules suggested by ATB, behaves too conservatively and misses opportunities to send beacons.

Behavior in highly dynamic scenarios

DynB was developed to be stable under heavy network congestion, and to be able to react quickly to density changes. It actually follows a very aggressive model for channel access. To validate the theoretical concept, we discuss the results of another experiment. This was planned as an extreme case of topology dynamics: Two disconnected clusters of 100 nodes each, both fully meshed, meet for 5 s. In practice, such a rapid increase in neighbors happens, for example, if a vehicle drives on a narrow side street towards an intersection with a major road. While on that street, radio communication to other cars is scarcely possible because the signal cannot pass through the big buildings on both sides. When the vehicle enters the intersection, it instantly passes into communication range with a huge number of other vehicles. The same behavior can be observed when a vehicle is being driven behind some big trucks on a freeway. As soon as it overtakes them, it may experience a huge change in channel load caused by the vehicles in front of the trucks.

The impact of each algorithm was observed over the course of 30 s. For simulations that converge towards stable behavior (TRC at low node densities and DynB at any density), observations in the transient phase are discarded; for simulations that keep oscillating (TRC at higher node densities), the transient phase is fixed to 10 s. For the remaining 30 s, the mean value of the beacon interval and the channel busy ratio, as well as the 5th and 95th percentiles are calculated.

Figure 5.36 illustrates the behavior. The values of the channel busy ratio, the core metric of both algorithms, are plotted. TRC uses thresholds of the channel busy ratio to switch between states. As these measurements are averaged over time, this leads to a pronounced delay until the system can react to changes in network topology. In order to compensate for this, it needs to target an overall underutilization of the channel, as evidenced by the plots. This also lead to a substantial increase of the beacon interval, which makes the protocol less suitable for safety-critical warning messages.

DynB achieves its goal of adjusting the channel utilization more quickly and more smoothly. No over- or under-compensation for the change in network topology can be observed; the adaptation is almost instant. Thus, DynB can target much smaller beacon intervals, always keeping the channel busy ratio as close as possible to $b_t = 0.25$.

5.5 Geocasting

If a node knows its own position, the position of the destination, and the positions of all direct neighbors, it can select the neighbor that is closest to the destination for the next hop towards the destination. The alternative is to send a message not to a specific node located at a geographic coordinate but to any node in a given area. This is referred to as geocasting (Navas & Imielinski 1997).

The term geocasting had already been introduced in the context of MANETs (Navas & Imielinski 1997; Ko & Vaidya 2002). It essentially refers to broadcast-based geographic routing, which applies to all applications in ad-hoc networks but also in vehicular networks.

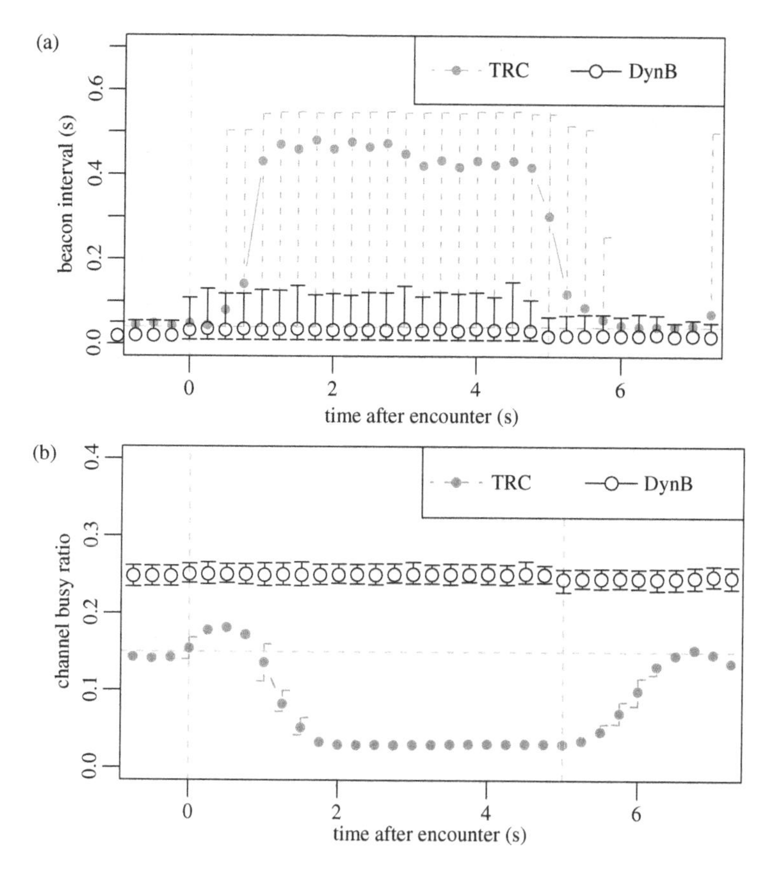

Figure 5.36 Channel capacity management as performed by the DynB and TRC, showing the mean and the 5th and 95th percentiles. (a) Adaptation of the beacon interval when the two clusters meet. (b) The impact on the channel busy ratio when the two clusters meet. © [2013] IEEE. Reprinted, with permission, from Sommer *et al.* (2013).

In the following, we study the application of geocasting to VANETs. In particular, we first look at the standardization efforts to make geographic routing as introduced in the framework of GPSR applicable for vehicular networks. In the second part, we study the TO-GO approach, which improves the concept of geocasting using digital maps, which are assumed to be anyway available to all vehicles equipped with IVC communication devices, as well as two-hop neighborhood information.

We need to emphasize that this is a very narrow look at possible approaches. Most importantly, we will return to this topic later in this chapter when discussing DTN concepts and particularly the DV-CAST protocol (see Section 5.7.1 on page 219).

5.5.1 ETSI GeoNetworking

Applying the lessons learned from the use of geographic routing in MANETs and early adoptions to VANETs, the ETSI incorporated this into its set of standards for IVC. The resulting standard was named GeoNetworking (ETSI 2010b).

GeoNetworking was designed to provide wireless communication among vehicles as well as among vehicles and infrastructure elements. The protocol suite supports a fully distributed and self-organizing ad-hoc networking concept. Infrastructure such as RSUs can be integrated if available, but is not necessarily required. According to the standardization body, GeoNetworking is well suited for highly mobile network nodes and frequent changes in the network topology.

The GeoNetworking approach supports a variety of applications, including road traffic safety, where multiple vehicles might have to be informed about a critical situation, as well as unicast communication for infotainment applications.

Strictly following the concepts of geographic routing, GeoNetworking provides two, strongly coupled functions: geographic addressing and geographic forwarding. Messages can be directed either to a specific node identified by its geographic location or to multiple nodes within a certain geographic area.

Similarly to GPSR, for each packet received the destination address is evaluated. If the receiving node is not the node addressed, it forwards the message greedily according to its local view of the network topology. Thus, each node has to maintain not only a set of its direct neighbors but also as much of the neighboring topology as possible. This information can be learned on the fly whenever the node receives (or overhears) messages.

For the forwarding process, so-called GeoAdhoc routers are defined. A location service is used to identify both GeoAdhoc routers and the location of a destination node. Thanks to the location service, addressing can be realized on the basis of either node identity or a geographic position.

In addition to routing to a specific location, GeoNetworking – just like GPSR and GHT – also supports routing of messages to a geographic area. In the vehicular networking domain, this is for example useful when sending traffic notifications or road safety warnings to all nodes that are within a certain safety boundary.

According to the standard, geographic routing comprises the following three forwarding schemes (ETSI 2010b): GeoUnicast, GeoBroadcast, and topologically scoped broadcast. All the schemes are to be used in conjunction with congestion-control techniques defined in the context of DCC (see Section 5.4.2 on page 185).

GeoUnicast

GeoUnicast is the direct application of geographic routing as initially proposed by GPSR. A packet is delivered to a specific node's location by means of unicasting the data via multiple hops. The location service mentioned can be used to determine the node's location information. Each relaying node forwards the message according to its local topology information until the packet reaches the destination. The concept is depicted in Figure 5.37.

GeoBroadcast

In contrast to GeoUnicast, GeoBroadcast is used to forward a message to a defined geographic destination area. Geographic routing is used to forward the message

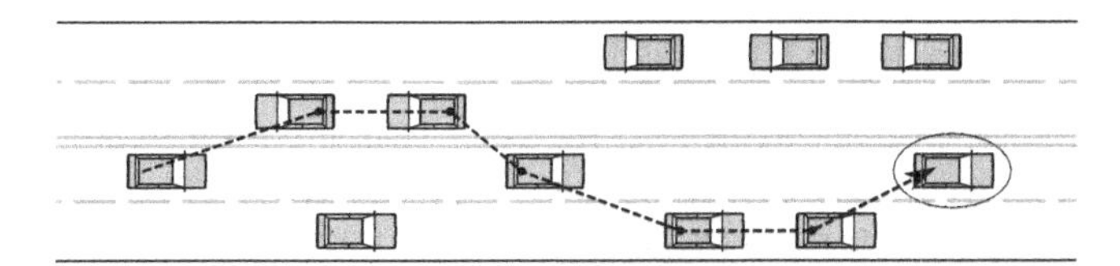

Figure 5.37 GeoUnicast allows one to send a message to a specific vehicle using geographic routing.

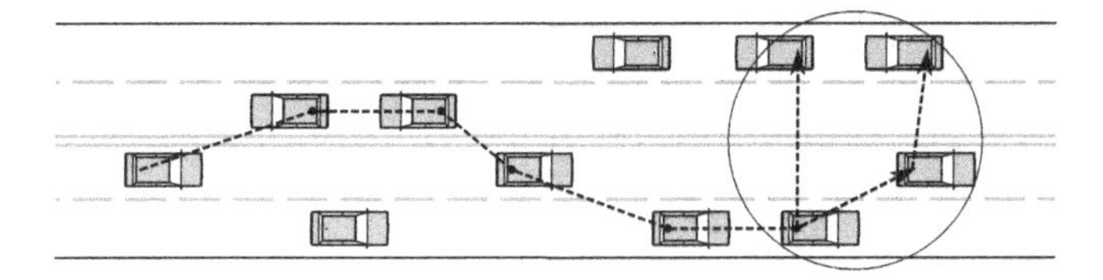

Figure 5.38 GeoBroadcast allows one to send a message to all vehicles in a certain geographic area.

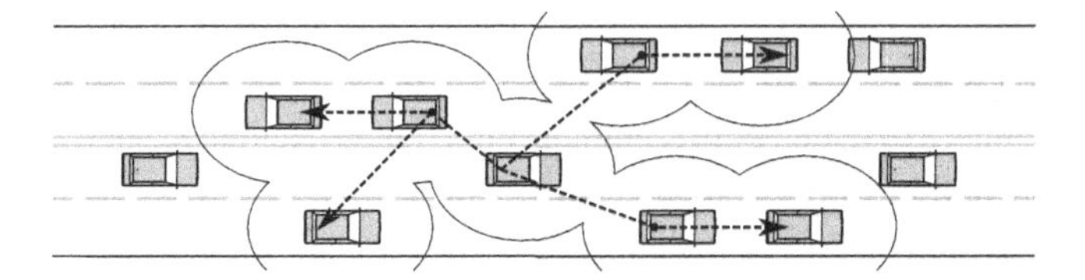

Figure 5.39 Topologically scoped broadcast can be regarded as scoped flooding to all vehicles in the n-hop neighborhood.

hop-by-hop until it reaches the first vehicles in the given area. Figure 5.38 illustrates this behavior.

The vehicle in the destination area is then responsible for re-broadcasting the packet to all other vehicles that are located in this destination area. Again, local topology information is used to determine all the relevant destination nodes.

The ETSI also defined a routing primitive GeoAnycast, which is essentially similar to GeoBroadcast, but the first vehicle in the destination area that receives the message does not have to forward the message any further.

Topologically scoped broadcast

Topologically scoped broadcast (Figure 5.39) essentially defines a way to support flooding constrained to a certain area. In particular, the re-broadcasting is limited by means of a TTL value and thus defines all nodes in the n-hop neighborhood. Since all nodes that receive a message with a positive TTL value are potential forwarders, broadcast storm suppression techniques have to be used to compensate for the potentially high overhead (see Section 3.4.2 on page 102).

Single hop broadcast is a specific case of topologically scoped broadcast, which is used to send packets only to the one-hop neighborhood.

5.5.2 Decentralized environmental notification messages

Decentralized environmental notification messages (DENMs) (ETSI 2013c) are generated like CAMs (see Section 5.3.2 on page 172) by the local ITS application and then broadcast to all vehicles in communication range. Unlike the periodic character of CAMs, messages generated and exchanged using the *decentralized environmental notification service* are inherently event-triggered. One of the key applications relying on this service is road hazard warning. The service uses DENMs to exchange information about road hazards or abnormal traffic conditions. One of the examples that is frequently referred to in the literature is a warning about an accident or the electronic brake light application.

DENM protocol

DENMs are generated like CAMs by the local ITS application and then broadcast to all vehicles in communication range. The standard leaves it open to the implementation and the use case whether and how messages are treated in the network. So, the protocol allows one to repeat the DENM as long as the event that is being described is still active. In addition, a DENM may be forwarded by another car, e.g., by means of geocasting, to inform other vehicles that are not in communication range about the event. The standard even allows one to transmit the event to a centralized TIS.

Thus, each event is unambiguously identified by a so-called action ID and the event may target a certain geographic destination area. Furthermore, the event is identified by means of an originating vehicle's ID, a geographic position, a detection time, and a duration. These attributes may change over time and over space.

In order to generate and update events, the following DENMs have been specified.

- New DENM: This message is sent by a vehicle that first detects a specific event. It includes all the relevant attributes describing the event.
- Update DENM: After an event has been created, it can be updated to provide additional information about the evolution of an event.
- Cancellation DENM: This message is used to indicate the termination of an event (must be sent from the originator of the event).
- Negation DENM: Similarly to the cancellation message, it too terminates an event, but the sender need not be the originator of the event.

DENM message format

Similarly to CAMs, the message format of DENMs is organized in the form of different containers that follow the DENM protocol header. The full message structure is depicted in Figure 5.40. The following containers have been specified.

- Management container: This container includes all the information related to DENM management including the event ID, the reference time, and the DENM type. In addition, information such as the position and duration of the event is included.

Table 5.6 Selected cause codes used in DENMs

Code	Description	Sub-code	Event
1	Traffic condition	2	Traffic jam slowly increasing
		4	Traffic jam strongly increasing
		6	Traffic jam slowly decreasing
		8	Traffic jam strongly decreasing
2	Accident	8	Assistance requested
3	Road works	5	Street cleaning
13	Wrong-way driving	2	Vehicle in wrong direction
92	Post-crash	1	Accident without e-Call triggered
97	Collision risk	1	Longitudinal collision risk
		2	Crossing collision risk

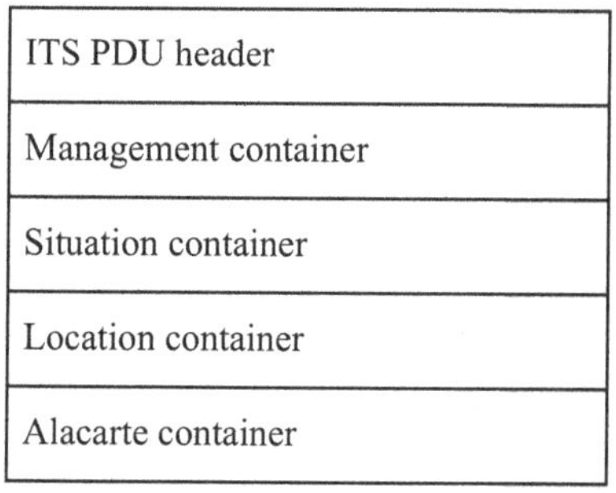

ITS PDU header
Management container
Situation container
Location container
Alacarte container

Figure 5.40 The general structure of a DENM.

- Situation container: The event type encoded by means of *cause codes* is provided in this container. For some codes, a direct mapping to Transport Protocol Expert Group (TPEG) codes is available. Selected cause codes are listed in Table 5.6.
- Location container: The event is described in terms of its heading and speed. The container also includes information about the road class.
- Alacarte container: This container summarizes all relevant information not included in the other containers. Examples include the lane number, temperature, impact of an expected crash, etc.

5.5.3 Topology-assisted geo-opportunistic routing

The idea behind topology-assisted geo-opportunistic routing (TO-GO) is to incorporate topology-assisted geographic routing with opportunistic forwarding (Lee *et al.* 2009, 2010b). The routing protocol exploits the broadcast nature of the wireless communication medium to analyze simultaneous packet receptions and to opportunistically forward the packet via a subset of the neighbors which have received the packet correctly. That is, TO-GO provides broadcast storm suppression together with geocasting that takes into account digital map information. In the following, we explore the characteristics of

TO-GO, which initiated a new era of information dissemination techniques in vehicular networks using geocasting.

Topology-assisted geographic routing

Geographic routing as we have discussed it in the scope of GPSR follows the assumption that all possible directions have an equal likelihood of available potential forwarder nodes. In vehicular networks, this is clearly not the case, because the positions of vehicles are constrained by the road network topology. Topology-assisted geographic routing makes use of digital map information to improve the routing performance in VANETs.

The first approaches towards topology-assisted routing date back to the early days of vehicular networking research (Lochert *et al.* 2005; Lee *et al.* 2007). The idea was to take advantage of the fact that an urban map naturally forms a planar graph where a junction (or intersection) is a node and a road segment is an edge in the graph. In the resulting geocasting concepts, junctions are the only places where a routing decision has to be taken. Packets are forwarded greedily along a street from one junction to the next one. Routing voids in the geographic routing concept are addressed by using GPSR's face-routing mechanism that is performed over the road topology.

TO-GO further improved the situation using an improved lens-shaped forwarding (Lee *et al.* 2010b). Unlike typical geographic routing schemes, a focus is created along the street to circumvent unnecessary directionality towards the side of a street. Figure 5.41 outlines the principles of this mechanism.

Overall, TO-GO used two algorithms to optimize geocasting in urban environments:

- The *next-hop prediction* algorithm is used to determine the packet's target node; and
- the *forwarding set selection* algorithm helps identifying a set of candidate forwarding nodes.

Before we study the design of TO-GO in more detail, we include a brief excursion to Bloom filters, which are used by TO-GO for the forwarding set estimation.

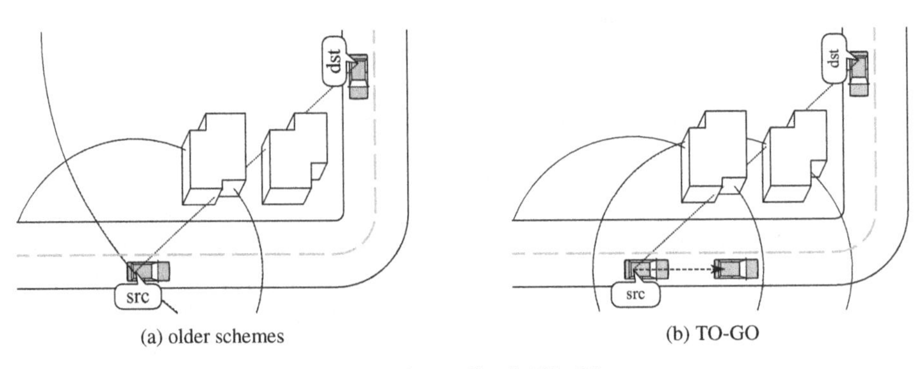

Figure 5.41 The lens-shaped area used for forwarding in TO-GO.

An excursion to Bloom filters

The Bloom filter addresses the problem of testing a series of messages one-by-one for membership in a given set of messages (Bloom 1970). Using conventional hash algorithms, the time required to identify a message as a nonmember of a given set scales with the number of messages in the set. Similarly, the space required to store all members scales with the same value. In many applications, both time and space are restricted but a certain uncertainty might be allowed, i.e., some false-positive rate can be tolerated.

Bloom (1970) proposed a new hash-coding method that provides exactly these features. The computational factors considered are the size of the hash area (space), the time required to identify a message as a nonmember of the given set (reject time), and an allowable error frequency. Bloom showed that allowing a small number of test messages to be falsely identified as members of the given set permits a much smaller hash area to be used without increasing the reject time. Thus, the use of Bloom filters is mainly intended for applications in which a large amount of data is involved and the storage of all values using conventional techniques is not feasible.

A Bloom filter is a hash area represented in form of a bit field X of size N. Each bit in X is individually addressable as $X_0 \dots X_{N-1}$ with addresses 0 to $N-1$. It is assumed that all bits in the hash area are first set to 0. We further need a set of hash functions $h(y)$ that is able to code a message y into a set of M distinct bit addresses. More formally, this means $h_i(y) = \{0, 1\} \; \forall i \in [0 \dots M-1]$.

Inserting a message into the Bloom filter X is realized as setting all bits in X to one for which $h_i(y)$ is 1:

$$\forall i \in [0 \dots M-1] : X_i \leftarrow h_i(y) \tag{5.21}$$

Intuitively, this means that the number of 1 bits in the Bloom filter increases continuously with each message inserted. Some bits for a new message y' may have been set to 1 already; others are explicitly set to 1. Figure 5.42 illustrates this procedure. In our example, three bits addressed as $[1, 5, 7]$ are set to 1 according to the hash function for the message $y =$ "foo".

In order to test whether a message y' is part of the Bloom filter, the same hash function set $h(y')$ is used to generate the respective 1 bits for the message. Formally, the test for $y' \in X$ is performed bitwise:

$$\forall i \in [0 \dots M-1], h_i(y') = 1 : X_i \stackrel{?}{=} 1 \tag{5.22}$$

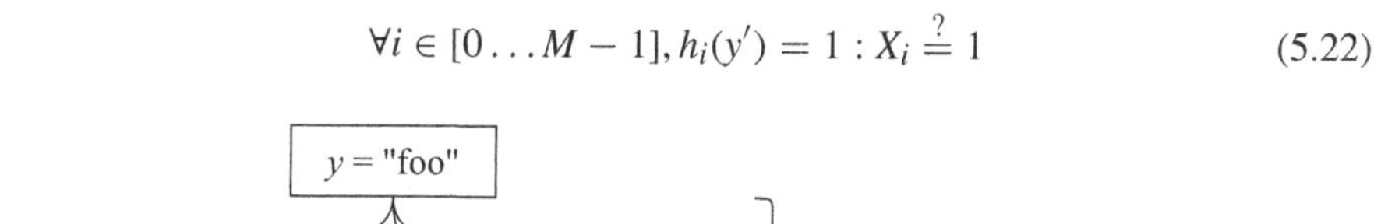

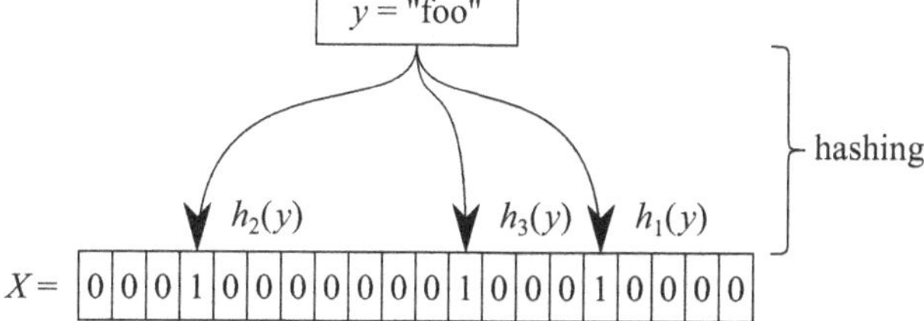

Figure 5.42 Adding new messages to a Bloom filter.

Thus, the message is accepted only if all bits identified by $h(y')$ are set to 1 in our Bloom filter X. If any of the bits is 0, the message is rejected. Please note that this is a probabilistic test. It can be said correctly whether a message was *not* part of the Bloom filter. Yet, false positives are possible.

Intuitively, it can be seen that for larger N the expected fraction of errors is smaller. Thus, depending on the expected number of input messages and the hash function set h used, N can be set to a value that keeps the number of false positives small enough for the envisioned application.

TO-GO's forwarding scheme

TO-GO uses hello messages not only to discover all neighboring vehicles that are in radio communication range but also to identify the *furthest neighbor(s)* and their locations in each direction of the urban map. Typically, there are exactly two such neighbors except for intersection nodes. This information is required in order to support junction forwarding prediction both in greedy mode and in recovery mode. The beacon also contains a Bloom filter of all neighbors the node has. Thus, upon receiving a beacon, a node would have a neighbor list that contains the neighbors from which it has received a beacon, every neighbor's furthest neighbor, and a Bloom filter of all neighbors the direct neighbors can see.

TO-GO uses this information to predict the target node, which is either the furthest neighbor on the same street or a junction node. The next-hop prediction algorithm is depicted in Figure 5.43. First, the node identifies the farthest node seen N towards the destination. Then, it finds the farthest node J towards the destination that is currently on the junction. Finally, it locates the farthest node N_J towards the destination as seen by J. If both N and N_J are on the same street, greedy forwarding is used. If, however, this is not possible, relaying via node J is performed.

After finding the target node, the current forwarding node must determine which nodes will be in a forwarding set. Essentially, all nodes in the forwarding set will compete to relay the message, but only one node should actually act as a relay. Thus,

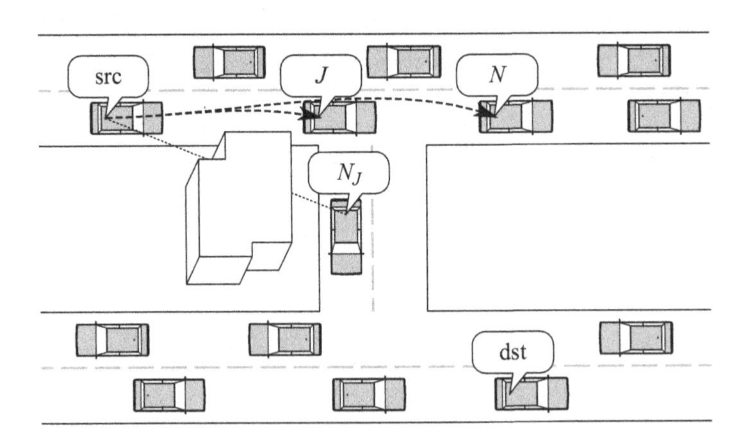

Figure 5.43 Geo-assisted forwarding using TO-GO.

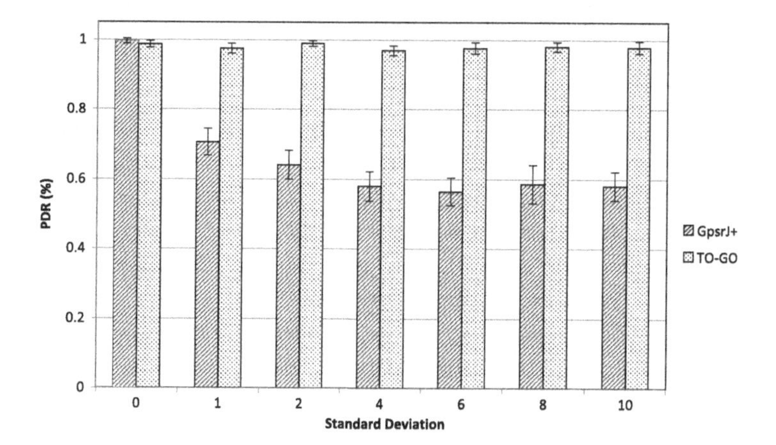

Figure 5.44 The performance of TO-GO compared with geographic routing in an urban environment. The PDR for increasing signal shadowing is plotted as the standard deviation σ. © [2010] IEEE. Reprinted, with permission, from Lee *et al.* (2010b).

all nodes in the forwarding set must be able to communicate with each other. As shown by Lee *et al.* (2010b), finding the optimal solution is *NP complete*. Therefore, TO-GO uses the following approximation, which makes use of the Bloom filter information received in the hello messages. The idea is to choose the neighbor M which has the maximum number of neighbors. Then, all those nodes that (a) can hear M, (b) are heard by M, and (c) are heard by all current members in the forwarding set are included in the forwarding set.

Knowing the resulting forwarding set, the message is finally multicast to all nodes in this forwarding set. The recipients compete to relay the message. Here, distance-based broadcast storm suppression is used, i.e., the node that would be able to make most progress will relay first. Other nodes suppress the re-broadcast if they are able to overhear the message. The algorithm scales with $\mathcal{O}(n^2)$, where n is the number of neighbors.

As can be seen in Figure 5.44, TO-GO clearly outperforms classical geographic routing. In this experiment, the packet delivery ratio (PDR) was measured in an urban environment for increasing radio signal shadowing. TO-GO was compared with classical geographic routing. The specific GpsrJ+ algorithm is an improved version of GPSR that already makes use of topology-assisted routing. Compared with this algorithm, TO-GO is able to maintain a PDR of almost 100%.

5.6 Infrastructure support

We started a discussion on the need for infrastructure support in Section 3.2.3 on page 65. Communication between vehicles is possible given two conditions.

- *Density:* The number of vehicles within communication range is above a certain threshold.

- *Penetration:* The percentage of vehicles equipped with the communication technology is high enough.

If both conditions are fulfilled, vehicles can exchange safety and non-safety messages without major issues and can even forward or relay the packets to other cars in the form of multi-hop communication. The protocols we study in this chapter can directly be applied as long as there is sufficient capacity on the wireless communication network.

If, however, one of the two conditions does not hold, the vehicular network faces major problems. This can be simply communication gaps, e.g., areas where no (equipped) vehicle can forward a message to other cars, or application failures, e.g., warning messages are not received or sent since there is no communication device to be deployed.

This situation can be observed in all scenarios. On a freeway, vehicles might not be able to exchange messages about congestion or even worse about an accident ahead of the cars. In a suburban or urban environment, vehicles might no longer be able to follow a green-light optimal speed advisory (GLOSA) or might not be able to help each other during intersection approaches.

These effects can already become visible in low-density scenarios like at night or in the presence of obstacles blocking the signal propagation. Pure vehicle-to-vehicle (V2V) communication can therefore lead to inefficient information transfer between the vehicles.

To overcome these problems, infrastructure elements can be used to relay messages or even to fully participate in the vehicular network by storing and forwarding road-traffic-related events. In the following, we study selected approaches for infrastructure elements including RSUs and even parked vehicles.

5.6.1 Roadside units

The most straightforward approach to roadside infrastructure has been the use of WiFi APs (Ott & Kutscher 2005). With the progress towards ITS, AP-like systems have been considered to participate in the vehicular network. So-called roadside units (RSUs) can be directly integrated in the IVC system (Lochert *et al.* 2008; Sommer *et al.* 2011c). Options regarding where to place RSUs include positions along a freeway, intersections in urban environments, or even almost random-looking locations such as bus stops that are connected to some backbone network. Most RSUs are therefore used to help connect groups of cars (or even single ones) that are disconnected due to there being a low vehicle density or too low a penetration rate.

Motivation and problem statement

RSUs are static components, placed on the road sides, which may communicate directly with the vehicles by means of wireless communication. In addition, RSUs are assumed to be connected via a backbone network to help speed up the dissemination of information between groups of cars. Essentially, the task of these static devices is to store and to relay information. In addition these units can be connected with one or multiple TICs, which manage and distribute traffic information in a more intelligent manner.

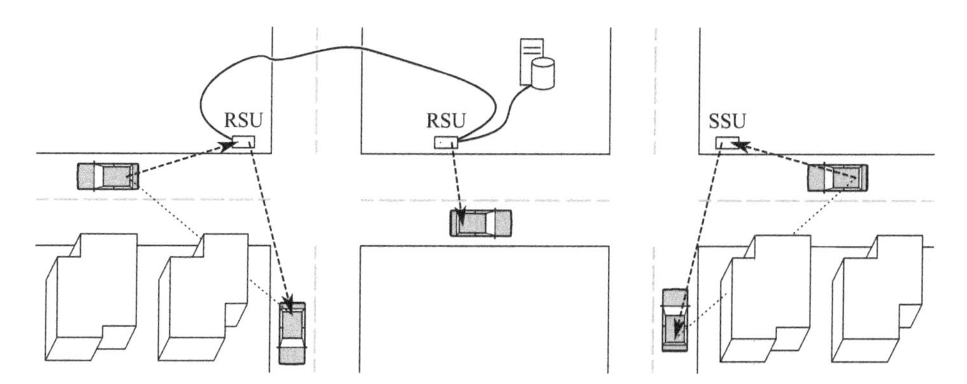

Figure 5.45 An example of RSUs used to connect vehicles that are not within communication range via a backbone network (left) and of an SSU bridging the communication gap due to signal shadowing.

Figure 5.45 outlines the concept (on the left). Shown are vehicles exchanging messages via RSUs and the backbone network connecting these. Also the communication gap between two close-by vehicles that are not able to communicate with each other due to some buildings blocking the radio communication (dotted lines) is bridged.

RSUs of course also help to exchange safety-critical warning messages such as CAMs or DENMs. They may relay a message, but also forward a message to an RSU that is geographically better suited for sending the specific message. Similarly, RSUs integrated with traffic lights may help in emergency situations.

A second type of infrastructure elements are stationary support units (SSUs) (Lochert *et al.* 2008; Sommer *et al.* 2011c). Operating in the vehicular network like an RSU, SSUs are assumed not to be connected via a backbone network. It is therefore much cheaper to deploy such units, which can be operated by a solar panel and a battery as a standalone system. An example is depicted in Figure 5.45 (on the right). Such simple units may have a substantial impact on the speed at which information is exchanged in vehicular networks – again, helping to bridge communication gaps in sparse vehicle distributions at night or for low penetration rates during the initial deployment of IVC systems. In the following, we concentrate on RSUs. Most of the discussion will also hold for SSUs.

Since RSUs and SSUs typically use the same protocol as the vehicles participating in the vehicular network (or at least a similar one), the main question to be answered is not protocol-specific but rather concerns the optimal placement of RSUs and the resulting impact. One of the first works studying optimality criteria for RSU placement was published by Lochert *et al.* (2008).

Later, this problem was reformulated using the coverage problem that is well understood in the area of wireless sensor networks (WSNs). Using coverage as the main metric, RSU deployment has been studied for infotainment and road traffic efficiency applications (Salvo *et al.* 2012). Meanwhile, several deployment options have been discussed in the literature (Aslam *et al.* 2012; Lee & Kim 2010; Barrachina *et al.* 2012).

In field operational tests (FOTs), the vehicular networking community tries to validate these concepts and also to shed light into other metrics such as the influence on road traffic efficiency applications (Gozálvez *et al.* 2012).

The main question is therefore that of the optimal distribution of RSUs or SSUs. In general, this must still be considered a challenge in ITSs. In order to outline the complexity, we will present and discuss selected distribution algorithms in the following, as have, for example, been described by Barrachina *et al.* (2012).

Obstacle-aware distribution

In order to increase coverage, the obvious starting point is to prevent areas being covered by more than one RSU. The additional coverage – this is called *k*-coverage in the scope of WSNs (Zhang & Hou 2005; Dietrich & Dressler 2009) – does not bring any benefit, since we can assume a rather high reliability of the RSUs and that there are no significant energy constraints. Thus, a placement policy whereby new RSUs are positioned so that the radio communication ranges do not overlap appears to be leading to optimized results. Yet, especially in suburban environments, the truth is much more complex. Cellular network planning already tells us that objects obstructing the radio signal must be incorporated into the planning process.

Obstacle-aware distribution (OAD) is a technique that takes the attenuation of the radio signal caused by obstacles like buildings into account (Ferula 2013). The objective is, again, to prevent areas being covered by more than one RSU. RSUs are placed only at intersections, since this positioning has been shown to be optimal in terms of radio connectivity (Lochert *et al.* 2008; Gozálvez *et al.* 2012; Barrachina *et al.* 2012).

We will compare two versions of the algorithm later in this section: a random positioning and a priority-based positioning. Using the random policy, intersections are selected in a random way. No additional information about the typical traffic flows on the street network is necessary. For the priority-based technique, all intersections are sorted according to the average road traffic on that intersection. OAD now selects the intersections at which to place an RSU by checking how many side streets of an intersection are already covered by another RSU. For this, it either randomly chooses an intersection to start with or starts from the top of the sorted list in the priority-based case. If the estimated coverage is below a certain threshold, the intersection is picked for placement of an RSU; and the algorithm continues until a pre-defined number of RSUs has been distributed.

More formally, the algorithm works as follows (Ferula 2013).

- For the priority-based technique, sort the list of intersections according to the average traffic on that intersection
- Until a pre-defined number of RSUs has been distributed:
  - Randomly choose an intersection/pick the intersection from the top of the list
  - Assess the coverage of that intersection
  - If the coverage is below a threshold level, deploy an RSU at this intersection

Density-based approach (D-RSU)

A different concept called D-RSU or the density-based approach was developed by Barrachina *et al.* (2012). The basic idea of this solution is to deploy more resources in areas with lower average road traffic. The underlying assumption is that in high-vehicle-density areas the communication between the vehicles is, on average, to be expected to be very good. Thus, additional RSUs might not provide the expected benefit.

The situation is different in zones with low traffic density. Here, it is necessary to place more RSUs in order to increase the reliability of the transmission of IVC messages.

According to the authors, RSUs should be deployed using an inverse proportion to the expected vehicle density. If, for example, in a specific city area the expected traffic density is about 60%, only about 40% of the RSUs should be placed in this part.

To measure the expected vehicle density, again the average occupancy of the intersections has been chosen. Using the maximum density $D_{\max}$, this value is inverted for each intersection i according to $D_{\text{inv}}[i] = D_{\max} - D[i] + 1$.

Afterwards an intersection is selected according to a weighted random-selection algorithm. The implementation of this algorithm was kept very simple:

- Until a pre-defined number of RSUs has been distributed:
  - Select a random number r in $\left[0, \sum_{\forall i} D[i]\right)$; initialize $v = 0$
  - For all intersections j in the list:
    - Calculate $v = v + D_{\text{inv}}[j]$
    - If $v > r$, intersection j will be used to deploy an RSU

Minimum cost distribution

A final deployment option is a minimum cost distribution (MCD) policy (Ferula 2013). In general, the development and maintenance of RSUs are very expensive. This is frequently discussed in the context of traffic lights, since not only the costs for the RSU itself but also the costs for backbone network connectivity and continuous maintenance need to be considered. Hence, the use of already existing infrastructure may be beneficial. The minimum cost deployment policy takes this into account and tries to install RSUs primarily at positions that already have an existing backbone network connection.

A type of infrastructure that has already been introduced in many cities is the intermodal transport control system. This service is primarily used for local public transportation and serves different tasks. An example is the dynamic passenger information which is intended to inform users of public transport services about departure times at bus stations. The communication between these units can be realized with different technologies, including FM or digital radio systems, cellular networks, and wired backbone networks. These communication links can, in theory, be reused for interconnecting RSUs.

The concept for the minimum cost policy is therefore to deploy RSUs exactly at these locations. We can easily assume them to be bus stops. The deployment policy can then use either the OAD or the D-RSU technique to select the most suitable bus stops.

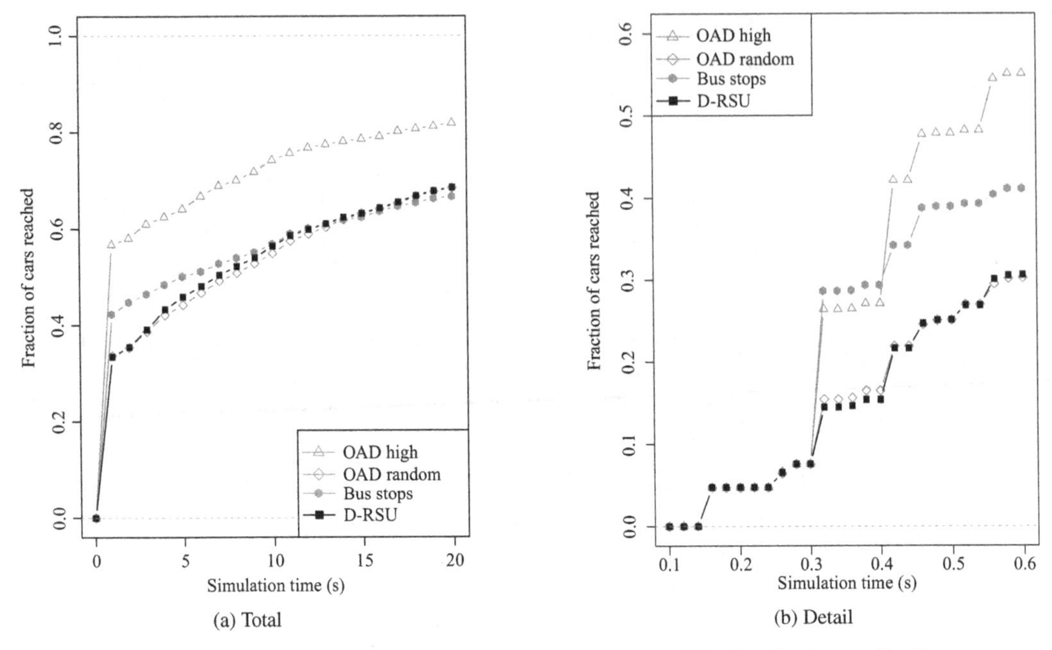

Figure 5.46 The performance gain using RSUs; an eCDF of the time it takes to distribute a specific message in the city of Innsbruck, Austria is plotted.

Performance impact of RSUs

In order to gain more insights on the advantage of RSUs, we briefly look at some results comparing OAD, D-RSU, and MCD (Ferula 2013). The simulation study was performed for different numbers of RSUs. The experiment was done for the city of Innsbruck, Austria using the Veins simulation framework. As a key metric, the time it takes to spread a specific message to vehicles potentially interested in the message was used.

Figure 5.46 shows the results in the form of an eCDF for low-density traffic (about 25 vehicles/km^2). As can be seen in Figure 5.46(a), irrespective of the RSU deployment strategy, a substantial gain in information spread can be achieved (38%–58% of the vehicles are informed almost immediately by means of the wired backbone connection and RSUs relaying the information). Over time, D-RSU and MCD perform almost equally well, being able to reach about 60% of all vehicles in the simulation after 20 s. Please note that it is theoretically impossible to reach 100% of the vehicles because some might be located in smaller streets in a more remote area of the city. OAD, on the other hand, allows one to spread the information even further, reaching about 80% of all vehicles after 20 s. This is because OAD was designed primarily for low-density scenarios.

Figure 5.46(b) reveals more details about the protocol's behavior in the first 0.6 s. During the initial 300 ms, the RSUs simply have to receive the initial message and to forward it via the backbone network to the other RSUs. Only after 300–500 ms does the impact of RSUs and the respective deployment strategy become visible.

5.6.2 Parked vehicles

Even though there are plans to equip major freeways with RSUs and also to consider placing RSUs at major intersections in urban environments, e.g., in projects aiming to provide intelligent traffic lights communicating with vehicles for GLOSA or adaptive traffic-light cycles, the most common criticism is that RSUs are simply too expensive in terms of deployment and maintenance. This also holds for other, less complex infrastructure elements such as SSUs. Thus, it seems likely that many parts of our road networks will not be covered by infrastructure elements in the near future. Given the two problems discussed in the last section, low vehicle density and low penetration rate, this will be especially critical in suburban environments. Considering the changing vehicle density at day and night, both traffic efficiency and cooperative-awareness applications will likely have to rely solely on other communication technologies.

Another possibility is to reconsider the capabilities of vehicles that are currently parked, i.e., not being used, but are equipped with IVC devices. Those cars could act as temporary RSUs. Parked vehicles are often located at exactly the right locations and at the right time: They are parked when the vehicle density is low (e.g., at night), but not parked when there are many cars on the streets already providing rather good connectivity (e.g., during the rush hour).

In the following, we study the concept of parked cars and the potential benefit of those devices. The first applications considered are cooperative awareness (Eckhoff *et al.* 2011c) and information downloading (Liu *et al.* 2011), both of which greatly benefit from support by parked vehicles.

Motivation

As we have discussed already in Section 3.2.3 on page 65, there are several studies available clearly showing the availability of vehicles parked on the streets in major cities. One of the most detailed studies was done for the city of Montreal, Canada (Morency & Trépanier 2008). It revealed some very interesting insights. During the study, which was done in 2003, out of 61 000 daily parking events, 69.2% of all parked cars were parked on streets while only 27.1% were parked on outside parking lots. A minority of 3.7% was parked in interior parking facilities. The study furthermore showed that parked vehicles were distributed throughout the whole city, which means that there is a high probability that a parking car is within transmission range of a moving car. Thus, almost all parked vehicles can, in theory, participate in the communication between vehicles, following the same approach as proposed for RSUs.

The same study revealed that, on average, the duration of one parking event was about 7 h (Morency & Trépanier 2008). Other studies found that, on average, a vehicle is parked for 23 h a day (Litman 2006). We therefore conclude that the use of parked cars as relays in vehicular networks can prove to be very helpful in supporting the message exchange – at any given time, most cars are parked; of these, most are parked on streets.

Considering the capabilities of parked vehicles in comparison with RSUs, a key advantage of vehicles is that they are energy autonomous: As a vehicle moves, its battery is continuously recharged. Of course, parked vehicles do not have this virtually

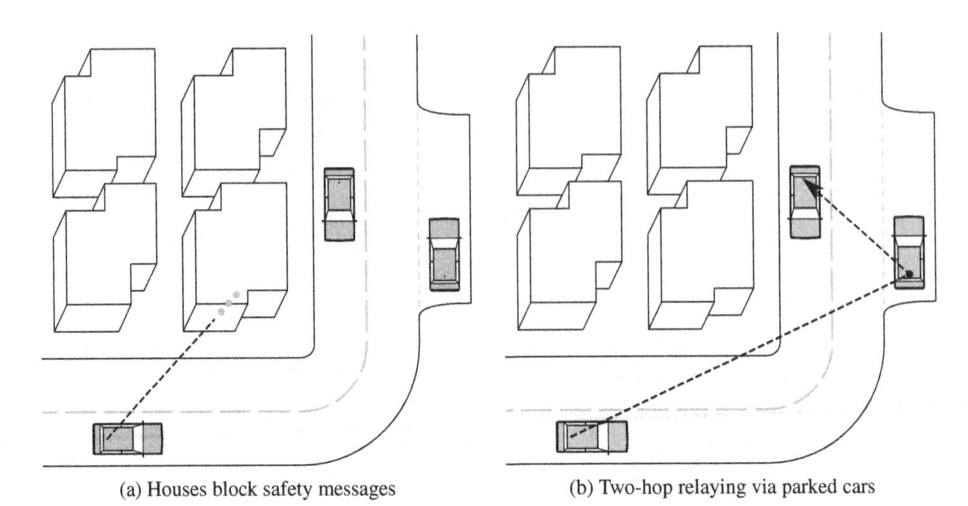

Figure 5.47 Utilization of parked cars as relay nodes can increase cooperative awareness in vehicular safety applications.

unlimited supply of power, since their batteries do not recharge while their engines are turned off. The battery drain due to their use as relays in the wireless communication network, however, is very low. Modern cars already come pre-equipped with dedicated electronics to keep certain devices powered on when the vehicle is not driving – and cutting power to these devices when the battery charge drops below a certain point. We discuss this in a little more detail in the next section.

Another benefit of parked cars is their parking position itself. Being located alongside the street and often near obstacles, they offer a promising possibility to relay messages exchanged by driving cars in order to bypass obstacles. This idea is shown in Figure 5.47. Therefore, conceptually, parked vehicles represent a set of dynamic SSUs, participating in the vehicular network, e.g., to enhance the performance of safety applications. We see a major benefit in the ubiquitous availability of such parked cars in comparison with RSUs and SSUs.

Given that the technical impact is very low, this might even become a motivation for mandated use, again, given the fact that road traffic safety can be substantially improved without the need to deploy an unreasonably (due to cost) high number of RSUs.

This leaves the question of how likely it is that parked vehicles will be able to participate in a vehicular network (Sommer *et al.* 2014a). From a technological point of view, parked vehicles are just like their moving counterparts. If we can expect moving vehicles to be equipped with IEEE 802.11p radios, we can expect this also to be the case for parked vehicles. Obviously, only a subset of moving and parked vehicles will be ITS-ready in the early days of IVC. Furthermore, increasing availability and market penetration of ETSI ITS- and IEEE WAVE-equipped cars can be predicted. The US Department of Transportation (US DOT) is evaluating the scalability, security, and interoperability of WAVE devices and applications in its *Connected Vehicle Safety Pilot Program*, aiming to jumpstart commercialization of the automotive and consumer electronics. The output of the study serves also to support National Highway Traffic Safety Administration (NHTSA) 2013 rulemaking, such as whether to make WAVE mandatory.[1]

Following this discussion, we argue that users will most probably be willing to join for the following reasons.

- The availability of communication via parked cars can substantially improve cooperative awareness and, therefore, vehicular safety.
- The success of social networks and of crowd-sourcing activities demonstrates the general willingness to share information for mutual benefit.

Impact of relaying on battery drain

For investigating potential communication strategies for parked vehicles, we assume that parked cars are virtually energy autonomous. In order to motivate this assumption, we investigated the energy needed for providing relay services (Sommer *et al.* 2014a).

A typical IEEE 802.11p OBU should not drain more than 1 W on average, which is a very generous upper limit. Now, if we consider a small car's battery providing 480–840 W h (Badescu 2003), we can run the system for approximately 20 days, fully draining the battery. Further assuming that we allow the system to use at most 10% of the battery's capacity, we can still use the system for two days.

Since most cars are parked for less than two days, the battery can be recharged while the car is in use and will then last for another two days for relaying wireless communication messages. Furthermore, we can say that the use of such a relay system for a parking time of less than one day would have no critical impact on the usability of the vehicle.

When considering large cars or hybrid cars, these numbers will be even better. For example, the battery of a Tesla Roadster has a capacity of 53 000 W h, providing energy for several years of constant radio transmission. Still, it is obligatory that the OBU of a parked vehicles does not discharge the battery to the extent that the car cannot be started again. The remaining power must always be sufficient for the ignition and other mandatory functions of the vehicle. Basically, there are two possible ways to overcome this problem. Either the on-board device knows about the battery level and can switch itself off accordingly, or the IEEE 802.11p device is equipped with a dedicated battery that is also recharged when the car moves again.

Using parked cars

The application range for using parked vehicles is rather large. It ranges from improving safety applications such as cooperative-awareness messaging (Eckhoff *et al.* 2011c; Sommer *et al.* 2014a) or intersection collision warning systems (Joerer *et al.* 2014) to traffic efficiency and even entertainment applications (Liu *et al.* 2011, 2012; Malandrino *et al.* 2012)).

[1] US DOT, The Connected Vehicle – Next Generation ITS, see http://www.itsa.org/industryforums/connectedvehicle.

The main idea behind all of these concepts is the same. Parked vehicles are treated like SSUs and directly participate in the vehicular network by relaying messages or storing and forwarding them according to the application-layer logic at later times. We will study this approach in more detail in this section.

Another concept is to treat clusters of parked vehicles as a distributed virtual RSU. This RSU can even be connected to the Internet (or some other backbone network) by means of a WiFi AP that is within communication range of one of the parked vehicles in this cluster or by means of a cellular 3G or 4G network uplink. It has been shown that this substantially improves the situation for content-download applications in general (Liu *et al.* 2011, 2012; Malandrino *et al.* 2012).

Let's concentrate on cooperative-awareness applications in the following. Assuming that each moving car periodically emits beacon messages containing its position and speed, parked nodes will of course be able to overhear these messages. If the parked vehicle acts like an SSU and re-broadcasts this beacon message so that other moving cars (which might be unaware of the original broadcast due to shadowing) will then pick up the beacon, the level of awareness can be significantly increased (Eckhoff *et al.* 2011c; Sommer *et al.* 2014a).

To obtain an upper bound for the safety benefit obtainable by utilizing parked vehicles as relay nodes, a simple relaying system has been created. The objective was to cope with the broadcast storm problem and to keep channel load low. Therefore, the relaying of messages was limited to two-hop transmissions, i.e., a maximum of one relay node. In the resulting system, moving vehicles generate safety beacons with a TTL value of 1. When another node receives one of these beacons, it decreases the TTL to 0 and re-transmits it. Packets with a TTL of 0 are never re-broadcast.

In a final system, a carefully designed relaying algorithm needs to be deployed in order to keep channel usage low but still ensure a benefit close to the potential upper bound. Possible solutions include the restriction of relaying to only special nodes, for example, nodes that are parked close to intersections (Benslimane 2004). Furthermore, a relaying node could be able to autonomously assess whether packet relaying helps improve cooperative awareness for nearby nodes by observing neighborship relations, including movement information such as speed or direction (Tonguz *et al.* 2007). Also, evaluating current channel conditions in order to determine whether a packet should be relayed seems to be a promising approach (Sommer *et al.* 2011c, 2013).

Performance gain for cooperative awareness

Investigating the performance gain to be expected on bringing parked vehicles into the picture, we compare the gain derived by use of parked vehicles with that of RSUs. The aforementioned simple relaying strategy has been used for this experiment in order to gain more detailed insights into situations in which RSU and parked-car relaying help to increase cooperative awareness. Sommer *et al.* (2014a) investigated the beneficial effects in extensive simulative studies parametrizing both node density and the amount of stationary nodes.

The metric used in this study is the general benefit, i.e., the additional amount of cars reached relative to the number of cars reached with moving-vehicle relaying only.

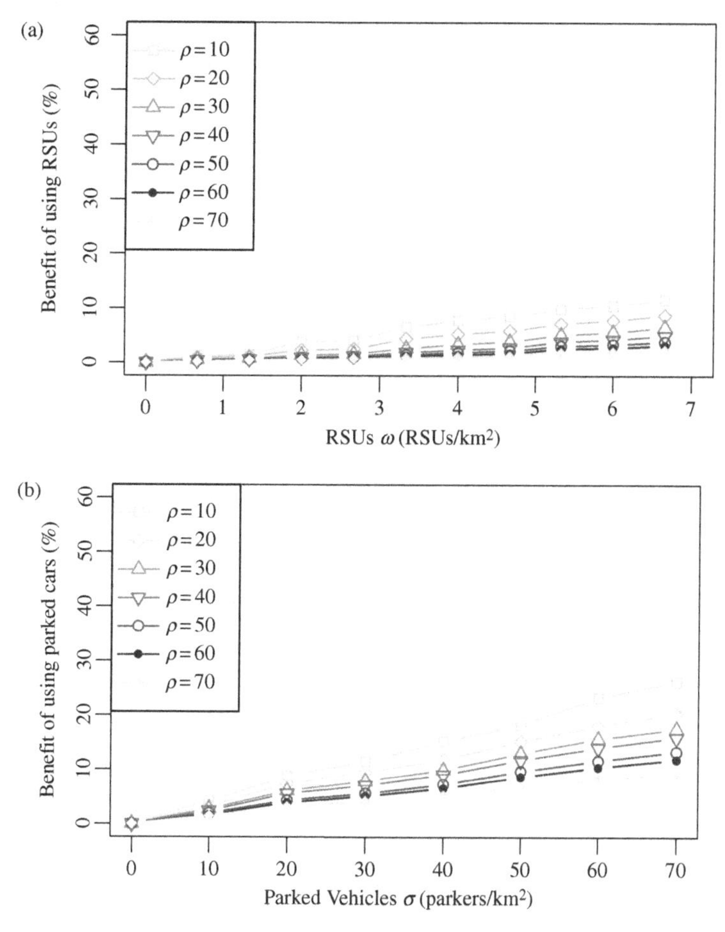

Figure 5.48 The increase in message-delivery success using (a) RSUs and (b) parked cars in a suburban scenario (the city of Ingolstadt, Germany). © [2014] IEEE. Reprinted, with permission, from Sommer *et al.* (2014a).

Two scenarios have been evaluated. The first was a suburban environment of the city of Ingolstadt, Germany, where parked cars were placed exactly to match realistic distributions. For the second scenario, a typical Manhattan-grid layout was used.

In all setups, we observe that the benefit of either parked cars or RSUs is higher when the node density ρ of moving vehicles equipped with IEEE 802.11p devices is lower (see Figures 5.48 and 5.49). With more driving vehicles in the network, the probability of a vehicle previously being unreachable due to a blocked radio signal but reachable through an intermediate driving vehicle increases – reducing the benefit of infrastructure elements such as parking vehicles and RSUs.

On comparing the performance of parked-car and RSU relaying in a suburban context (Figures 5.48(a) and 5.48(b)), we observe that even with RSUs placed at the most frequented junctions with a density of $\omega \approx 6.7\,\text{RSUs/km}^2$ the benefit does not exceed

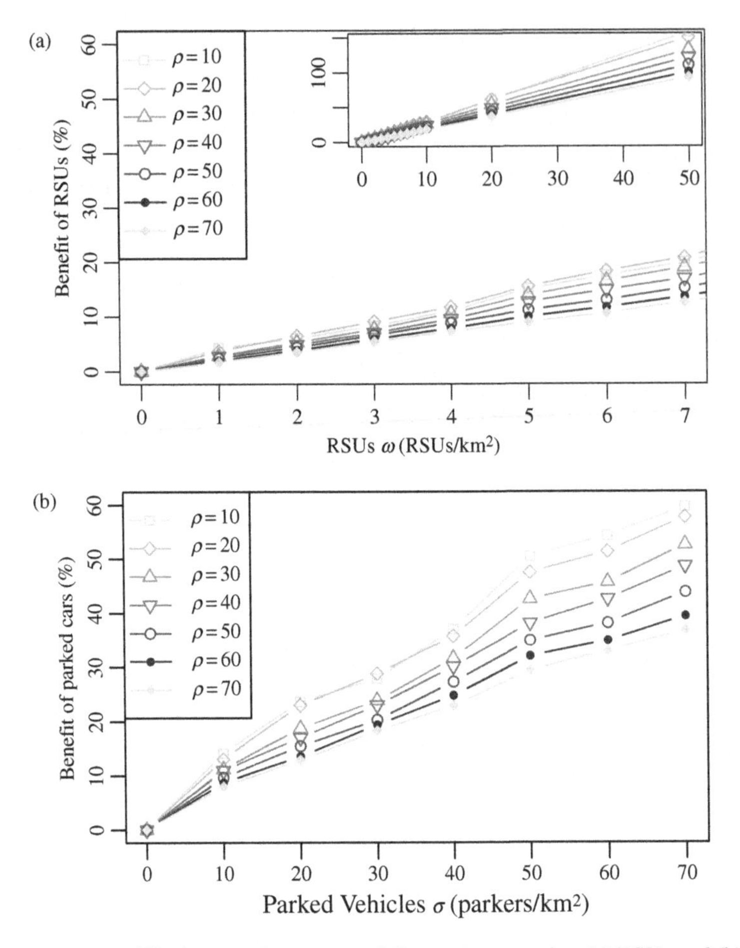

Figure 5.49 The increase in message-delivery success using (a) RSUs and (b) parked cars in a Manhattan-grid scenario. © [2014] IEEE. Reprinted, with permission, from Sommer *et al.* (2014a).

10%. By utilizing the readily available parked cars in the area we obtained peak values of up to some 25%.

In this scenario, the amount of RSUs deployed is evidently lower than the number of parked vehicles; however, it is unlikely that in a suburb the density of RSUs deployed exceeds the number of simulated nodes, though for the parked vehicles a density as high as $\sigma = 70\,\text{parkers/km}^2$ is considered to be realistic.

On studying the same experiment for the Manhattan-grid scenario, we would expect an even more substantial gain. Radio signal shadowing caused by buildings is much more pronounced here, thus the benefit obtained from relaying by infrastructure nodes is higher than that in the suburban area (see Figures 5.49(a) and 5.49(b)). Furthermore, compared with the suburban setup the benefit has changed in favor of the use of parked vehicles rather than RSUs. For example, it takes only about $\sigma = 20\,\text{parkers/km}^2$ to reach the same amount of vehicles as with $\omega = 7\,\text{RSUs/km}^2$. Equipping nearly every junction with a roadside unit (RSU) (Figure 5.49(a), top right) would lead to an almost perfect coverage, improving cooperative awareness by up to 150%, but at unrealistically high cost.

The results presented demonstrate the potential gain for cooperative-awareness applications when using RSUs or making parked cars take over this role. Depending on the scenario, use of RSUs can increase the amount of reachable hosts. Yet, parked cars can boost this gain at negligible cost.

Relaying benefit in the day/night cycle

The remaining question is: to what extent does the typical day/night cycle have an influence on the communication between the vehicles? Typically, vehicles are parked for a certain period, used to travel to work or for shopping, and parked again (at these places), before eventually being driven home again. Therefore, we examine the effect of moving vehicles becoming parked ones and vice versa. For this experiment, the total amount of vehicles $\rho + \sigma$ was considered invariant, but the ratio ρ/σ was varied (Sommer *et al.* 2014a).

Figure 5.50 depicts the results. The benefit of using parked vehicles for the communication between moving vehicles compared with not making use of the parked cars is shown. Again, the results are plotted for a typical suburban scenario (the city of Ingolstadt, Germany) and the more extreme case of a Manhattan grid. In both scenarios, the trend is clear. With more vehicles becoming parked ones, i.e., the density of moving vehicles dropping significantly, the cooperative-awareness application becomes less effective. However, if parked vehicles are used as relays, the quality of the application remains almost stable.

Cooperative awareness is particularly important at night, when lighting conditions might be bad yet drivers are more inclined to drive faster on the now almost empty streets. With the help of parked cars, vehicles can experience the same level of cooperative awareness at night as they would if there were still many more moving vehicles on the street. For both day and night, we can conclude that with parked cars we can achieve the same level of cooperative awareness in sparsely populated areas as we would have in those with many moving vehicles.

5.7 DTN and peer-to-peer networks

In the previous sections, we studied ad-hoc routing, geographic routing, and beaconing-based information dissemination techniques. All these concepts require a certain minimum of connectivity between the vehicles. This is not necessarily always the case in vehicular networks. Therefore, techniques to manage these intermittent connections are needed.

Of course, this problem had already been identified in the early days of vehicular networking research. Thus, many of the concepts that we studied in the previous sections particularly focus on this issue as well. The motivation for providing a separate section just on the problem of spontaneous or intermittent connections is to highlight two main

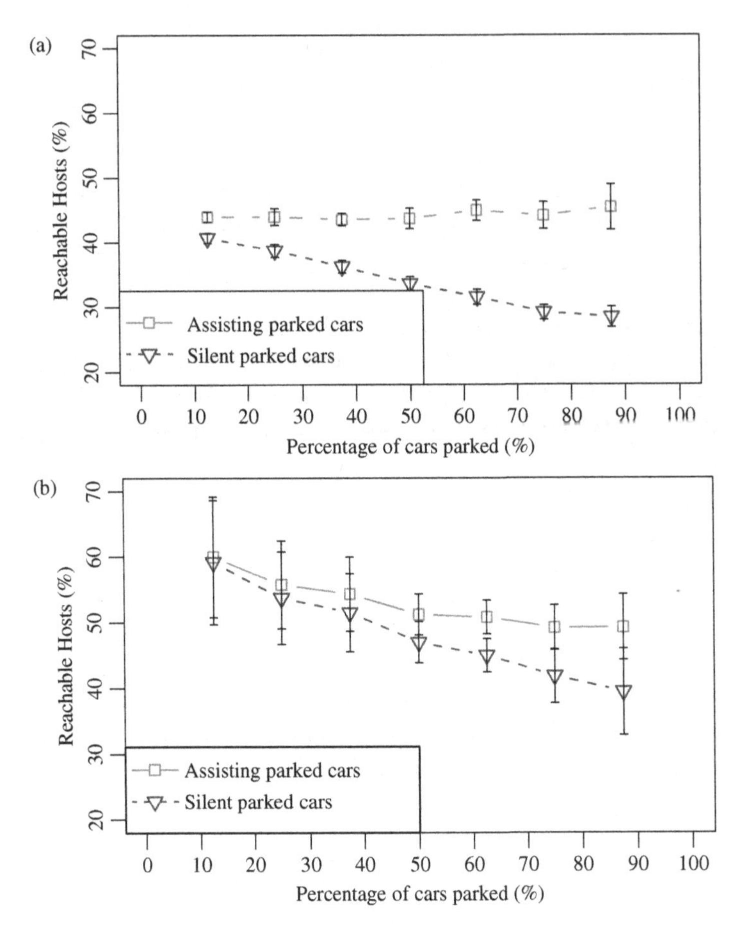

Figure 5.50 The percentage of hosts reached within the safety range in a typical day–night scenario. (a) Manhattan grid and (b) a realistic scenario. © [2014] IEEE. Reprinted, with permission, from Sommer *et al.* (2014a).

principles in networking that help one to overcome the connectivity problem also in vehicular networks.

We focus on two approaches in our discussion.

- Delay/disruption-tolerant network (DTN) – In the DTN research field, the main focus is on exchanging information between mobile devices such as mobile phones (Hui *et al.* 2005; Zhu *et al.* 2013; Cao & Sun 2013; Khabbaz *et al.* 2012). Users participating in the network experience spontaneous connectivity to other users according to their social relationships and their mobility pattern. Investigating DTN solutions in detail would require yet another textbook. We therefore investigate selected examples that highlight how to exploit these spontaneous connections for efficient information dissemination in VANETs.
- Peer-to-peer networks – The storage and management of information in massively distributed systems has been investigated in the scope of peer-to-peer

networks (Steinmetz & Wehrle 2005; Lua *et al.* 2005; Ranjan *et al.* 2008). Peer-to-peer routing algorithms, which were originally used for distributed information storage in the Internet, have later successfully been used in many areas. We have already explained how distributed hash tables (DHTs) can be integrated with geographic or virtual-coordinate-based routing. Again, the entire area of peer-to-peer networks has already been the key focus of several textbooks. We show how peer-to-peer concepts can be applied to the problem of content downloading and to a distributed TIS in vehicular environments.

5.7.1 Distributed vehicular broadcast

The idea behind distributed vehicular broadcast (DV-CAST) is to make the protocol scenario-aware (Tonguz *et al.* 2010). In particular, it is important to know whether broadcasting as discussed in Section 5.4 on page 174 will be able to deliver a message, or whether DTN concepts need to be employed to overcome network fragmentation. DV-CAST therefore supports two operation modes.

- Multi-hop broadcast: If the network is connected, multi-hop beaconing with broadcast suppression is performed to deliver the message to all vehicles in the surroundings.
- Store–carry–forward: If the network is fragmented, message-ferrying techniques as proposed in the framework of DTNs are used to carry the message until it can be delivered to appropriate destinations.

In the following, we outline the algorithm used by DV-CAST in more detail and also look at the resulting performance gain. Please note that the DV-CAST concept was specifically designed for use in freeway scenarios. This becomes obvious on looking at the algorithm design.

Algorithm

The basic algorithm of DV-CAST is as follows. Nodes periodically send hello beacons containing their position, speed, etc., as also proposed in the ETSI CAM beaconing. Using this information, each node maintains three independent neighbor tables.

- *Same direction, ahead* – all vehicles that are driving in the same direction with current positions ahead of the vehicle under consideration.
- *Same direction, behind* – as above, but including all vehicles with positions behind the current vehicle.
- *Opposite direction* – all vehicles, irrespective of their position, that are driving in the opposite direction.

Each message to be sent via DV-CAST uses geographic addressing and contains a source location as well as a destination region of interest (ROI). Messages are forwarded using geocasting, i.e., multi-hop broadcasting that is geographically constrained towards the destination ROI. Without additional help, such a geocast will quickly stop if no further relays are within communication range.

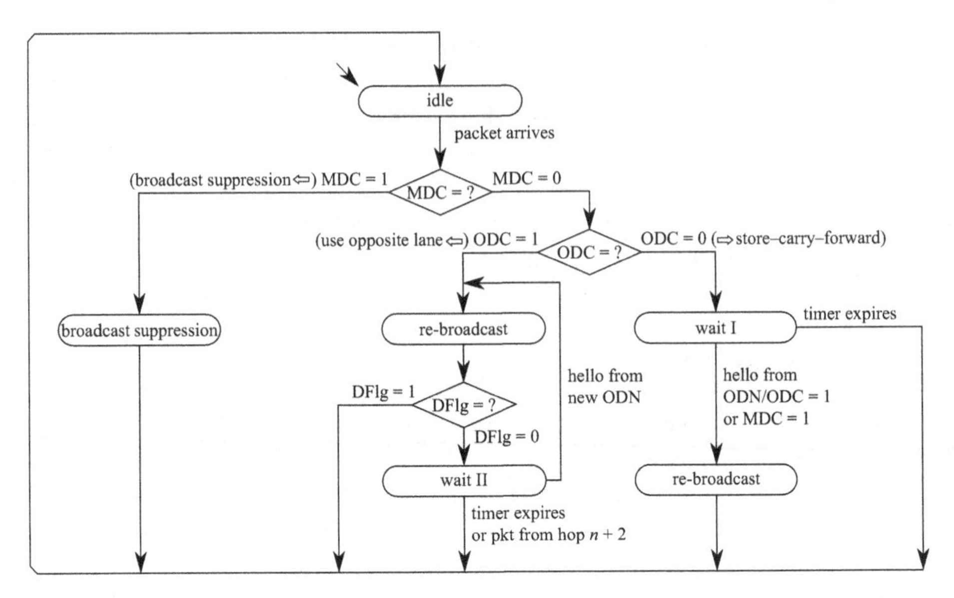

Figure 5.51 The DV-CAST algorithm used to decide between geocasting (broadcast suppression) and store–carry–forward (re-broadcast to a vehicle traveling in the opposite direction).

The algorithm used by DV-CAST is outlined in Figure 5.51. For each message, three flags need to be evaluated (Tonguz *et al.* 2010).

- Destination flag (DFlg): This flag is set if the vehicle processing the message is located within the message's target ROI. Thus, it is (one of) the intended recipient(s) of the message.
- Message direction connectivity (MDC): If there is any vehicle driving in the same direction that is closer to the destination ROI, this flag is set. Intuitively, this flag characterizes a car at the back of a cluster or group of vehicles on a freeway.
- Opposite direction connectivity (ODC): This flag determines whether a car is connected to at least one vehicle driving in the opposite direction. This car could, if necessary, be used as a message ferry to the next group of cars following the vehicles but not within radio communication range.

As can be seen in Figure 5.51, MDC is checked first for each arriving message. If connectivity towards the destination ROI is given, the geocast is continued, i.e., broadcast-suppression techniques are employed as mentioned by the authors of DV-CAST. This represents the case of a well-connected neighborhood.

One exception is the case when the vehicle is the last in the group, i.e., MDC is not set, but the vehicle is in the target ROI of the message. In this case, independently of the ODC flag, the message is re-broadcast to make sure that all neighbors in the ROI also receive the message.

In the case of a sparsely connected neighborhood, i.e., MDC is not set, ODC is checked. If it is set, i.e., when there are vehicles connected to the current one that are driving in the opposite direction, the message is re-broadcast to make sure that it is

Table 5.7 DV-CAST decision matrix

MDC	ODC	DFlg	Scenario	Actions
1	0/1	1	Well connected	Broadcast suppression
1	0/1	0	Well connected	Relay using geocast with broadcast suppression
0	1	1	Sparsely connected	Re-broadcast and assume that a vehicle driving in the opposite direction will help to relay the message
0	1	0	Sparsely connected	Re-broadcast and carry and forward to the first neighbor traveling in the opposite direction or in the message direction encountered
0	0	0/1	Totally disconnected	Wait and forward to the first neighbor traveling in the opposite direction or in the message direction encountered

received by one of those vehicles. The receiver simply stores and carries the message until it finds another vehicle driving in the original direction. This is characterized by the two WAIT functions. "WAIT I" is used if there is no vehicle within direct communication range. In this situation, the vehicle acts as a message ferry until either a vehicle traveling in the same direction (MDC) or one traveling in the opposite direction (ODC) is detected. This also represents a totally disconnected network. "WAIT II" is used to describe the state of the vehicle waiting for another opportunity to deliver the message to one traveling in the opposite direction.

Table 5.7 summarizes the different scenarios and associated actions to be taken by DV-CAST.

Performance

Let's have a brief look at the performance impact of such a mixed broadcast and DTN approach. The performance of DV-CAST has been assessed in several studies considering its reliability, efficiency, and scalability. We want to concentrate on a single metric only, namely the broadcast success rate. It describes the percentage of instances in which the protocol can successfully deliver the message to the nodes in the target region, i.e., it represents a measure of the reliability of the protocol.

In Figure 5.52, the broadcast success rate is plotted over distance. In this experiment, Tonguz *et al.* (2010) aimed to understand DV-CAST's ability to reach the ROI even at greater distances. Figure 5.52 compares DV-CAST with simple beaconing (listed as broadcast in the figure). Also, five different vehicle densities were considered.

As can be seen, simple beaconing is not able to forward the message over larger distances. This becomes even more evident for more sparsely distributed vehicles. In contrast, DV-CAST is able to maintain a certain broadcast success rate even for larger distances. This is a result of the store–carry–forward scheme used by DV-CAST to overcome the problem of sparsely connected or even not connected situations.

In highly connected scenarios (even assuming a small but well-connected cluster of vehicles), this comparison is a bit unfair for both protocols. Broadcast suppression

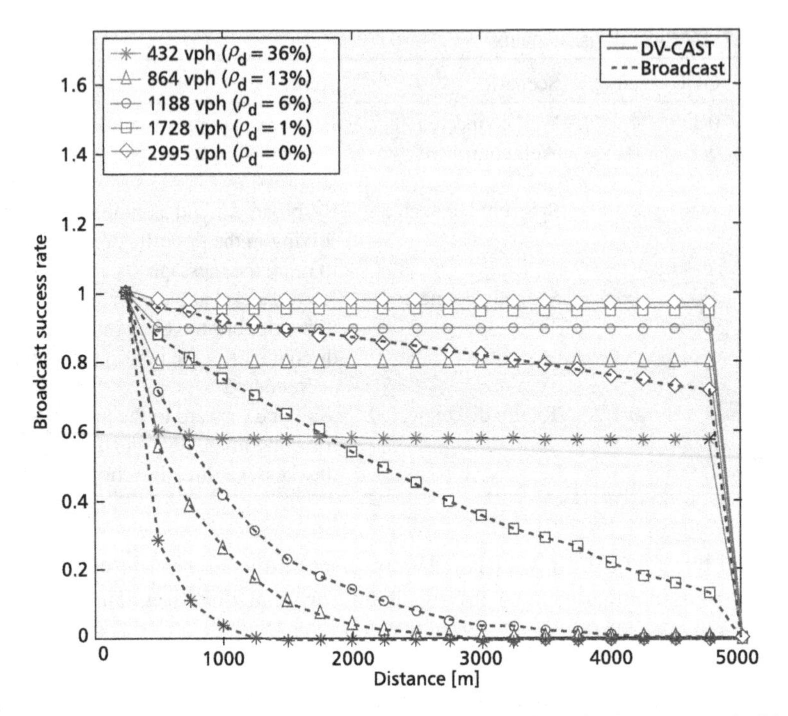

Figure 5.52 The broadcast success rate over distance of DV-CAST compared with simple beaconing for various numbers of vehicles per hour (vph). © [2010] IEEE. Reprinted, with permission, from Tonguz *et al.* (2010).

as performed by DV-CAST is not able to handle such high load scenarios; neither is simple broadcasting. DV-CAST can, however, be combined with the adaptive beaconing solutions discussed in Section 5.4 on page 174 to benefit from congestion-control techniques.

5.7.2 MobTorrent

A first concept to be studied that also heavily borrows from concepts known in the peer-to-peer networking domain is MobTorrent (Chen & Chan 2009). MobTorrent is able to provide download capabilities from intermittently connected roadside WiFi APs. The idea is to inform one or multiple APs about a download so that these APs can fetch the content in advance in order to provide chunks of data whenever the vehicle comes into communication range with these APs. This procedure is partly inspired by BitTorrent, a popular peer-to-peer-based content-downloading system in the Internet (Qiu & Srikant 2004; Izal *et al.* 2004).

The key feature in MobTorrent is the scheduling algorithm that replicates the prefetched data on mobile helpers so that the total amount of data transferred and the average data rate to the mobile clients are maximized. These mobile helpers are used to help vehicles download content on demand. This concept relies on opportunistic forwarding schemes in DTNs.

Because of these two concepts, torrent-like data chunks downloaded from the Internet and DTN-like helper systems complementing each other, we selected MobTorrent as an example for content downloading in vehicular networks. In the following, we explore the architecture and communication principles of MobTorrent.

Architecture

MobTorrent supports a number of components that act together in the content-downloading process. In particular, we have to distinguish the following roles and systems.

- *Mobile clients* are the vehicles that require help to download content from the Internet.
- *Roadside APs* provide Internet access to the mobile clients. They are assumed to be distributed at the roadside, being provided, e.g., by coffee shops, taxi stops, or traffic lights.
- *AP discovery servers* provide location information about available APs.
- *Mobile helpers* are those vehicles that are currently not downloading but willing to offer their bandwidth to help peer vehicles to download content.

MobTorrent also supports prefetching of content by so-called forerunners, i.e., vehicles driving in the same direction but arriving earlier at the AP. These could help download the content by using the ad-hoc network between themselves and the following client to send the received block over a few hops to the mobile client.

Figure 5.53 outlines the download process of MobTorrent.

At time t_1, the vehicle announces its interest to download a certain file. This must be sufficiently large to minimize the overall delay – small files can be immediately downloaded without the help of MobTorrent. For this announcement, the mobile client first acquires a list of APs on its route through the road network. Depending on the file size, and on the available APs and their locations, the mobile client selects a set of adequate APs and contacts them to request prefetching of data blocks or chunks.

At time t_2, the selected APs start downloading the requested blocks from the Internet and store them for later retrieval by the mobile client or a mobile helper.

At time t_3, the mobile client passes by one of the APs. It immediately starts downloading the prefetched blocks (if they are available) using its WiFi connection to the AP. In our example, a mobile helper passes by another AP that provides another block of prefetched data. As the mobile helper is driving towards the mobile client, it downloads this block to store and carry it to the mobile client.

Finally, at time t_4, the two vehicles meet and the mobile client can download the second block from the mobile helper, thus completing the download even before encountering the second AP.

Scheduling and performance

Scheduling of the transmissions, i.e., solving the question of which AP should prefetch which data block, can be realized according to different optimization criteria. First, the data transfer rate to the mobile client can be maximized. This is done by maximizing the

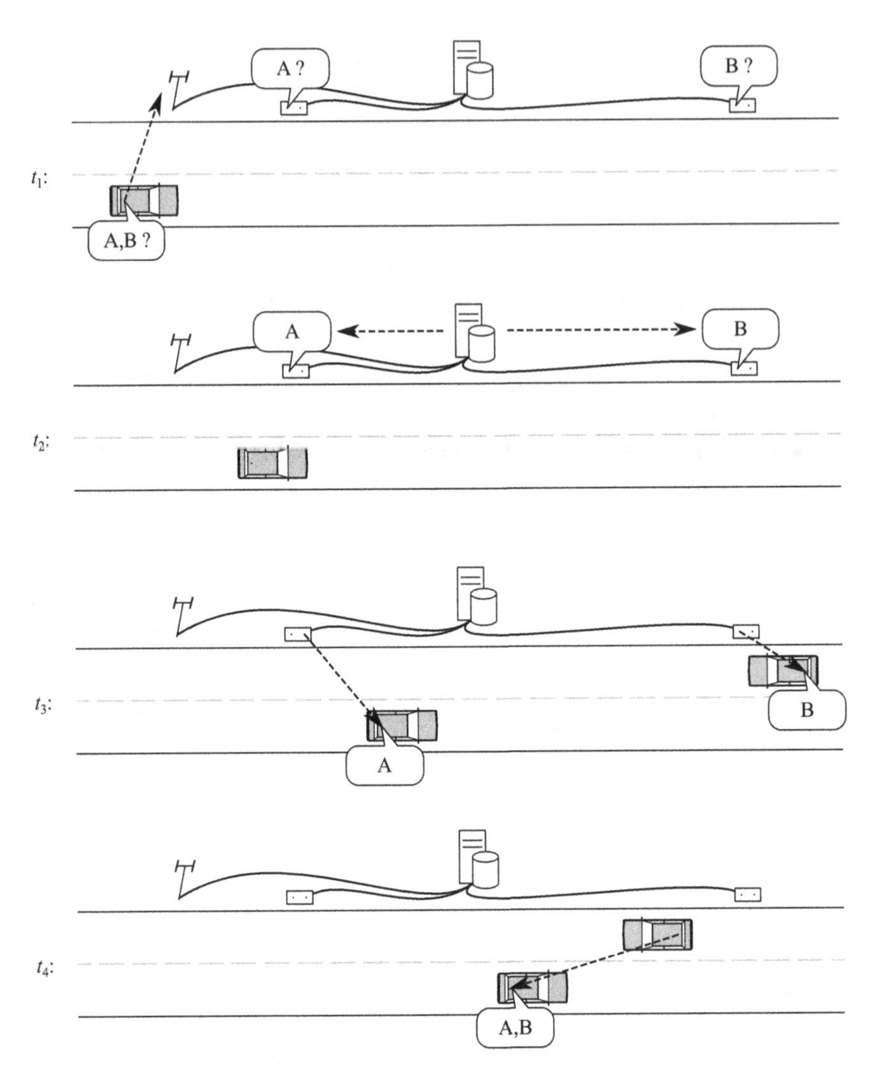

Figure 5.53 Control and data flow for MobTorrent downloads.

data transfer from each AP to the mobile client. In this scheme, relaying from mobile helpers (or forerunners) cannot be planned.

Second, the delay can be minimized. This objective can be achieved by fully exploiting all contacts between the vehicle, APs, and mobile helpers. Please note that this bound is loose, since it is possible that some contacts between the mobile client and a relay cannot be fully utilized because the relay does not carry enough data.

Figure 5.54 shows some selected performance results. In their experiments, Chen & Chan (2009) assessed the download rate of MobTorrent for different numbers of vehicles, both in a testbed and in trace-driven simulation. In particular, the bus network of the National University of Singapore was used. The graph compares four algorithms. The one-hop case serves as a baseline. Here, only APs send data to the

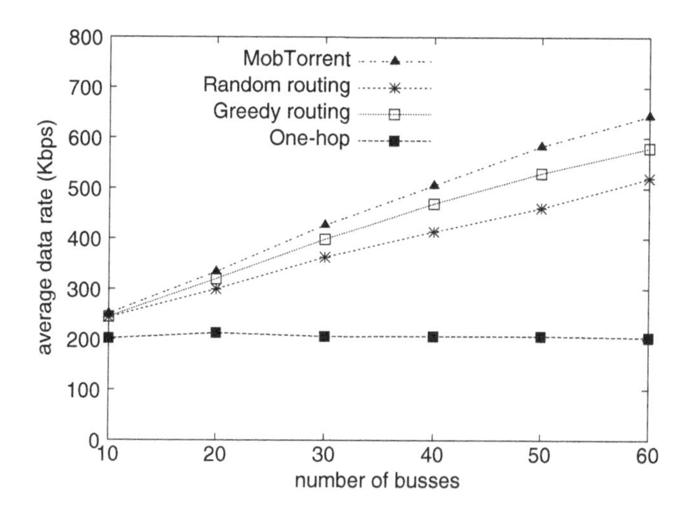

Figure 5.54 The download performance of MobTorrent for different configurations. © [2009] IEEE. Reprinted, with permission, from Chen & Chan (2009).

mobile client. Random and greedy routing perform replication schemes to the different APs either randomly or greedily according to the expected traveling time of the vehicle. MobTorrent finally makes full use of mobile helpers and forerunners to download the requested content.

As can be seen in Figure 5.54, the average data rate is independent of the number of vehicles if APs alone are used for the download (the one-hop case). Greedy and random prefetching of data both lead to significant improvements and also help exploit the available number of helper vehicles. The careful scheduling of data blocks as proposed by MobTorrent improves the situation even further.

5.7.3 PeerTIS

The next example we discuss is the peer-to-peer traffic information system (PeerTIS) (Rybicki *et al.* 2009). The idea is to make full use of a 3G or 4G network connection for information exchange between participating vehicles. Let's return to the requirements of a typical TIS: Traffic information must be current, with a high spatial resolution.

Another observation needs to be considered in order to design the PeerTIS system: A TIS does not rely on random access to the stored information. Instead, there is a strong spatial correlation between the location of a vehicle and the (multiple) queries the vehicle may address to the TIS database. Assuming a centralized TIS, a vehicle planning a route will most likely query information segment by segment along the planned route. Thus, the queries "follow" the road network's topology.

PeerTIS database

PeerTIS relies on peer-to-peer data-management concepts. All items of information are stored in a strictly distributed manner, i.e., each participating vehicle becomes responsible for a subset of all of the traffic information.

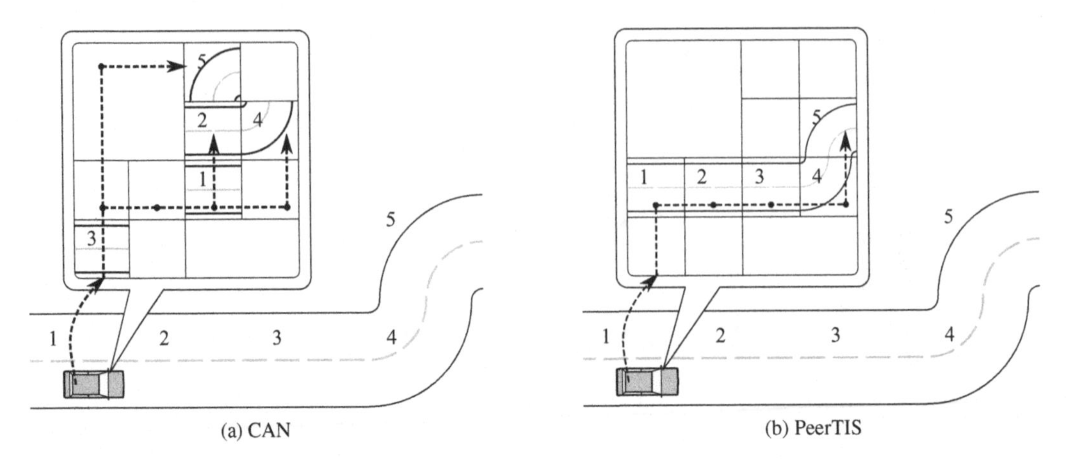

Figure 5.55 Data access with hashing (CAN) and exploiting the geographic information of the digital map (PeerTIS).

Each information block contains as data at least the road segment, the mean speed on this road segment, the current time, and a user ID. Vehicles may create such information blocks or update them with more current sensor information. Overall, this guarantees a high spatial resolution and a high precision. However, we still have to discuss the query time in such a distributed system.

Let's start with the database used by PeerTIS. Inspired by the DHT used by the content-addressable network (CAN) peer-to-peer protocol (Ratnasamy *et al.* 2001), PeerTIS also makes use of such structures. The difference is that PeerTIS explicitly dispenses with hashing. Instead, it exploits the structures given by the underlying digital map of the road network.

Figure 5.55 illustrates these concepts. As shown in Figure 5.55(a), the classical hash-based storage system would lead to the situation that information about consecutive road segments is stored at random clients. This is a result of one of the most important properties of the hash functions used: The system must create a pseudo-random number derived from the input data, i.e., the road segment. Querying data for a planned route will obviously lead to a substantial overhead.

PeerTIS makes use of the underlying map. It stores traffic information at vehicles that manage a certain part of the digital map as shown in Figure 5.55(b). The most innovative aspect of PeerTIS is how the map is organized and how vehicles become responsible for specific tiles.

The joining process as organized by PeerTIS works as follows. A vehicle joining the PeerTIS network sends a request to a vehicle that is already an active PeerTIS client. This can be done either by querying local neighbors, e.g., using WiFi or IEEE 802.11p, or by means of an Internet-based repository. The car receiving the request splits the tile of the map it is responsible for into two subtiles, keeps responsibility for one half and hands over all stored data as well as the responsibility for the second half to the newly joining vehicle. The process is bootstrapped by defining a first vehicle as the root, i.e., making it responsible for the entire map.

Obviously, this process helps to distribute the storage load to all participants. Moreover, load is spatially balanced: Areas of higher vehicle density are, in turn, served by more vehicles. Yet, it remains unclear how an efficient query can be realized. We explore this part in the next subsection.

Queries to the distributed TIS

Actions in the PeerTIS system include the following steps:

1 Join the PeerTIS network
2 Calculate a naïve route (and some alternatives)
3 Query current TIS data for these routes
4 Recalculate the routes and pick the best alternative
5 Periodically check for updates

The most important observation is, as has already been mentioned, that the queries of segments are not random but predictable, i.e., segment by segment following the topological information of the road network. This leads to the situation in which first the information related to the beginning of a road segment is queried, then the information for the next road segment, and so forth. Translated to our database, this means that one tile of the map is queried, then a tile that is just next to the previous one, and so forth. This procedure is depicted in Figure 5.56.

In order to speed up the process, adequate pointers are needed, namely pointers that are similar to the finger tables in DHTs. For our map, this can easily be realized by storing the address information of all the vehicles responsible for any tile that directly connects to the locally stored tile of the map.

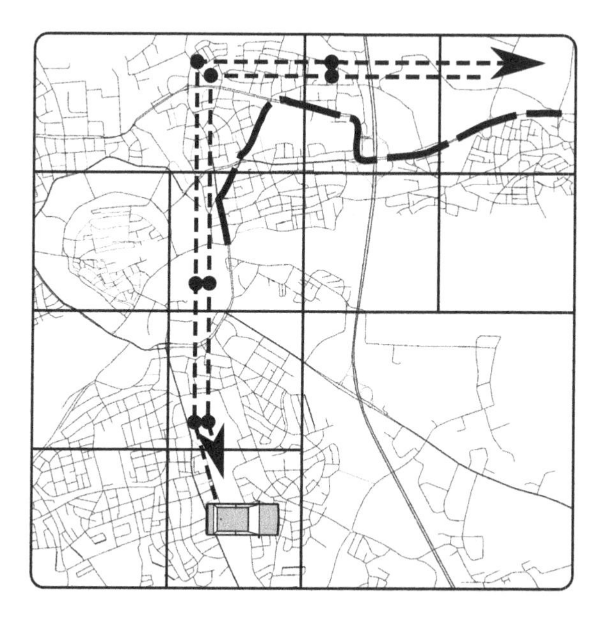

Figure 5.56 The query procedure of PeerTIS.

Let's investigate the query process shown in Figure 5.56 in more detail. Our vehicle is about to take the route in the network that is emphasized in the picture. It therefore wants to query the traffic information for the starting point of the tour first. Since the address of the vehicle responsible for this tile is not yet known, the vehicle sends the query in the appropriate direction, i.e., to a vehicle that is responsible for a tile geographically closest to the starting point of the trip. In our example, the selected tile north of the originating vehicle already directly connects to the tile including the starting point. The query can therefore directly be relayed to the appropriate vehicle.

In general, this procedure would already allow the stepwise query process of all tiles that the planned route touches. The communication can, however, be further optimized by relaying the query along the intended path and collecting all related traffic information before sending the reply back. This is outlined by the arrows in Figure 5.56.

Open challenges of PeerTIS include the need to define adequate caching algorithms to reduce the number of necessary queries, and data replication to prevent loss of information if a vehicle involuntarily leaves the network (e.g., because it is simply shut down or there is no further 3G network connectivity). Furthermore, the underlying assumption is that one relies on a single digital map. In practice, available map data will be heterogeneous, either provided by different organizations or just reflecting different versions of the same map.

6 Performance evaluation

The methodology chosen greatly influences the quality of performance evaluation. Field operational tests (FOTs) are being conducted in many areas and provide some very helpful first results. Yet, large-scale experimentation is conceptually complicated or even infeasible – just consider the possible number of parameter configurations that need to be tested. Thus, simulation is in most cases the method of choice for performance studies. This chapter studies how such simulations should be performed in the field of inter-vehicle communication (IVC) and which tools are available to aid researchers. One of the key concerns is the correct and realistic modeling of the vehicles' mobility. Besides that, the "correct" choice of the scenario has a strong influence on the expressiveness, the validity, and the comparability of simulation experiments. This chapter also studies the impact of radio signal propagation models, the influence of the human driver's behavior, and suitable metrics for finally assessing the performance.

This chapter is organized as follows.

- Performance measurements (Section 6.1) – In this section, we discuss strategies and techniques for performance evaluation of vehicular networking applications and protocols in general. In particular, we outline recent measurement campaigns, including the tools used, and give a high-level overview on simulation techniques.
- Simulation tools (Section 6.2) – This section covers typically used tools for simulating vehicular networks. We introduce the necessary network simulation as well as road mobility simulation and summarize integrated frameworks used for more holistic approaches to simulation of vehicular networks.
- Scenarios, models, and metrics (Section 6.3) – We study the scenarios, models, and metrics needed in this section. It makes a huge difference whether the protocols have been studied in the correct environment and using models accurately representing the realistic behavior of wireless communication channels and vehicular networking protocols. We summarize this discussion with an overview on metrics that can (and should) be used for the final performance assessment.

6.1 Performance measurements

In this first section, we explore potential ways for assessing the performance of vehicular networking protocols and applications. In particular, we start with a brief overview of methodologies that can be used in this field. The main focus of this section, however, is to introduce ongoing FOTs and the scope of simulation techniques representing the current state of the art in performance measurements. We will see that both methodologies have their specific advantages but also significant limitations when it comes to evaluating protocols and applications on a large scale and with a very high degree of realism.

6.1.1 Concepts and strategies

In general, we have three methodologies at hand to assess the performance of a system: analysis, simulation, and experiments (Law 2006). It is not feasible to make a general recommendation regarding which of the three will be the best for a certain application scenario. Yet, it is definitely possible to select a methodology on the basis of empirical evidence or lessons learned in the more than 20 years of research in vehicular networking.

Figure 6.1 outlines the selection strategy. Experimenting with the actual system has many advantages. Most importantly, one can exactly see how the system behaves in a certain context without making unrealistic assumptions and without simplifications that may lead to an incomplete analysis of the system performance. On the other hand, experiments may be very expensive and time consuming, even for small-scale testing of selected system properties. This is mainly due to the fact that the system needs to be embedded in the context of interest, which, in the case of vehicular networks, may easily lead to situations in which several devices need to be investigated in combination and the system performance may be substantially impacted by the environment, e.g., looking at radio signal propagation in line-of-sight (LOS) or non-line-of-sight (NLOS) scenarios. We study ongoing experiments in a little more detail in Section 6.1.2, where we focus on both small-scale testing and large-scale FOTs.

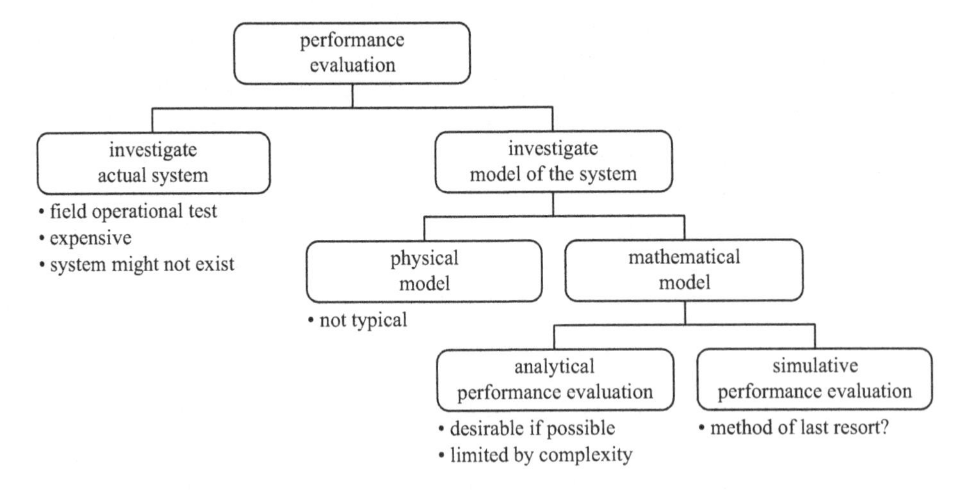

Figure 6.1 Analysis, experiment, simulation – the quest to select the best-fitting approach.

If experiments with the actual system are not possible, the remaining option is to experiment with a model of the system. This can either be a physical model, which is not typical in the context of vehicular networks (actually, this is not typical for any networking applications at all), or a mathematical model of the system. In many situations, it would be desired to derive an analytical solution for the stationary behavior of the system given by the mathematical model. However, models easily get too complex, rendering such a solution not feasible.

The remaining option – and we specifically emphasize this in discussing the selection process for choosing the right methodology – is simulation, seemingly the method of last resort. Simulation has a long tradition in computer networking, so we can rely on a huge basis of background information for doing simulations right. This, unfortunately, does not prevent researchers making the same mistakes over and over again. In Section 6.1.3 on page 243, we explore the feasibility of using simulation techniques in the area of vehicular networks in more detail, including all the potential pitfalls.

Simulation has so far been the key approach when it comes to performance measurements and system evaluation in our field. Therefore, after a brief look at FOTs, we devote the remaining sections in this chapter to simulation as a tool, investigating tools as well as the scenarios, models, and metrics used.

6.1.2 Field operational tests

When it comes to experimental studies of vehicular networking solutions, we have to distinguish between two general types of experimental setups, each as important as the other and both complementing the many simulation studies.

- *Small-scale testing* primarily helps one to acquire a better understanding of certain functionalities of a more complex vehicular networking system. First of all, small-scale testing is often feasible and not too expensive. In many cases, only a few components of the communication system are needed. Depending on the test scenario, these will have to be deployed in a few cars used for the experiment. The primary use is to *validate* simulation models or to *develop* new models based on empirical data.
- *Field operational tests (FOTs)* are considered the queen among all tests. Vehicles equipped with a rather complete set of communication devices and on-board units (OBUs) are used on a larger scale to reproduce a scenario that is at least similar to the typical use of the vehicular networking applications under consideration. In practice, several hundred vehicles are needed even for smaller experiments. Obviously, such FOTs quickly become extremely expensive and also require careful management of all the resources involved.

In the following, we discuss selected experiments used for performance measurements of vehicular networking solutions both on a small scale and on a large scale. We start with a brief overview of the measurement equipment used since no definitive commercially used IEEE 802.11p receiver is available, as yet.

Measurement equipment

The measurement equipment that has been used for FOTs can be categorized into three groups of devices.

- Integrated field test devices – These are commercial products designed especially for FOTs. In most cases, a field-programmable gate array (FPGA) is used to implement the physical layer and the lower MAC-layer protocol standards.
- Standard WiFi chipsets – Since IEEE 802.11p requires only marginal modifications compared with IEEE 802.11a, standard chipsets can be used to tune into the specific dedicated short-range communication (DSRC) channels and to change bandwidth and timings.
- Software-defined radio (SDR) – This option provides the most flexibility in terms of modifications to the lower-layer protocol behavior, but requires the reimplementation of all of the protocols needed.

Integrated field test devices

Prototypes for ETSI ITS and IEEE WAVE radios have been developed by an increasing number of companies. Examples include the DENSO Wireless Safety Unit (WSU), the NEC LinkBird-MX, and the Cohda Wireless MK2 and MK3. Many others are now available, and their number is increasing.

The common characteristic of all these devices is that at least the lower-layer functionality of the ETSI ITS or IEEE WAVE protocol stack has been implemented and tested by the respective manufacturers. Changes and updates are possible in the form of software updates since the lower-layer functionality is usually implemented in an FPGA. Meanwhile, even fully ETSI-compliant protocol stacks supporting cooperative awareness messages (CAMs) and decentralized environmental notification messages (DENMs) as well as GeoNetworking are available.

The resulting setup using FOTs is depicted in Figure 6.2. The shark-fin antenna assembly installed on the roof is an early prototype including FM radio, GPS, 3G, and IEEE 802.11p antennas. The car is further outfitted with an omnidirectional antenna mounted next to it in order to compare results from the different antenna systems. In addition, an additional GPS that is used to log position information with each transmission can be seen.

The advantage of these integrated field test devices is that all the functionality needed is immediately available for experiments. Also, one can expect to rely on a fully standard compliant protocol implementation. The most critical disadvantage is that in many cases no access to internals of the device is available, which would be useful, e.g., to modify certain parameters in the protocol stack.

Standard WiFi chipsets

On the other side of the spectrum, adapted stacks for the popular Atheros chipset AR5414A-B2B (which is, e.g., used in the Unex DCMA-86P2, pictured in Figure 6.3) have been developed in several projects. This allows one to use cheap hardware solutions, which are fully compliant to ETSI ITS or IEEE WAVE standards. Also,

Figure 6.2 The experimental setup for FOTs. A standard shark-fin antenna assembly, an omnidirectional antenna, and a GPS receiver on the roof are shown, as well as the installation of the DENSO WSU IEEE 802.11p radio in the trunk (inset). © [2011] IEEE. Reprinted, with permission, from Sommer *et al.* (2011a).

Figure 6.3 Unex DCMA-86P2 (Mini PCI card) mounted on a PC Engines Alix system board.

this hardware has been tested in great detail. On the other hand, these solutions are bound to the capabilities of the chipset, and have not fully been validated in field tests when it comes to the use of ETSI ITS or IEEE WAVE.

This is actually the most critical aspect. The implementation of IEEE 802.11p (PHY and lower-layer MAC) can be assumed to be working correctly according to the standard, but all higher-layer functionality needs to be implemented by the end user. Thus, no fully featured ETSI ITS or IEEE WAVE implementation can be expected.

Figure 6.4 Ettus Research USRP N210 SDR hardware platform

Software-defined radio

Most recently, the third option has become one of the most popular ones that also seems to be a perfect candidate for future implementations to be used in the car market. Experimental research on wireless communication protocols frequently requires full access to all protocol layers, down to and including the physical layer. Software-defined radio (SDR) hardware platforms, together with real-time signal-processing frameworks, offer a basis to implement transceivers that can allow such experimentation and sophisticated measurements.

Bloessl *et al.* (2013c) presented the first fully functional open-source simulation and experimentation framework for IEEE 802.11p for SDR systems. The system has been implemented for GNU Radio, a real-time signal processing framework, and fitted for use with the widely used USRP N210 from Ettus Research, pictured in Figure 6.4. The core of the software framework is a modular orthogonal frequency-division multiplexing (OFDM) transceiver, which supports all four modulation schemes BPSK, QPSK, QAM-16, and QAM-64 (Bloessl *et al.* 2013a, 2013b). The system is implemented fully in software and runs on a typical laptop computer.

Most importantly, such SDR-based IEEE 802.11p transceivers can be regarded as a first step towards SDR-based systems to be deployed in real cars. We believe that this will be a necessary step since the product cycle of typical cars is very long and installed systems for IVC need to be able to be changed by means of software updates only (Bloessl *et al.* 2013c). With such a solution, even PHY and MAC protocol updates become feasible.

Small-scale testing

Let's start with small-scale testing activities. As mentioned before, these are mainly used for validation of simulation models, or to develop new models on the basis of empirical data. It is of course impossible to discuss all such activities as of today – and also not within the scope of our discussion. We have selected a few of these activities to outline both the scope and the aims of these experiments. We also want to introduce some of the most important aspects of the experiment design.

Please note that such small-scale experiments give very important insights into the behavior of selected parts of the vehicular networking system but not into the application performance as such – large-scale testing is needed for this. The examples we selected span radio signal modeling, in particular shadowing by buildings and other cars, to fast fading caused by multipath transmissions.

Radio signal shadowing

The choice of the path-loss model to be used heavily influences all typical communication metrics including the radio's transmission range and the packet error rates. These, however, are crucial when one wants to determine, e.g., neighbor counts in a realistic way. A major aspect in the path-loss models is the impact of obstacles.

Classical stochastic models might lead to severe deviation from realistic behavior of single communications. Thus, several measurement campaigns have been performed to develop new radio signal shadowing models on the basis of empirical data. By analysis of the results of extensive experiments using vehicles equipped with IEEE 802.11p radios, the shortcomings of traditional, non-ray-tracing models in capturing the effects that predictable shadowing can have on vehicular ad-hoc network (VANET) applications have been outlined, and the results also helped in the development of completely new, but still computationally inexpensive, simulation models.

First, we concentrate on radio signal shadowing caused by buildings. Several measurement campaigns have been conducted to develop new simulation models on the basis of empirical data (Sommer *et al.* 2011a, Mangel *et al.* 2011b).

Sommer *et al.* (2011a) performed an extensive series of experiments in a wide range of scenarios, gathering log data from continuous IEEE 802.11p transmissions between cars. The radio used was part of the DENSO WSU platform, mounted in the trunk of an Audi A4.

Two cars were equipped with these devices for this small-scale experiment. One car was configured to broadcast its current position at intervals of 200 ms by sending WAVE short messages (WSMs) on the control channel (CCH), i.e., at 5.89 GHz. The other car was configured to receive the messages. For each packet, its timestamp and sender position, as well as the receiver position and the dBm value of the received signal strength (RSS), were collected.

For building the model from the experimental data, the measurement results were then correlated with the position and 2.5D shape of buildings (i.e., their outline and height). For this, OpenStreetMap geodata and satellite imagery were used. Figure 6.5 shows a sample of the log data plotted as an overlay on top of the road network and building outlines using a custom application that is based on the OpenLayers API. The transmissions were visualized by drawing the line of sight corresponding to each, using color coding to indicate the attenuation it experienced (Sommer *et al.* 2011a).

In this way, it was possible to verify the accuracy of data and to associate each recorded RSS value with the number of exterior walls intersected by the line of sight

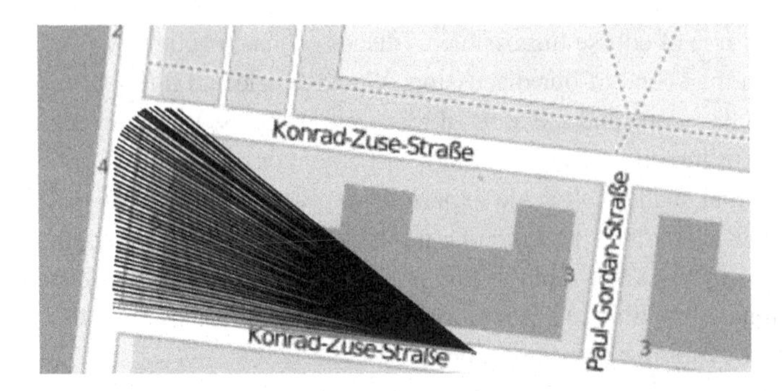

Figure 6.5 Use of geodata about the road network and building geometry for the data correlation and verification step. Each line indicates one successful transmission between sending and receiving car, with a line's brightness representing the measured RSS value. © [2011] IEEE. Reprinted, with permission, from Sommer *et al.* (2011a).

between sender and receiver and the length of this intersection. In a second step, the plausibility of RSS measurements was validated by comparing results from the unobstructed scenario with expected values from an analytical model, which was based on the simple free-space path-loss model.

The second example also focuses on radio channel modeling. The key question addressed in this experiment was as follows: Which physical effects have a major impact on signal propagation and, therefore, have to be considered in simulations?

Segata *et al.* (2013) investigated this question in a freeway scenario. The goal was to understand, from experimental evidence, the effects of shadowing in NLOS scenarios due to different vehicle types obstructing the line of sight for two vehicles driving on a freeway. One application where this study is of particular interest is platooning (Fernandes & Nunes 2012, Segata *et al.* 2012).

This experiment involved multiple vehicles, two for performing the radio communication and a variety of other vehicles driving in between and obstructing the LOS of the radio communication. Overall, experiments were performed assuming either perfect LOS conditions or NLOS conditions with obstacles of different types in between, in particular a car, a van, and a truck. The obstructing vehicle was driven either by one of the authors (car, van) or by volunteers such as helpful truck drivers.

Figure 6.6 shows the measurement scenario as well as the placement of the radio antenna and the GPS antenna on the rooftops. While driving, data frames sent back and forth between the two vehicles at different distances were recorded, while logging information such as signal power and GPS position.

A sketch of the measurement scenario is shown in Figure 6.7. The radio signal as received by the second car is transmitted in a multi-path manner. The direct LOS is obstructed by the truck, so the signal on this ray is expected to be very weak. Other rays bouncing off different objects along the freeway will arrive slightly delayed but likely with a higher signal strength.

All the experiments were repeated for different distances between the two communicating vehicles (80 m, 120 m, 160 m, and 200 m). The minimum distance was 80 m,

Figure 6.6 Measurement on a freeway to determine the impact of radio signal shadowing and fading under LOS and NLOS conditions. © [2013] IEEE. Reprinted, with permission, from Segata *et al.* (2013).

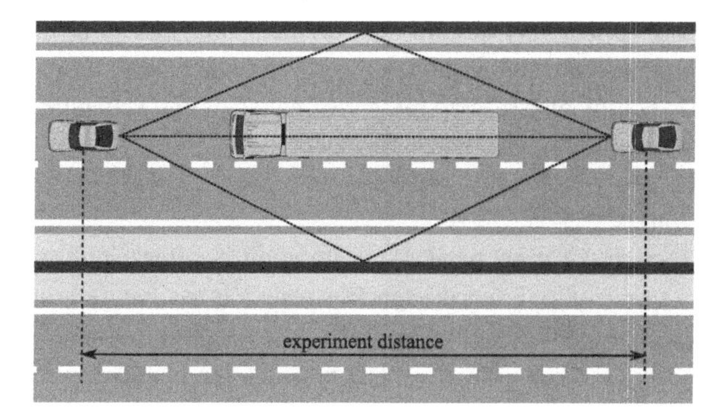

Figure 6.7 A sketch of the scenario with a truck as obstacle. © [2013] IEEE. Reprinted, with permission, from Segata *et al.* (2013).

which is quite close, but still permits a truck to drive in between while maintaining a safe distance. The majority of these tests had to be aborted for two reasons. With large distances, the first problem comes from the fact that on a public freeway it is impossible to prevent other vehicles interfering with the experiment. The second is due to the topology of the road, which is never really straight, so as soon as it bends slightly, the vehicles get in mutual LOS.

The study revealed that one must consider not only the average value of the RSS values because the received power spans over a particular range (Segata *et al.* 2013). Sample results of this study are shown in Figure 6.8. We therefore investigate the individual distributions of received power per experiment and compare them by means of kernel density estimates using a Gaussian smoothing kernel, as illustrated in Figure 6.8. Existing vehicle shadowing models can be extended to reproduce the effects analyzed.

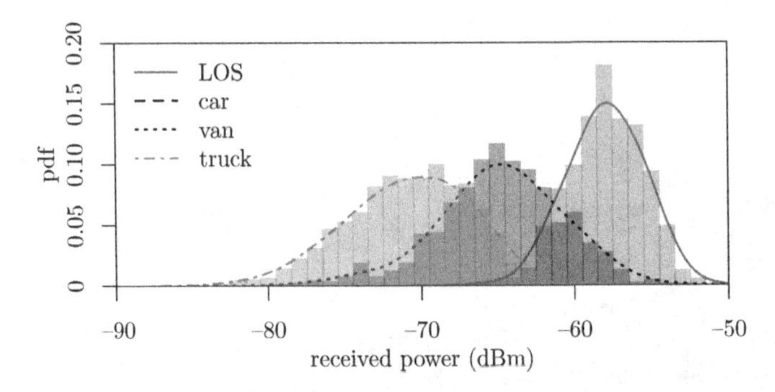

Figure 6.8 A histogram of received power for a subset of the 80-m experiments with overlaid estimated probability density functions. Note that results from the car experiment almost coincide with LOS and are thus not plotted. © [2013] IEEE. Reprinted, with permission, from Segata *et al.* (2013).

We explore the resulting empirical models of both experimental studies in detail in Section 6.3.2 on page 279. As we have seen, this rather simple experiment has already allowed researchers to collect very helpful measurements that can be further analyzed and used for modeling the radio signal shadowing in vehicular environments. Of course, this experiment does not allow assessing the performance of any vehicular networking application in detail as only a single sender was involved.

Platooning experiments

As mentioned in Section 3.1.3 on page 48, platooning is often cited as one of the most visionary ITS application. Even though ITS research has focused on platooning since the early 1980s, it is still an active research area touching on many challenging problems. This is further motivated by the many benefits that platooning could provide once deployed.

The first initiative addressing platooning applications was the Partners for Advanced Transit and Highways (PATH) project (Shladover 2006). PATH is the only ITS research program to make a comprehensive long-term investment in research on platooning, named automated highway system (AHS) in this context. While the national program was being planned, PATH had already received substantial research funding support from the California Department of Transportation (Caltrans).

From a research point of view, platooning is extremely challenging, insofar as it involves several research fields including sensors, control theory, and communications. Any controller designed for supporting platooning will need frequent and up-to-date information about vehicles in the platoon in order to avoid instabilities that might lead to collisions. This is where the networking community comes into play: A platooning system requires an information update frequency of at least 10 Hz (Ploeg *et al.* 2011). Whether such communications requirements can be satisfied by the plain ETSI ITS or IEEE WAVE stack is still unclear, and further work is needed in order to complete the system.

Cooperative adaptive cruise control (CACC) controllers have been investigated ever since the beginning by the pioneering PATH project (Shladover 2006), but they are still under continuous improvement either by academic researchers (Ploeg *et al.* 2011) or by car manufacturers, such as in the European Safe Road Trains for the Environment (SARTRE) project (Bergenhem *et al.* 2010). The idea in SARTRE is that platoons form autonomously and can travel on public motorways mixed with human-driven vehicles.

As can be seen from our discussion, intensive testing is needed in order to validate developed concepts and algorithms. This has been done on a larger scale in PATH and SARTRE.

Let's briefly focus on the SARTRE demonstrator (Jootel 2012). The system as it has been used for testing includes both trucks and cars within the platoon. The lead system was always a bigger truck driven by an experienced driver. The platoon size was limited to six vehicles in total. Typically, one or two trucks led the platoon, followed by smaller vehicles, including cars of different types, with different body styles and different engines. Like in realistic use, the vehicles therefore had a wide range of characteristics, which provides more of a challenge to the developers. A sample test setup is shown in Figure 6.9.

As an example from all the experiments, let's study the fuel reduction achieved. This is one of the key motivations for platooning applications and most certainly the key element from a business point of view.

Figure 6.10 outlines the results as presented in the final project report of SARTRE (Jootel 2012). The fuel reduction is plotted for each vehicle in the platoon and for different distances between the vehicles. For safety reasons, the gap in front of a light vehicle has not been decreased to distances below 8 m. As can be seen, the reduction is substantial for both of the trucks involved (up to 16%). Even the leading truck benefits because of the changed aerodynamics. This is a great incentive for finding such a leader. For the light vehicles, the gain is in the range 5%–15%. Together with the advantage that the cars are fully autonomous while in the platoon, i.e., the driver can concentrate on other things or simply relax, this outcome provides substantial motivation for further investigations and research in platooning.

Figure 6.9 A platooning experiment as conducted in the context of the EU SARTRE project.

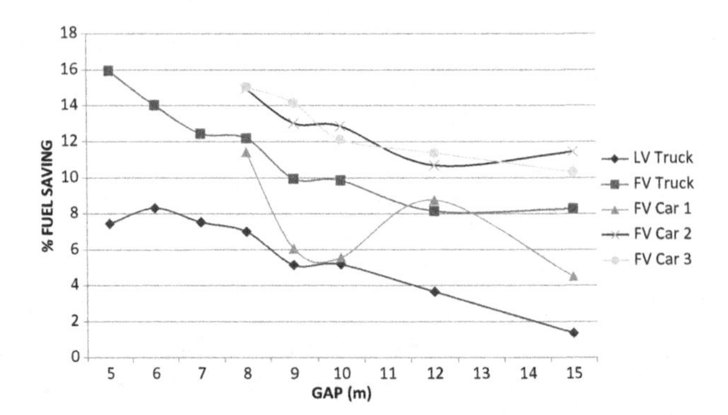

Figure 6.10 Reduction of fuel consumption as measured for all six vehicles in the platoon.

To conclude, we can say that such smaller-scale testing has already provided extremely valuable insights into the behavior of the system. Interoperability with other platoons and also with completely different applications, however, cannot be assessed in such an experiment. Large-scale FOTs are needed in order to learn about these issues.

Large-scale field operational tests (FOTs)

The need for large-scale FOTs has been identified by the research community as well as governments and the automotive industry. In 2009, the German sim^{TD} project started to investigate the applicability of vehicle-to-X (V2X) communication in a field trial. Only a few years later, the US Department of Transportation (US DOT) started a similar project of even larger scale, mainly focusing on an area around Ann Arbor, MI. Both FOTs provided new insights into how supposedly simple communication protocols behave in reality.

sim^{TD}

In the sim^{TD} project (Safe and Intelligent Mobility – Test Field Germany), partners from the automotive domain, telecommunication domain, federal state government, and research institutes worked together in order to deploy and test the results of previous research projects in a large-scale metropolitan field trial in the Frankfurt Rhine–Main region (Stübing *et al.* 2010). The objective was to test V2X technology for the first time in a large-scale experiment. The project aimed at equipping several hundred vehicles with V2X technology, allowing one to gain new insights into the protocols' behavior at that scale.

Eighteen partners, jointly funded by the German Federal Ministry of Economics and Technology (Bundesministerium für Wirtschaft und Technologie, BMWi), the Federal Ministry of Education and Research (Bundesministerium für Bildung und Forschung, BMBF), and the Federal Ministry of Transport, Building and Urban Development (Bundesministerium für Verkehr, Bau und Stadtentwicklung, BMVBS), worked together to implement and validate twenty-one applications by a large fleet of vehicles and road-side stations (Weiß 2011, Eckhoff *et al.* 2011a). Up to 100 vehicles were supposed to be driven by hired drivers, complemented by up to 300 cars of a free-flow fleet (taxis, ambulance cars, police cars, etc.).

Insights obtained in sim^{TD} were expected to contribute towards the harmonization and standardization of vehicular networking, which they did. According to Stübing *et al.* (2010), sim^{TD} addressed the following objectives:

- increased road safety and improved efficiency of existing traffic systems using V2X communication;
- definition and validation of a roll-out scenario for identified applications, especially with respect to scientific questions;
- consolidation of car-to-X applications for traffic efficiency, vehicle safety, and value-added services;
- specification of test and validation metrics and methods in each phase of the overall system development in order to enable measurements and evaluations of the results;
- consolidation and harmonization of requirements with respect to feasibility and performance.

Figure 6.11 outlines the system architecture used. In our context, all the components on the left-hand side are of interest. The right-hand side shows the centralized ITS stations providing and aggregating information for all participants.

Most important is that all cars are equipped with multiple radio technologies from IEEE 802.11p and WiFi to cellular technologies that are based on 3G and 4G networks. The cars are allowed to exchange information directly as well as via pre-deployed roadside units (RSUs). In addition, all cars are equipped with human–machine interface (HMI) solutions.

Among other things, the focus of sim^{TD} was on the application class of traffic information systems (TISs). In particular, a self-organizing traffic information system (SOTIS)-like protocol operating independently of infrastructure and of pre-installed digital maps was investigated (Wischhof *et al.* 2003). Since a high market penetration rate is crucial for the performance of a decentralized TIS, it is desirable to create sensibly priced systems. By enabling the vehicles to dynamically create map information themselves, it is foreseen that one could achieve a fully self-organizing system working even at low penetration rates.

US DOT

The US Department of Transportation (US DOT) initiated a large-scale FOT in 2010 as a follow-up to a previous vehicle-to-vehicle (V2V) project. The main goal was to study the interoperability of IVC technology as well as the behavior of standardized protocols on a larger scale.

Interoperability was defined in this context as the ability for V2V safety systems to successfully function across any, and all, equipped vehicles regardless of make, model, and model year. This is of course a very ambitious objective, given that only a few

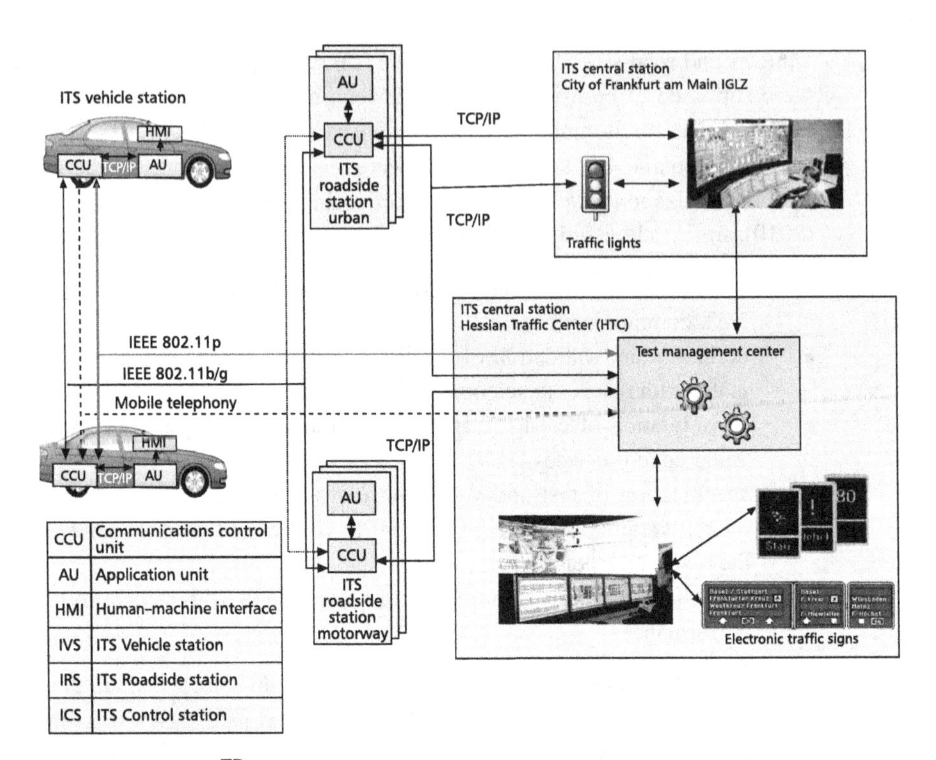

Figure 6.11 The sim^{TD} system architecture. © [2010] IEEE. Reprinted, with permission, from Stübing *et al.* (2010).

WAVE radio systems are on the market and all of them need to be considered prototypes (see page 232).

The US DOT V2V interoperability project was planned in two phases.

- Phase I: 30-month project between January 2010 and June 2012
- Phase II: 21-month extension through March 2014

It was a collaborative effort between eight automotive original equipment manufacturers (OEMs) and the US DOT. Essentially, all major automakers were part of the project, which was coordinated by the National Highway Traffic Safety Administration (NHTSA).

The FOT was executed mainly for light vehicles in the area of Ann Arbor, MI. The test objectives focused on technical issues regarding the use of 5.9-GHz ITS related to interoperability, scalability, security, and reliability. The cars were equipped with four different OBUs from various suppliers. In addition, major concerns were also to study the drivers, behavior, and driver acceptance.

In order to provide necessary inputs for standardization bodies, tests included different ways to disseminate information using IEEE WAVE. The most important message type considered was the basic safety message (BSM). In the scope of the FOT, the following protocol options were experimented with: fixed-rate beaconing, adaptive transmit power, and adaptive transmit rate. This is in line with the ongoing European standardization by the European Telecommunications Standards Institute (ETSI).

On the basis of the findings in the FOT, the US federal government announced in February 2014 that the NHTSA plans to begin working on a regulatory proposal that would require the installation of V2V devices in new vehicles in a future year.[1] This NHTSA announcement coincides with the final standardization of higher-layer networking protocols in Europe by the ETSI. We are thus now entering an era that might change the game in road traffic management.

6.1.3 Simulation techniques

The main focus of the remainder of this chapter will be on simulation tools, models, and metrics. In this section, we explore simulation techniques from a higher level, outlining the state of the art in simulation of vehicular networking applications and protocols.

In this context, we first have to address the question of what distinguishes vehicular networks from any other type of network. In other words, why are existing simulation techniques, which have been developed and validated over decades of networking research, no longer fully suitable or might at least lead to results and observations that might strongly differ from a realistic system behavior?

In particular, we need to discuss a few primary aspects that distinguish simulation of vehicular networks from that of other wireless networks.

- Simulation of vehicles' motion, which is unique for road traffic compared with, let's say, human mobility models, has to be reproduced in a realistic way. Simple random-waypoint models as are frequently used for simulating mobile ad-hoc networks (MANETs) are no longer useful. Microscopic simulation of the movement of vehicles needs to consider car following, lane change, intersection models, and other factors.
- The models and even the metrics are different. First, models that reflect human behavior need to be considered, the (non-) reactions to traffic bulletins being a prominent example. Second, radio communication between vehicles is often constrained by buildings and other vehicles. Furthermore, classical networking and communications metrics might not tell the full story about a certain VANET application. Metrics such as the travel time and the CO_2 emission need to be considered as well.
- The comparability and reproducibility of simulation studies is a topic that is heavily discussed in the community. In general, all these questions need to be addressed for any simulation-based study of networking solutions. Lessons should have been learned from the many years of experience, yet we see that there are still many unresolved issues and potential pitfalls, so we decided to address these topics in this textbook as well.

[1] See http://www.nhtsa.gov/About+NHTSA/Press+Releases/USDOT+to+Move+Forward+with+Vehicle-to-Vehicle+Communication+Technology+for+Light+Vehicles.

Modeling vehicle mobility

Mobility modeling in the vehicular networking domain has been one of the most controversial problems discussed. Owing to the lack of a way to accurately describe the mobility of vehicles on our road network, early approaches used simplifications that severely impacted the measurement results obtained.

At the same time, transportation and traffic science was already able to provide better mobility models, classified as macroscopic, mesoscopic, and microscopic models, according to the granularity with which traffic flows are examined. It took a long time for the vehicular networking community to implement these models in their simulation approaches.

In the scope of this textbook, we mainly concentrate on microscopic mobility models that accurately model the behavior of each car individually. Macroscopic models, however, are of almost equal interest when it comes to large-scale simulation that is focusing not on individual vehicles but on entire flows of road traffic. We study this in more detail in Section 6.2.2 on page 259.

In the following, we concentrate on the historical evolution of mobility models used in simulations of VANET protocols and applications (Sommer & Dressler 2008). Figure 6.12 illustrates the change in the models used.

Early approaches relied on relatively simple models using random node movement. These models have frequently been used in simulation of MANETs, and substantial expertise in their integration into network simulation tools has been acquired. However, because such mobility models do not realistically reflect car movements on our roads, the search for more realistic but also more complex solutions started. As an intermediate step, traces were recorded, e.g., from taxis or other public transportation, and then used as an input for network simulation. By incorporating expertise from the road traffic engineering community and adding road traffic microsimulation, i.e., the mobility simulation of individual cars taking into consideration the road network, other vehicles, and traffic rules, these mobility traces can also be generated on demand. Recent advances

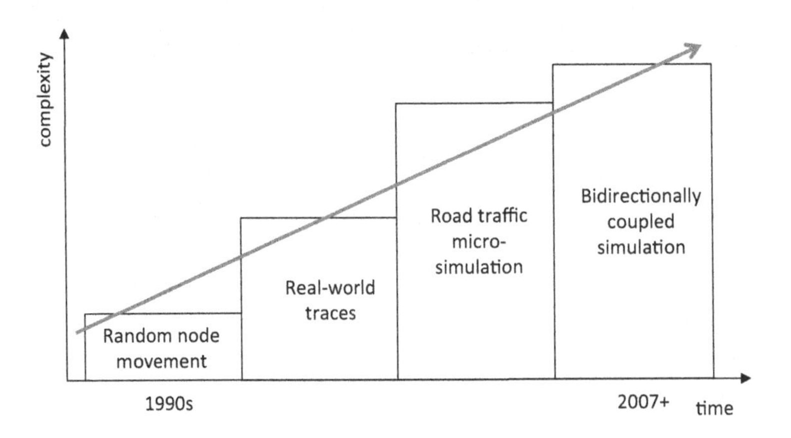

Figure 6.12 The historical evolution of mobility modeling strategies and techniques in VANET research.

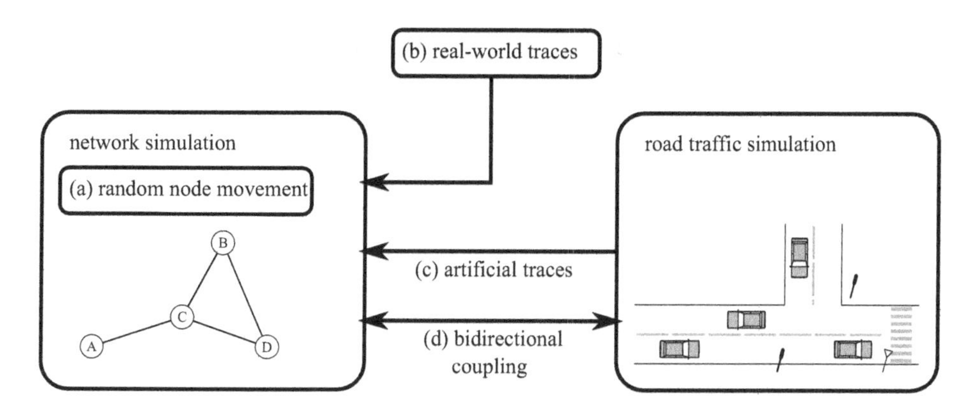

Figure 6.13 Mobility modeling techniques for simulation of VANET protocols and applications.

are based on tightly coupled road traffic microsimulation and network simulation. It is obvious that the inherent complexity of such mobility models is strongly increasing.

Figure 6.13 shows an architectural view of mobility modeling in VANETs. All four of the aforementioned simulation approaches are included in the figure, labeled (a) to (d). In the following, we briefly introduce these concepts and outline their key advantages and weaknesses. We closely follow the discussion by Sommer & Dressler (2008).

Random node movement

All ad-hoc networking research started by bringing node mobility into play for the first time. Assuming unconstrained node movement in a completely random manner, the *random-waypoint* mobility model has frequently served as the mobility model of choice (Johnson & Maltz 1996). Following an early ETSI recommendation, this was later replaced by the *Manhattan-grid* model (ETSI 1997). Here, mobile nodes should move along a grid of possible paths – assuming a downtown Manhattan scenario.

It has been shown that such random node movements provide vastly different results from more sophisticated vehicular mobility models, sometimes not even reaching a steady state (Yoon *et al.* 2003). Still, derivatives of them have been in use ever since. The root cause is that it is much easier to use these models rather than considering, for example, inertia or constraining vehicles to pre-defined roads.

Real-world mobility traces

Compared with the use of random mobility models, the modeling of node mobility on the basis of sets of pre-recorded real-world mobility traces was a major step towards realistic vehicle simulation. Traces have been recorded in many projects around the globe. The most complete traces result from public transportation such as city buses (Jetcheva *et al.* 2003). One of the most popular examples is the Shanghai taxi trace (Huang *et al.* 2007). In general, the idea is to continuously log GPS information while the vehicles are being driven. Many of these mobility traces were then post-processed and made available to the research community.

The use of these traces in event-based simulation is straightforward. Applying the concept of trace-driven simulation, each position update, i.e., each GPS record, is reflected in the simulation as a single event at exactly the time it was recorded. Of course, the absolute time can be shifted as long as the relative time between two records is maintained. With each event, the vehicle is instantly moved to the new position. Such trace-based mobility models result, of course, in most realistic vehicle movement in network simulation. However, their use is limited since only a few traces are available and even those do not reflect all of the vehicles on the road. Furthermore, changing just a single parameter such as the vehicle density and keeping all other parameters unchanged is infeasible in reasonably large scenarios.

Microscopic mobility simulation

Instead of using recorded traces with all their drawbacks, the use of vehicle mobility simulators promises to lead to the same results but without inheriting their limitations. Here, the realism of node movement is constrained only by the complexity of the mobility simulator used. Approaches range in complexity from simple, collision-free node movements (Saha & Johnson 2004), to the use of a fully featured mobility simulator based on multi-agent systems (Naumov *et al.* 2006), and the use of common mobility models from the field of transportation and traffic science (Krajzewicz *et al.* 2012).

Artificial mobility models have the advantage of providing simulations with very realistic mobility traces while at the same time allowing the mobility parameters to be freely adjusted in order to examine their influence on a simulation's outcome. Still, many simulation experiments remain impossible in the VANET framework. Not only does the node mobility influence the network connectivity, and hence network traffic, but also, in many real-world VANET scenarios, the network traffic influences the node mobility. The best example is a TIS: If the application works properly, vehicles will be informed about traffic congestion and may have a chance to plan a detour. This, however, instantly invalidates the pre-recorded or pre-calculated trace.

Bidirectionally coupled simulation

In all these cases, the loop between road traffic simulation and network traffic simulation needs to be closed. This, however, requires intensive cooperation among the different simulation tools. The term *bidirectionally coupled simulation* has been coined to highlight this interaction. Starting with early simulators (Sommer *et al.* 2007, Piorkowski *et al.* 2007), the community eventually decided to integrate all these efforts into a single interface, the traffic control interface (TraCI) (Wegener *et al.* 2008). This interface is now supported by all major IVC simulation frameworks (Wegener *et al.* 2008, Sommer *et al.* 2011b, Rondinone *et al.* 2013).

It has been shown that this integrated or bidirectionally coupled simulation not only provides more detailed insights into effects on (and of) network traffic, but at the same time has only a negligible impact on the runtime of simulations.

As illustrated in Figure 6.13, two inter-dependent processes are running concurrently, namely the network simulator and the road traffic simulator. Several parameters are stored in both worlds. Both the network and the road traffic simulator need to share data like the position and speed of the simulated vehicles, and potentially even about the road network topology. Other data like properties of the radio signals and the state of the vehicle controller are local to the network simulator and the road traffic simulator, respectively.

The integration of the two simulators is based on the aforementioned TraCI interface. The simulation time is managed by the network simulation side, and movement updates about the simulated vehicles as well as control commands steering the road planning and even fine-grained actions of the cars within the road traffic simulator are exchanged at regular intervals.

Thus, bidirectionally coupled simulation of VANETs generally consists of two alternating phases (Sommer & Dressler 2008).

- While the network simulation is running, it sends parameter changes to the road traffic simulation, altering driver behavior or road attributes, and influencing vehicles' routing decisions.
- At regular intervals controlled by the network simulator, the road traffic simulation performs traffic computations that are based on these new parameters and sends vehicle movement updates to the network simulation.

Integrated IVC frameworks have been developed to enable widespread use of these techniques. Examples include TraNS (Piorkowski *et al.* 2007), Vehicles In Network Simulation (Veins) (Sommer *et al.* 2007, 2011b), and iTetris (Rondinone *et al.* 2013). Researchers from the network simulation community can now directly build on the work of researchers from the transportation and traffic science community – and vice versa. We will explore selected frameworks in more detail in Section 6.2.3 on page 262.

IVC-specific models and metrics

We will get back to the most important models and metrics that need to be considered for simulating IVC protocols and applications in Section 6.3.2 on page 274 and Section 6.3.4 on page 290, respectively, including a more detailed discussion of the examples shown in this section. Here, we want to discuss examples for models and metrics that are not available in classical network simulators. These examples include the impact of human drivers' behavior, radio signal shadowing and fading, and metrics that are not related to networking such as the CO_2 emission and the fuel consumption of the vehicles involved. Please note that the examples selected serve only to identify the need for such models and metrics. Depending on the application, there might be even more relevant ones.

Human drivers' behavior

The need to include the human drivers' behavior in the simulation of vehicular networking applications has been identified as a very important aspect (Dressler & Sommer 2010, Dressler *et al.* 2011b). This is reflected by the interaction between the vehicle and the driver – at least as long as we are not talking about autonomous cars. Decision making might be strongly influenced by experiences and need not necessarily reflect what is proposed by the system as the technically optimal solution. In fact, human drivers'

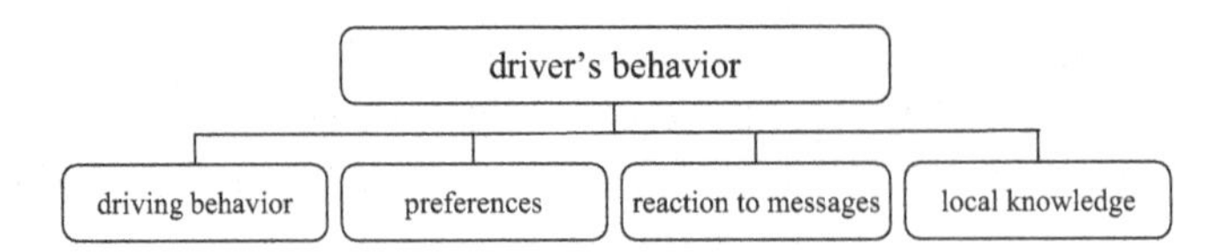

Figure 6.14 Driver behavior submodels.

behavior impacts a system not only on a microscopic level (as studied by adequate car-following models), but also on a macroscopic level (impacting route planning and route changes): Depending on the driver's knowledge and several additional aspects, either the recommendations of the IVC based information system are considered, or no action is taken.

The impact of individual human drivers' behavior on overall road traffic has been a topic of interest ever since the early days of traffic information systems. For example, König *et al.* (1994) developed driver behavior models using AI techniques for the driver's route planning. The authors considered four submodels that influence the driver's behavior as shown in Figure 6.14.

In particular, the following aspects need to be considered.

- The driving behavior, which is determined by factors such as aggressiveness, age, and gender, influences the microscopic mobility of the vehicle. This is usually modeled within the mobility models as discussed before.
- The preferences of drivers influence both the route selected (a factor that is integrated into navigation systems today) and the motivation to accommodate changes to this route.
- Most important in our case is the specific reaction to received messages, which are communicated to the driver via (heads-up) displays or other audiovisual interfaces. Depending on the situation and the experience of the driver, the reaction can range from completely ignoring the information to acting exactly as suggested by the technical system.
- Local knowledge is difficult to model and also somewhat related to the reaction to received messages. Essentially, the behavior will be different depending on whether the driver is at a well-known place or at some remote and completely unknown site.

Here, we primarily consider the reaction to received messages. The most comprehensive literature study of human factors has been the one conducted by Dingus *et al.* (1996). They provide guidelines for advanced traveler information systems and commercial vehicle operation. A very interesting aspect identified in this study is that human drivers tend to resist deviating from their present route to avoid congestion, i.e., they prefer to follow their "traditional" routes. This report also summarizes four driver classes.

Dingus *et al.* (1996) further studied a mix of all these classes to represent realistic driver behavior. They came up with a statistical distribution, which helps to implement these classes despite considering a statistical sample only, which represents typical

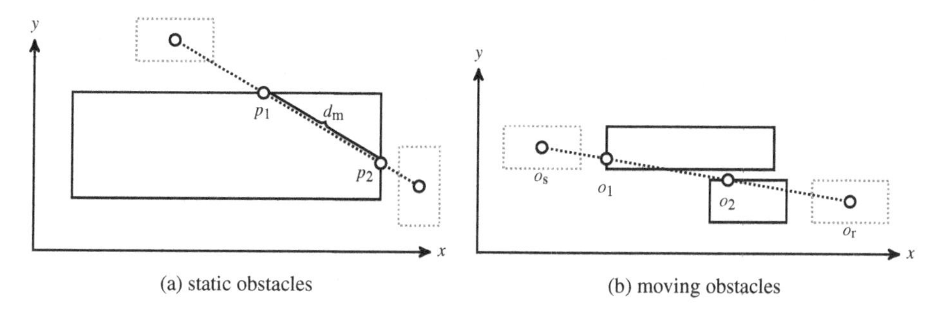

Figure 6.15 Calculation of line-of-sight intersection points with buildings and vehicles.

drivers in a very specific region or country. We study the resulting models in more detail in Section 6.3.3 on page 285.

Radio signal shadowing and fading

In any wireless network, thus, also in vehicular networks, the radio channel model must encompass path loss, shadowing, and fading (Mecklenbräuker *et al.* 2011, Segata *et al.* 2013). In network simulation, fading due to large-scale path loss, deterministic small-scale fading, and probabilistic loss effects is most commonly calculated as a sum of independent loss processes (Ahmed *et al.* 2010, Nagel & Eichler, 2008). The receive power can be expressed on the basis of these terms, the transmit power of the radio, and the transmit and receive antenna gains. We provide a detailed discussion of the exact models and their impact in the field of IVC performance evaluation in Section 6.3.2 on page 274. In the following, we introduce the most important IVC-specific channel characteristics.

Path-loss models compute the average attenuation of a signal in relation to the propagation distance. The simplest – and still the one which is most widely used in vehicular network simulations – is the *free-space model* (Sommer *et al.* 2012). Free-space propagation takes into account only the distance d and the wavelength λ. Empirical adaptations thereof aim to account for non-ideal channel conditions by introducing an additional environment-dependent path-loss exponent α. However, more realistic treatment of the path loss takes into account the fact that radio propagation will commonly suffer from at least one notable source of attenuation: constructive and destructive interference of a radio transmission with its own ground reflection.

A physically more correct approximation (Rappaport 2009) of path loss must therefore be based on the phase difference φ of two interfering rays, leading to a *two-ray interference* model. From the length difference of the direct line-of-sight propagation path and the length of the indirect, non-line-of-sight path via reflections on the ground, the phase difference of interfering rays can be derived (Sommer *et al.* 2012).

Shadowing models are used to reproduce the attenuation of a radio signal induced by obstacles, such as buildings or other vehicles blocking the direct line of sight (see Figure 6.15). This can be modeled either stochastically, in particular with a log-normal distribution (Mecklenbräuker *et al.* 2011), or geometrically (Giordano *et al.* 2010, Sommer *et al.* 2011a, Mangel *et al.* 2011a), taking into account the objects that the direct ray has to traverse. The latter approach requires at least a rough geometric description of the scenario, but modern simulators include means of specifying geometries, or are capable of importing real-world maps (Sommer *et al.* 2011b).

For studying the shadowing effects caused by buildings, in particular models that abstract from all reflection and diffraction effects are good candidates. As shown in Figure 6.15(a), by relying on building outlines only, shadowing effects can be calculated in a reproducible way, focusing on medium-scale reception dropouts only. Such models have been developed, for example, by Sommer *et al.* (2011a) and by Mangel *et al.* (2011b). Commonly available geodata, e.g., as provided by the OpenStreetMap project, can be used to appropriately model buildings and the respective radio signal shadowing in vehicular network simulation. Using empirical data, these models can be employed to quantitatively capture medium-scale path-loss effects in suburban environments.

Of course, depending on the effects that should be captured, a different level of granularity – from a highly complex, fully deterministic model to a simple, fully stochastic one – is applicable. There exists a whole body of literature on how to properly model radio obstructions in wireless networks.

For the calculation of the impact of moving vehicles on power loss, a similar technique can be employed (Boban *et al.* 2014, Sommer *et al.* 2013). For every signal transmission between a pair of vehicles o_s and o_r, the set of vehicles that intersect their direct line of sight needs to be identified.

As illustrated in Figure 6.15(b), all that is needed in the first step is to store all points where this line of sight enters the bounding box of a vehicle as $\{o_1, o_2, \ldots\}$. Similarly to the problem of finding obstructing buildings, this problem can be cast as a *red–blue intersection* problem, drastically reducing its computational complexity below that of a brute-force search.

Finally, *fading* models capture the fluctuations of the received signal power caused by multi-path effects. Like for shadowing, this effect can be reproduced geometrically by means of ray tracing (Schmitz & Wenig 2006, Maurer *et al.* 2005). This effort is, as can be expected, extremely expensive and time consuming. Thus, stochastic models have been derived to reproduce these effects. Examples are Rician, Rayleigh, and Nakagami fading (Mecklenbräuker *et al.* 2011). The choice of one of these models (and related parameters) depends on the simulated environment (freeway, highway, urban, rural, etc.).

A stochastic model that takes into account both shadowing and fading effects is the Suzuki distribution (Suzuki 1977). It combines log-normal and Rayleigh distributions. However, being completely stochastic, it is not able to exploit exact geometric representations of the scenarios, which might be crucial for simulating road-traffic safety applications.

Travel time and CO_2 emissions

Besides the many different models that are relevant for the simulation-based performance evaluation of vehicular networks, we also have to question the typical metrics used for assessing and comparing the performance of classical wireless networks. Such more classical metrics tell us about the capabilities of the communication system to send messages within certain time bounds or to deliver a certain fraction of messages in a reliable way. Yet, they are of limited use when it comes to assessing a vehicular networking application as a whole.

Metrics that might be relevant, and that are completely unrelated to the wireless communication system as such, include the following (this list is of course far from being complete and simply highlights selected metrics).

- Collision probability – Safety metrics are very complicated to deal with. Even though safety metrics such as "the number of crashes prevented" would be ideal to demonstrate the effectiveness of a safety application, these numbers are extremely difficult to get (Dressler *et al.* 2011a). We study the collision probability for intersection collision warning systems (ICWSs) later, in Section 6.3.4 on page 294.
- Travel time, average speed, and effective average speed – The travel time of the cars is frequently used as a more descriptive metric for efficiency applications. The travel time reveals the ability of the TIS to efficiently re-route cars in the case of congestion. The same information is also revealed when using the effective average speed of a vehicle (even the variance is of interest in order to see whether just a few vehicles experience longer travel times than the majority).
- Emissions and fuel consumption – When it comes to exploring the effectiveness of a vehicular networking application, the travel time or average speed shows only one side of the picture (Sommer *et al.* 2010c). These metrics become optimized even if longer routes have to be taken in order to prevent being stuck in traffic jams. At the same time, this detour might lead to substantially higher fuel consumption and emissions.

We explore these metrics in detail in Section 6.3.4 on page 298. In the following, we briefly study the possibility of measuring emission values (we use CO_2 just as an example) even in simulations.

Let's discuss the difference between travel time and emissions as metrics fundamental to the evaluation of vehicular networking applications. It should be noted that the travel time provides only measures of the microscopic behavior of individual cars and, thus, of the extent to which the individual driver benefits from the system. A completely different view would be to analyze the overall behavior, i.e., the ability of the system to smoothe entire traffic flows. This can be provided either by looking at the variance of vehicle speeds or, as a combined metric with the distance traveled and revealing further interesting aspects, by looking at the resulting emissions (frequent accelerations result in a sharp increase of CO_2 emissions) (Sommer *et al.* 2010c).

Very accurate modeling of the gas consumption and emissions is provided by the EMIT model, which has been calibrated for a wide range of different emissions including CO_2, CO, hydrocarbons (HC), and nitrogen oxides (NO_x) (Cappiello *et al.* 2002). The basic operation is depicted in Figure 6.16. Speed, acceleration, and the characteristics of the particular vehicle are used to calculate the gas consumption using an engine model. From these results, emissions after passing through a catalytic converter, which is assumed to have reached operating temperature, can be estimated very precisely.

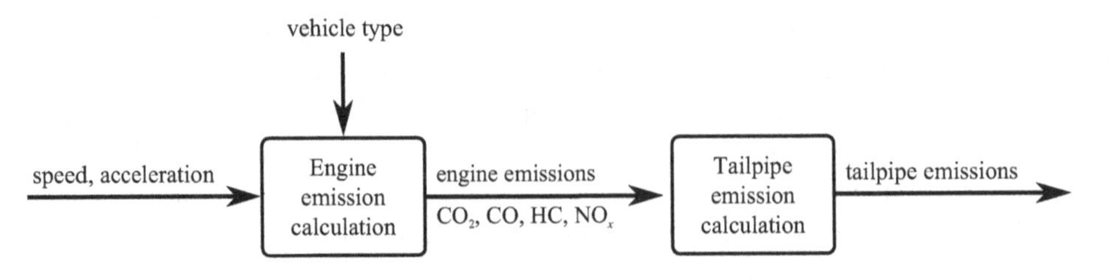

Figure 6.16 Two-stage emission calculation in the EMIT model.

This model has been integrated into the Veins simulation framework (Sommer *et al.* 2010c). In a proof-of-concept study, it was possible to show that the travel time and CO_2 emission do indeed reflect two different optimization criteria that cannot easily be treated in combination. We show more details about the model as well as selected results from the mentioned study in Section 6.3.4 on page 290.

Comparability and reproducibility

Before we can start digging into the details of simulation tools and models, we definitely need to discuss whether and how simulation experiments can be compared and reproduced by other groups. As a matter of fact, the credibility of network simulations, and therefore of vehicular network simulation as well, has been a constant topic of discussion.

Conducting sound simulations requires at least some familiarity with statistics. Knowing how to set up experiments and how to interpret result data is not as straightforward as it might seem at first glance, and complete textbooks, such as Law (2006), have been written on the topic.

Before we can even tackle the evaluation of simulation results, however, the much broader issue of safe software design, testing, and openness of code needs to be considered. Merali (2010) outlines five basic principles for scientific programming: documenting and saving development progress in a version-control system, tracking input data through all pre-processing steps, writing software that can be tested (and conducting these tests), and sharing code with peers.

Returning to the matter of conducting sound simulations, the work of Pawlikowski *et al.* (2002) investigated a wide body of literature in the field of network simulation, particularly evaluating the use of appropriate pseudo-random-number generators and the proper analysis of simulation output data. They were able to show that most of the simulations presented were not able to satisfy these two requirements. These findings, even though they were very disappointing for the simulation and modeling community as well as for the networking community, very positively influenced the way simulations are carried out.

The question of the comparability and reproducibility of simulation experiments must be considered unresolved. This would be essential since any scientific activity should be based on controlled and independently repeatable experiments (Pawlikowski *et al.* 2002). Unfortunately, we observe that, especially for vehicular network simulation studies, it turns out that generating reproducible and validated simulation results is extremely difficult (Joerer *et al.* 2012a,c). In the following, we report on a study carried out to identify the degree of reproducibility of simulation studies published between 2009 and 2011.

Literature survey

Joerer *et al.* (2012a) surveyed all of the relevant papers published between 2009 and 2011 which were presented at the leading IVC conferences. This amounted to a body of literature exceeding 1000 papers arising from 116 simulation studies focusing on IVC using short-range communication. The following conferences were selected.

- *ACM VANET* (Workshop on VehiculAr Inter-NETworking), which has been held annually in conjunction with ACM MobiCom since 2004. The workshop initially focused on VANET topics, but soon widened its scope to vehicular networking in general, recently including also topics related to long-range cellular systems.
- *IEEE VNC* (Vehicular Networking Conference), which is the youngest of the major vehicular networking centric events, and has taken place annually since 2009. This IEEE Communications Society conference focuses on vehicular networking in general and has a strong focus on IVC in particular.
- *IEEE VTC* (Vehicular Technology Conference), which is held semiannually (in spring and fall – referring to the seasons of the northern hemisphere) as a flagship conference of the IEEE Vehicular Technology Society, and has a long history that dates back to 1950. Considering only the last decade of vehicular networking research, the conference has focused on research topics regarding the physical layer and medium access.

Let's first look at the MAC protocols used. As shown by Eckhoff *et al.* (2012), it is important to use a fully featured IEEE 802.11p MAC model, especially at high node densities or when a high load is imposed on the wireless channel. As shown in Figure 6.17, the simulation studies reviewed used a wide variety of MAC protocols until the new standard was released, followed by a phase of quickly increasing consolidation.

The use of IEEE 802.11a or IEEE 802.11b is decreasing over the years. However, IEEE 802.11b without or with simple adaptations to common WiFi models to mimic the behavior of IEEE 802.11p (we call these 802.11p′) has been used very frequently, even though this approach can produce realistic results only in low-density scenarios (Eckhoff *et al.* 2012). Figure 6.17 also indicates that new MAC protocols have been proposed in only a few studies.

Most alarming is that in a relatively large number of simulation studies the authors indicated the use of 802.11 models, but did not state which one out of the current IEEE 802.11 family of standards was used (or whether they relied on just the IEEE 802.11 base standard published in 1997); or did not specify the MAC protocol used at all.

It has further been shown that the mobility model used in IVC simulations has a substantial influence on metrics like the number of unreachable nodes, the average path length, and topology changes (Saha & Johnson 2004). There is a clear benefit when using a dedicated road traffic simulator for these purposes (Sommer & Dressler

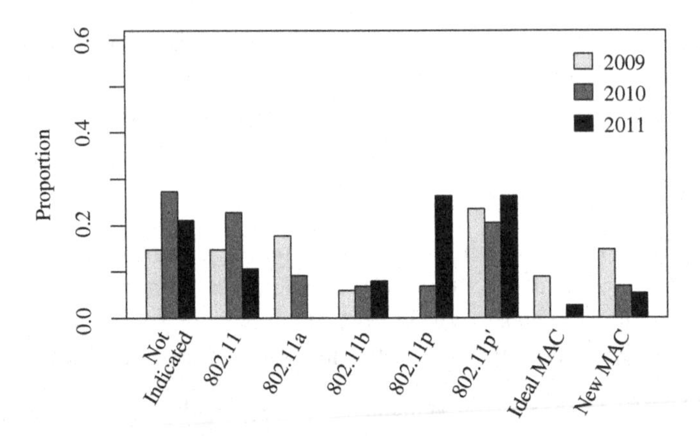

Figure 6.17 Distribution of MAC protocols. © [2012] IEEE. Reprinted, with permission, from Joerer *et al.* (2012c).

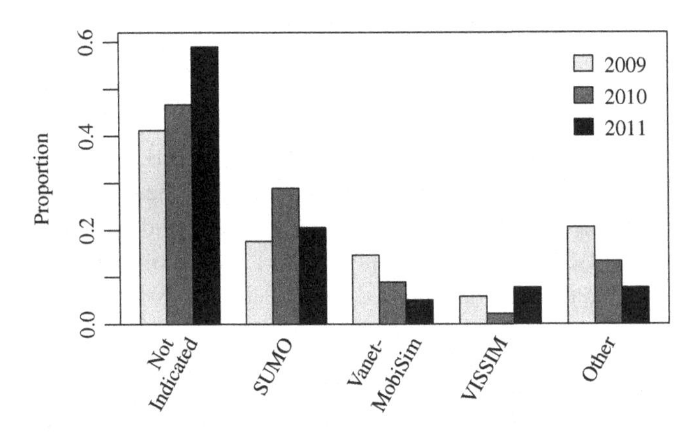

Figure 6.18 The distribution of road-traffic simulators. © [2012] IEEE. Reprinted, with permission, from Joerer *et al.* (2012c).

2008). As outlined before (see Section 6.1.3 on page 244), road traffic and network simulation need to be coupled bidirectionally if the IVC protocol studied may influence the behavior of the vehicles on the streets.

As shown in Figure 6.18, the most popular road traffic simulator, Simulation of Urban Mobility (SUMO), has been used in more than 20% of all papers. Other popular simulators include the dedicated vehicular network movement simulator VanetMobiSim and Vissim, which is a commercial tool.

Although the impact of accurate mobility modeling had already been shown in 2004 (Saha & Johnson 2004) and was fully confirmed in 2008 (Sommer & Dressler 2008), there is no observable positive trend towards applying realistic mobility models. On the contrary, the proportion of simulation studies in which the authors did not indicate whether a proper a road traffic simulator was employed has grown – from 40% in 2009 to almost 60% in 2011 (Joerer *et al.* 2012a).

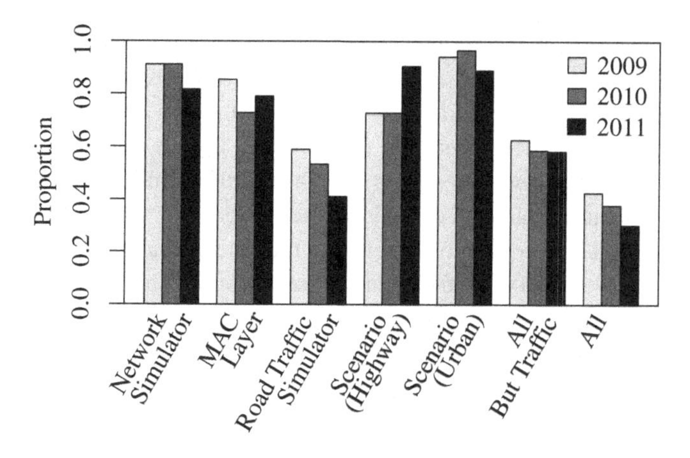

Figure 6.19 Trends in current IVC simulation studies. The prevalence of model descriptions by aspect and year. © [2012] IEEE. Reprinted, with permission, from Joerer *et al.* (2012c).

Comparing apples and oranges

As we have seen in the statistics presented, there is a clear trend towards using standardized protocols developed specifically for IVC (most prominently IEEE 802.11p) instead of relying on common WiFi variants. This, in combination with a consolidation of tools and models, allows one, in principle, to share setups and implementations for better comparability and reproducibility of simulation results.

Yet, Joerer *et al.* (2012a) observed that in a large number of cases the simulation setting and parameters are not fully clear. This includes precise information on the models and tools used as well as on the scenarios studied.

Figure 6.19 outlines these findings. It can be seen that, although the descriptions for individual factors (tools, models, and scenarios) are getting better, the overall quality of the description of simulation settings still requires improvement. On looking at the whole set of aspects, we found that only about one third of the publications specified *all* of them correctly. The results indicated as *all but traffic* summarize those publications taking into account all the listed categories but the road traffic simulator. As can be seen, the authors of only about 50% of the simulation studies reviewed properly mention the network simulator used, the MAC protocol employed, and the scenario studied.

6.2 Simulation tools

We now turn to an in-depth discussion of the available tools for conducting simulations of vehicular networks. Naturally, any simulation can always be written from scratch, implementing the aforementioned models and a tailored execution environment in a programming language of choice. However, not only does this put the burden of implementing (all of) the right models as well as that of model verification (whether the model is implemented correctly) on the developer but also proper execution of simulations depends on secondary aspects like correct implementation, seeding, use of pseudo-random-number generator (PRNG) streams, avoiding initialization bias, and the collection of statistically sound results. Moreover, the use of self-developed implementations hinders reproducibility and comparability of simulation results. Multiple studies, such as those by Pawlikowski *et al.* (2002), Kurkowski *et al.* (2005), and Joerer *et al.* (2012a), have repeatedly shown that some or all of these aspects are more often than not simply missed in studies. Simulation tools have therefore become an integral part of the simulative performance evaluation of vehicular networks.

Road traffic simulation tools mostly follow a time-stepped approach: At fixed time increments (e.g., every 100 ms of simulation time – not to be confused with wall-clock time, that is, real time, which commonly passes orders of magnitude slower), all vehicle movements and states, as well as the state of traffic-light controllers, will be recalculated.

Similarly, the most common type of network simulation tools in this domain, and the one we will focus on in the following, follows a discrete-event simulation (DES) approach. The central concept of such a simulator is an ordered queue of scheduled events. Each event is scheduled for a specific simulation time at which it will be triggered, giving the simulation a chance to react to the event. Such a reaction will typically change the state of the simulation and/or trigger new events to be scheduled in turn. After an event has been processed, the simulation will move on to processing the next event in the queue, advancing the "current simulation time" to that of the next event. Depending on the level of granularity desired in the simulation, an event might represent anything from a message being received by a vehicle, to an electromagnetic wave being picked up by an antenna, or a voltage level in a circuit changing from 0 V to 5 V.

6.2.1 Network simulation

We will now take a more detailed look at three simulators that are in widespread use: ns-3, OMNeT++, and JiST. With the exception of ns-3, all of the simulators presented in the following differentiate between the simulation engine and the simulation model library from which scenarios are assembled. Often, one model library will be maintained by the same company or community which also maintains the simulation engine. In the cases of OMNeT++ and JiST, these are the INET Framework and SWANS, respectively. Table 6.1 gives an overview of the discussed simulation tools.

The ns-3 network simulator

Out of the simulators we discuss, the network simulator ns-3 (Font *et al.* 2010) is the simulator with the longest historical record. It has its roots in NEST (Dupuy *et al.* 1989), a network simulation prototyping testbed written in C at Columbia University, which

Table 6.1 An overview of the simulation engines and model libraries discussed

Engine	Language	Default model library	Language
ns-2	C++	ns-2	Objective Tcl
ns-3	C++	ns-3	Python
OMNeT++	C++	INET	C++
JiST	Java	SWANS	Java

```
1  NodeContainer n;
2  n.Create(2);
3
4  InternetStackHelper().Install(n);
5
6  NetDeviceContainer d = CsmaHelper().Install(n);
7
8  Ipv4AddressHelper ipv4;
9  ipv4.SetBase("10.1.1.0", "255.255.255.0");
10 Ipv4InterfaceContainer i = ipv4.Assign(d);
11
12 UdpServerHelper server(4000);
13 server.Install(n.Get(1)).Start(Seconds(1.0));
14
15 UdpClientHelper client(i.GetAddress(1), 4000);
16 client.Install(n.Get(0)).Start(Seconds(2.0));
17
18 Simulator::Run();
```

Listing 6.1 Sample simulation script for ns-3 (adapted and shortened from documentation): Two nodes send or receive UDP data.

was later modified to create the REAL network simulator, a network simulator designed for testing mechanisms for congestion and flow control, and then rewritten again as the network simulator ns (in C++ and Tcl). A further rewrite in C++ and Objective Tcl with shadow objects yielded ns-2. The most recent total rewrite is ns-3, which relies on C++ and Python, and was written with the aim of being more scalable and having a more extensible structural and modular implementation. Listing 6.1 shows an example simulation script for ns-3.

The basic distribution of ns-3 is available under the GNU GPLv2 license, and it follows an object-oriented approach to modeling, making it easily extensible. Since ns-2 has often been criticized for being hard to learn and offering limited functionality and guidance for statistically sound simulations (Kurkowski *et al.* 2005), ns-3 is being distributed with a much updated user manual and multiple statistical frameworks are currently under development.

While there is no integrated development environment (IDE) or graphical execution environment available for ns-3, the simulator can record detailed traces that can be written to disk and, later, visualized using the included *nam* (short for network animator) tool or Wireshark. A core feature of ns-3 is its ability to seamlessly integrate – and be integrated into – testbeds: Simulations of ns-3 networks can serve as the local area network (LAN) or wireless network of virtual machines and ns-3 simulations can be run on top of physical network stacks.

Many of the available ns-2 models have already been rewritten for ns-3 and the ecosystem of available models is continuously growing, as the new simulator is increasingly commonly being used. Newer models have been written for ns-3 from the ground up, such as the iTetris suite of models for the simulation of ETSI ITS-G5 applications (Rondinone *et al.* 2013).

The OMNeT++ simulation environment

The goal of OMNeT++ (Varga & Hornig 2008) is to offer a complete simulation environment, shipping with an optional IDE for writing models, graphically assembling them into simulations, parametrizing the simulations, and running them. It equally well supports the manual writing of model code and network description files, or even a purely command-line-driven workflow. Similarly, simulations can run either in batch mode, or in a graphical user interface (GUI) that allows researchers to inspect and modify the simulation while it is running. Figure 6.20 shows screenshots of running IDE and GUI. OMNeT++ was further written with statistical correctness of simulations in mind, e.g., guiding users in setting up multiple independent replications of simulations with one or multiple independently seeded PRNG streams even in parallelized simulations, or detecting transient phases in simulations, and it integrates with the R project for statistical computing for analysis and plotting.

The development of OMNeT++ started in 1992. While OMNeT++ is open source, it is not free software. It is free only for academic use; for commercial use (here, the authors appeal to the integrity of users), a license is required. Yet, because of its modular and open approach to modeling, OMNeT++ today has a strong user community that tends to favor free and open-source software licenses when developing one's own modules. This has caused a thriving ecosystem of module libraries to spring up, each of which focuses on problems of a particular research domain.

Among the most popular of module libraries in current use is the *INET Framework*, which focuses on models of IP-based fixed and mobile hosts and wireless networking, and is maintained by the authors of OMNeT++. Other examples of module libraries are MiXiM for precise multi-channel multi-technology physical-layer modeling, OverSim for overlay network simulation, and the Veins framework for vehicular network simulation.

JiST/SWANS

JiST (Barr *et al.* 2005) development started comparatively late. It is a simulation engine that was written as a proof of concept for virtual-machine-based simulation, JiST being

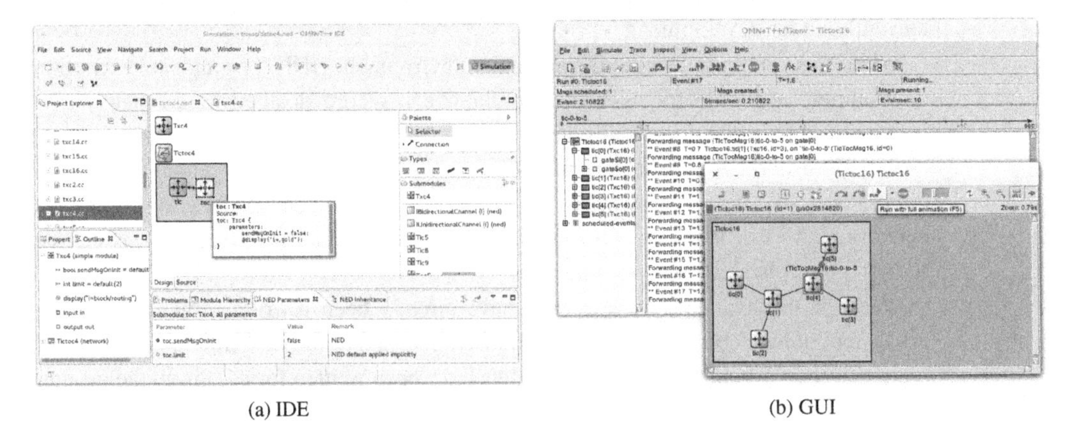

(a) IDE (b) GUI

Figure 6.20 Screenshots of the OMNeT++ IDE and a simulation running in GUI mode.

```
1  package jist.Hello;
2
3  import jist.runtime.JistAPI;
4
5  public class Hello implements JistAPI.Entity {
6
7    public static void main(String[] args) {
8      Hello h = new Hello();
9      h.myEvent();
10   }
11
12   public void myEvent() {
13     System.out.println(JistAPI.getTime());
14     JistAPI.sleep(1);
15     myEvent(); // schedule new event
16   }
17 }
```

Listing 6.2 Sample simulation in JiST (stepping simulation time by 1 s, printing 1, 2, 3, . . .).

short for *Java in Simulation Time*. It was developed as a basis for (and in tandem with) the SWANS wireless ad-hoc network simulator.

It is based around a *rewriter* component that modifies Java classes as they are being loaded by the Java runtime. The goal is to replace all references to wall-clock time with references to *simulation time*, a clock that progresses only on `sleep` calls. Simulation time advances independently for all nodes in a simulation, which are called *entities*. Consequently, method calls on other entities in the simulation are rewritten to be delayed until the simulation times of both entities have been brought into sync by jumping forward in time. This results in a classic DES system, but with little change to the semantics of the regular Java language. Listing 6.2 shows example code for a simple simulation in JiST.

Multiple independent extensions of JiST/SWANS exist, such as SWANS++, which aimed to continue development of SWANS, as well as bugfixes by Technion and extensions by Ulm University for IVC simulation.

6.2.2 Road traffic simulation

Just like with network-simulation tools, a wide variety of software exists for the express purpose of conducting road traffic simulations.

Depending on the desired granularity, such simulators need to implement a whole range of models. As illustrated in Figure 6.21, these simulators need to model not just motion constraints (both static constraints like speed limits and access restrictions as well as dynamic constraints like traffic lights and lane closings), but also traffic demand (when to generate which vehicle, where the driver wants to go, and how the driver chooses to get there). A more detailed taxonomy can be found in the literature, e.g., by Härri *et al.* (2009), as can a comprehensive overview of current road traffic simulators (Barceló 2010).

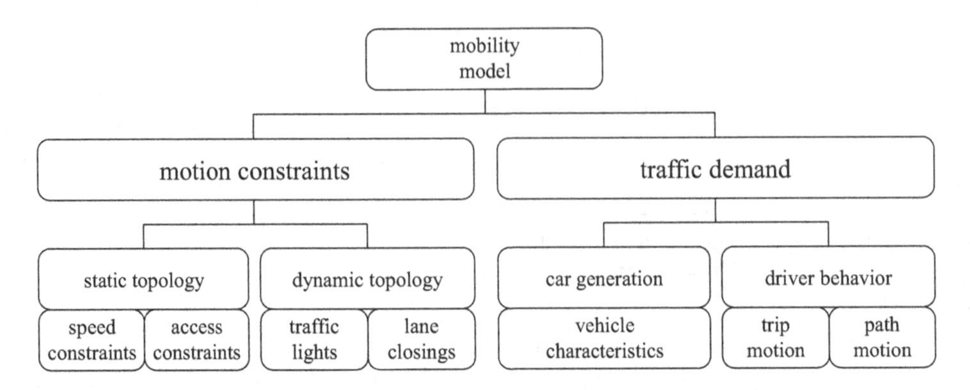

Figure 6.21 An overview of basic road traffic simulation concepts in increasing order of detail.

We now take a closer look at the top two simulators that are currently being used for the generation of synthetic vehicle movement: SUMO and Vissim.

The SUMO simulation environment

SUMO (short for *Simulation of Urban Mobility*) (Krajzewicz *et al.* 2012) is currently the most popular road traffic simulation toolkit for vehicular networks research (Joerer *et al.* 2012a). It is free and open-source software licensed under the *GNU general public license* (version 2 or later), easily extensible, and has been ported to a wide variety of software platforms.

The development of SUMO started in the year 2000. It has since then grown into a multi-modal simulation toolkit that includes a traffic simulation engine, tools for synthetic-network generation or import from public and private databases (such as OpenStreetMap or transport authorities), and tools for synthetic or trace-based demand generation (e.g., again, from transport authorities).

The microscopic traffic simulator was designed with scalability as a primary concern. This allows the simulation of city-scale networks in real time. Simulations can be run in batch mode or in a GUI, shown in Figure 6.22, which allows a researcher to interact with and modify the running simulation. Further, a choice of models for car movement is supported, such as the IDM (Treiber *et al.* 2000), the Kerner three-phase model (Kerner *et al.* 2008), and the Wiedemann model (Wiedemann 1974).

For easy extensibility and remote control of simulations, SUMO implements the aforementioned TraCI interface.

Vissim

Vissim (Lownes & Machemehl 2006, Fellendorf & Vortisch 2010) is a traffic simulation toolkit for Microsoft Windows that has been available commercially since 1994. Its microscopic traffic simulator is based on the Wiedemann car-following model (Wiedemann 1974), but with numerous additions and extensions, up to a multi-modal traffic simulation toolkit that can also simulate visibility and human perception and the movement of individual pedestrians.

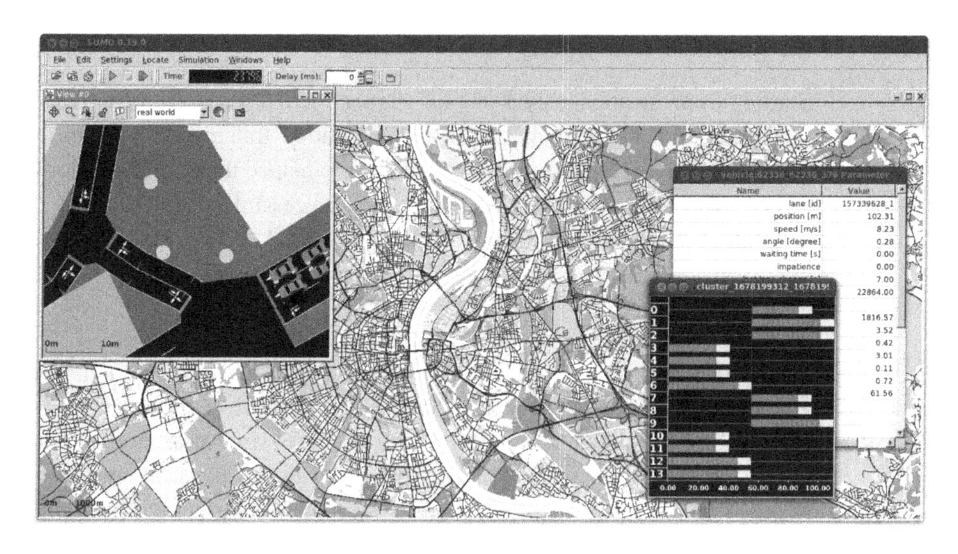

Figure 6.22 A screenshot of the SUMO microsimulation engine GUI.

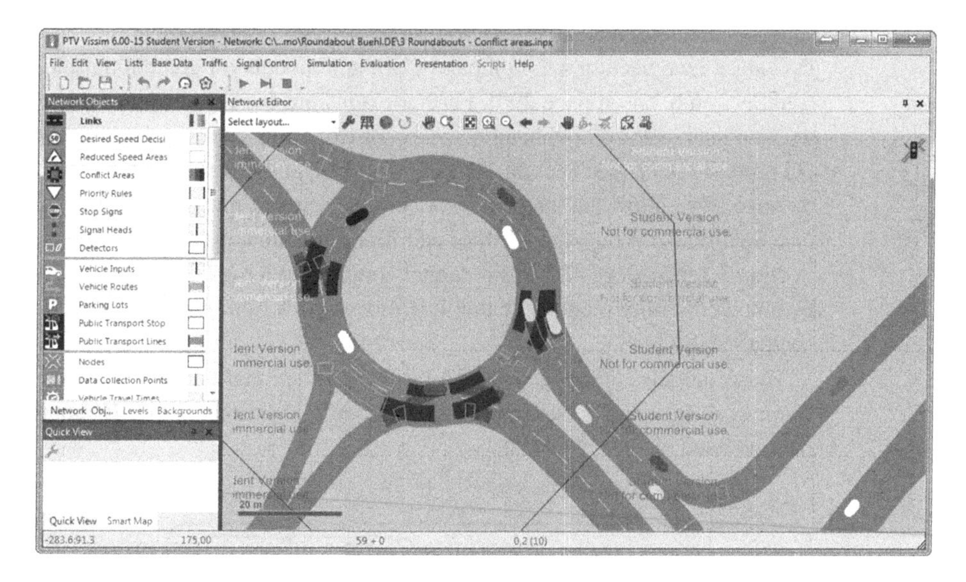

Figure 6.23 A screenshot of the Vissim IDE.

It supports graphically editing, modeling, and executing simulations in an IDE, shown in Figure 6.23. Other components are available for importing networks and demand modeling, graphical editing, and export of animations in 2D/3D videos or render meshes. Vissim can further be extended by user-created DLL libraries, e.g., with a custom driver model or for additional data logging. Further, a COM interface allows direct connection to the running simulation. Finally, Vissim can be used with related tools and Vistro for forecasting and optimization.

6.2.3 IVC simulation frameworks

Starting from the previously discussed simulation tools for network and road traffic simulation, several bidirectionally coupled simulation frameworks have been developed. Consolidation efforts in the past few years have helped these efforts to converge.

In the following, we discuss the three most successful examples of such coupled IVC simulation frameworks: Veins, iTetris, and VSimRTI.

Veins

The structure of the Veins (short for *Vehicles In Network Simulation*) simulation framework (Sommer *et al.* 2008, 2011b) is illustrated in Figure 6.24. It is based on the OMNeT++ simulation engine, which manages simulation execution control, data collection, and event scheduling. Starting from this, a physical-layer component that is based on the MiXiM (Köpke *et al.* 2008) model library provides a DES abstraction of electromagnetic waves to compute interference effects of multi-channel multi-technology transmissions.

On top of this, Veins provides a comprehensive set of IVC-centric simulation models, such as channel models (see Section 6.3.2 on page 274), protocol models of the IEEE IVC stack (see Section 4.2.3 on page 125), and models for computing selected metrics (see Section 6.3.4 on page 290). Further extensions by a diverse user community – the website lists over 100 publications using or extending Veins – simulate the ETSI ITS-G5 protocol stack (see Section 4.2.3 on page 128) and driver behavior (see Section 6.3.3 on page 285), allow the integration of cellular network model libraries such as for LTE (see Section 4.1.3 on page 113), or add new car-following models such as CACC (see Section 3.1.3 on page 48). Bidirectional coupling to a road traffic simulator is achieved by a node mobility model stub that interfaces with a running instance of a TraCI server such as SUMO.

These models, together with the OMNeT++ simulation environment, provide the framework in which custom simulation models of protocols or IVC applications can be integrated.

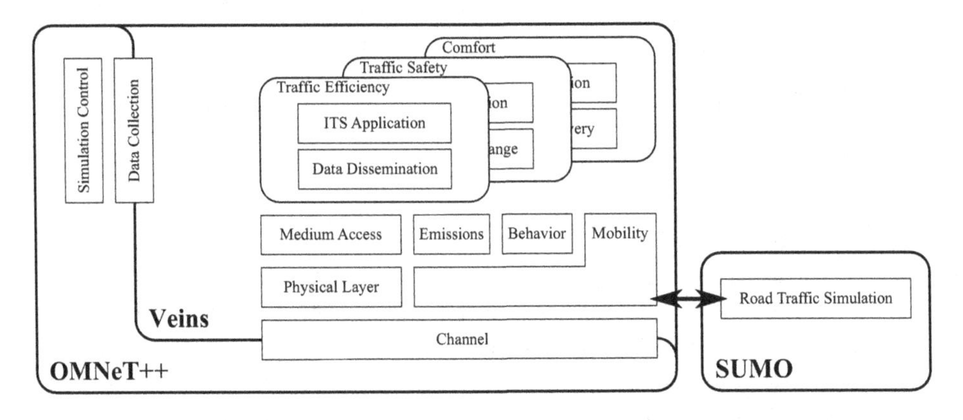

Figure 6.24 The Veins simulation framework.

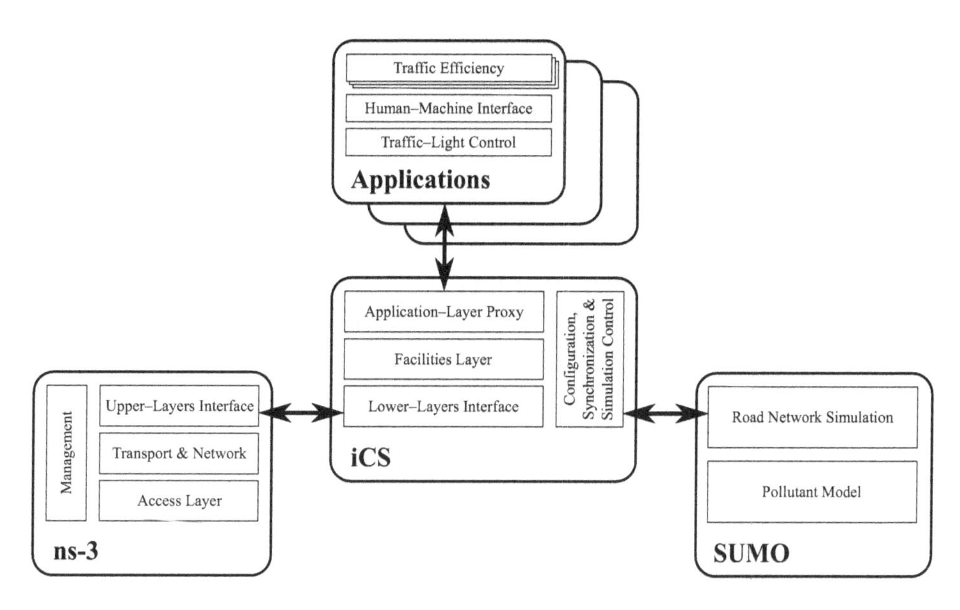

Figure 6.25 The iTetris simulation platform.

iTetris

iTetris (Härri *et al.* 2011, Rondinone *et al.* 2013), illustrated in Figure 6.25, is structured as four coupled function blocks. The central control system, called iCS, implements a subset of the ETSI ITS facilities layer, caches and forwards data, and interacts with dedicated simulation components. These are the following: a network simulator based on ns-3, a road traffic simulator based on SUMO, and an application built on top of an API that is specific to iTetris.

iTetris was developed in the context of the European Union FP7 research project *Integrated Wireless and Traffic Platform for Real-Time Road Traffic Management Solutions* running from 2008 to 2011 and released as free and open-source software. The project website lists over 30 related publications. Anyone can apply for access to the source code. In the scope of the project, the network simulator was extended with custom models of IVC-specific channel models and ETSI ITS layers in order to be able to send protocol-compliant messages and the road traffic simulator was extended with detailed emission models. These models have been fixed and integrated in more recent versions of ns-3 and SUMO.

The application manager connects to user-defined applications via an IP socket, communicating with them over an iTetris-specific protocol with reference implementations for Java, C++, and Python. This protocol allows applications to send messages or to instruct the iCS to subscribe to (and report back with updates from) the road traffic simulation, the network simulation, or the ETSI ITS facilities layer implemented in the iCS.

VSimRTI

VSimRTI (short for *V2X Simulation Runtime Infrastructure*) (Schünemann 2011, Naumann *et al.* 2009) goes one step further in decoupling simulation components. As illustrated in Figure 6.26, it is structured around a lightweight core.

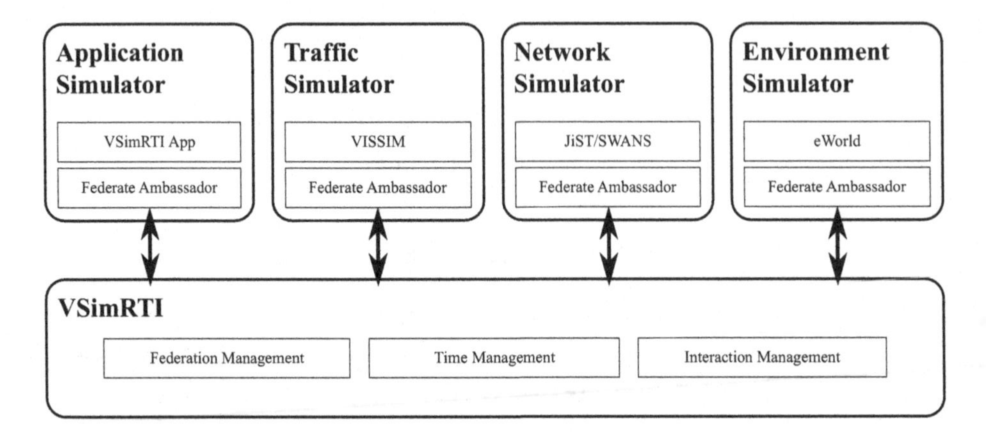

Figure 6.26 VSimRTI sample simulator coupling.

Its goal is to provide a generalized framework for coupling domain-specific simulators. For this, it implements an ambassador concept, which is inspired by fundamental concepts of the high-level architecture (HLA). VSimRTI provides the core, managing the synchronization of, interaction between, and lifecycle management for connected simulators.

The coupled simulators are envisioned to implement generic VSimRTI interfaces, making it easy to exchange one for the other. This approach also enables simulators to process just a subset of the complete simulation, e.g., two road traffic simulators to simulate two different cities. Further, optimistic synchronization techniques for high-performance simulations are planned to be supported for such simulators.

All coupling interfaces are completely language agnostic, and VSimRTI already provides interface components for many simulators: Vissim and SUMO as well as OMNeT++, ns-3, and JiST/SWANS are all supported. Reference implementations of domain-specific simulators for, e.g., cellular networking, are provided as well, and new components are continually being created. To date, the project website lists over 20 publications. A Java-based application simulator can execute IVC applications with VSimRTI-specific interfaces as well as close-to-life interfaces, e.g., for polling sensors or sending messages.

6.3 Scenarios, models, and metrics

Simulating an IVC protocol or application requires not only the correct tools, but most importantly a full set of scenario descriptions and implemented models capturing the behavior of the wireless communication, the protocols themselves, and even the behavior of the human driver. In this section, we study these components with a strong focus on IVC-specific aspects. In particular, we review typically used scenarios that, if not correctly chosen according to the intended application scenario, may have a strong influence on the resulting performance. The same holds for the simulation models used. Here, specific models for simulating the wireless channel and all the IVC-specific protocols are needed, as well as models identifying the behavior of the human driver. We conclude this section with a discussion of metrics that can (and should) be used for assessing the overall performance of IVC applications on a microscopic scale as well as on a larger scale.

6.3.1 Scenarios

The selection of the simulation scenario has possibly the biggest influence on the results. Depending on the vehicular networking protocol or application being studied, one has to very carefully select and parametrize the scenario configuration.

As an extreme example, let's consider a new protocol for information dissemination in emergency situations. The protocol has been designed to support ultra-low-latency communication to warn drivers (or even to support automated reactions) in the case of there having been an accident. If this protocol if tested in a sparse setup with only a few other vehicles in communication range on a single-lane freeway, the simulation results obtained will be unlikely to represent the most critical cases in terms of access to the wireless channel. On selecting an eight-lane freeway at rush-hour density, the protocol might completely fail due to interference and the congested channel.

In contrast to all the cellular networking technologies for which the Third Generation Partnership Project (3GPP) defined reference scenarios (also to provide comparability), the vehicular networking community has no such references yet. In the following, we discuss which aspects might be relevant in scenario definition. The final configuration is of course part of the more general performance-evaluation process.

Road network

The first step towards the scenario definition is the selection of the appropriate part of the road network. This step clearly depends on the IVC protocol or application under study as well as on the intended benchmark or research hypothesis.

Single street or freeway

The simplest scenario is a single street. In this scenario (see Figure 6.27(a)), cars can drive in an east–west direction on one or multiple lanes. If needed, lanes can be combined to support directions, i.e., cars coming from both directions. In some sense, this models a typical *freeway* scenario. This scenario is applicable for many IVC applications, from TISs to road traffic safety.

Even though this scenario sounds simplistic, it is not. Even for a single-lane street, realistic car-following models need to be applied. If multiple lanes are supported, appropriate lane-change models and overtaking maneuvers need to be in place.

The scenario becomes more interesting, especially when studying TIS applications, when a detour is possible. Such a scenario is depicted in Figure 6.27(b). Here, new models are needed in order to control vehicles leaving the freeway and entering on another ramp.

Starting from these simple scenarios, one can play around with more complex ones, not necessarily leading to new insights when it comes to studying vehicular networks in

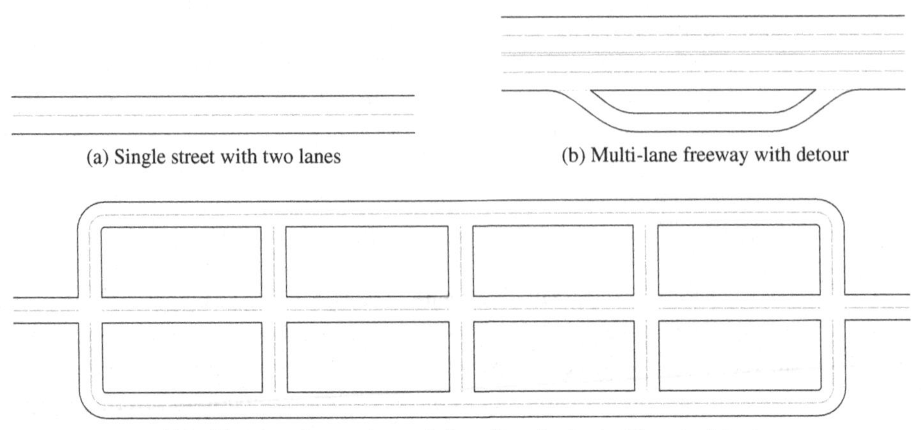

Figure 6.27 Simple single-street, freeway, and ladder scenarios.

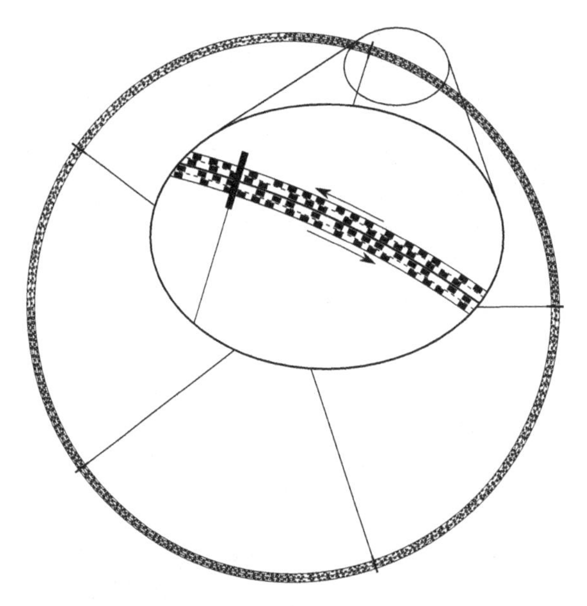

Figure 6.28 A circular street as a borderless scenario. © [2007] IEEE. Reprinted, with permission, from Sommer *et al.* (2007).

simulation. A scenario that is frequently used in the community is that of a major road with many possible detours placed in the form of a ladder (see Figure 6.27(c)).

The most challenging problem, however, is the generation of vehicles in such a scenario. The typical approach is to create vehicles and place them at one end of the street so that they can drive towards the other end. Yet, this can easily end up in border effects and simulation artifacts close to the end of the road.

This issue can be overcome by using a circular road as depicted in Figure 6.28. This represents the freeway scenario where both ends of the road are simply connected in the simulation environment. The circular street can of course also be extended with one or multiple detours. The key advantage of the scenario is that road traffic mobility can be simulated with hardly any side effects besides an initial setup phase when the street is initially filled with cars. On the other hand, the circular road needs to be long enough, or many applications that temporarily store information cannot easily be used in this scenario because the same simulated vehicle will pass the same road segment multiple times. Thus, if it has already, for example, collected traffic information regarding this segment, it will be able to teleport such information in time and space.

Single intersection

Especially for assessing safety-critical applications, the *single-intersection* scenario might be an excellent candidate. The idea is to simulate two streets that cross each other. Again, the scenario sounds simplistic, yet the level of complexity is much higher than that in the freeway scenario.

Figure 6.29 outlines the intersection scenario. As can be seen, two streets (running east–west and north–south) cross each other in the form of an intersection. Vehicles may pass the intersection obeying the traffic rules. If no traffic lights are in place, either of the streets is a preference road, or each car has to check for a vehicle from the right, or a four-way stop is implemented. This intersection management is very complex to model. If traffic lights are in place, the scenario becomes even more complex: traffic lights completely change the behavior of the vehicles, and, of course, traffic-light-switching programs (static or adaptive) need to be implemented.

This scenario brings into play further, very interesting, components. As is also depicted in Figure 6.29, buildings might be located at each corner. These buildings not only have an impact on the visual contact of cars approaching the intersection, but also may severely add to the path loss of the wireless communication. Thus, radio-shadowing models must be correctly implemented and configured to realistically simulate this scenario.

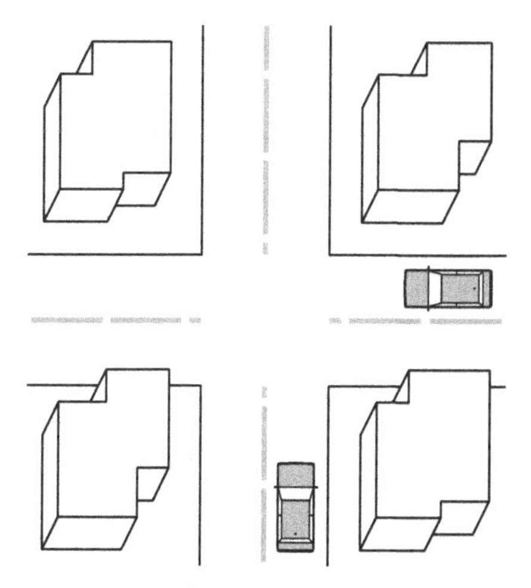

Figure 6.29 A schematic view of the X intersection scenario.

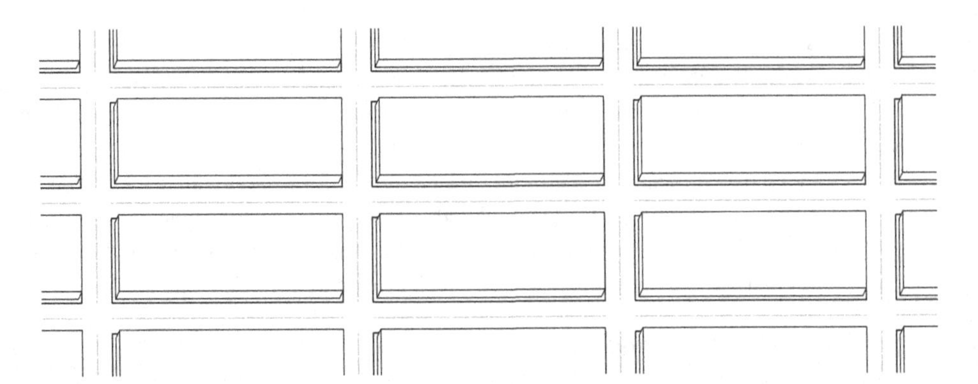

Figure 6.30 The Manhattan-grid scenario providing a rough approximation of an urban road network.

Manhattan grid

The two initial scenarios, a simple street and an intersection, can also be combined to create a first approximation of urban environments. In the literature, the *Manhattan grid* is one of the most frequently used scenarios. It combines an easy-to-create structure with a medium level of accuracy with respect to the degree of realism (assuming that any city looks like downtown New York).

The scenario is depicted in Figure 6.30. The parameters to be modified include the number of lanes per street as well as the number of streets in the horizontal and vertical directions. Besides these key parameters, the same conditions apply as already discussed for the single-street and intersection scenarios. Traffic lights can be added, as well as certain traffic rules. Buildings constitute a central component in any inner-city scenario, and should be carefully modeled and simulated.

Regarding the generation of vehicles, the circular (or, in this case, torus) approach can be realized again by virtually "connecting" each end of a street, i.e., creating a teleporter that removes a car as soon as it reaches the end and places it again at the beginning without changing its speed or other parameters. In general, the regulation of traffic flows is very complex, and the Manhattan-grid scenario is very sensitive in this aspect. If cars are just created somewhere, possibly at random locations, one might see a very unrealistic traffic pattern.

Realistic maps for urban scenarios

What remains to be done in order to provide the highest possible level of realism is to use real map information to prepare a scenario for an *urban* or *city environment.* Services like OpenStreetMap provide geodata free for public use. All the entries were created by volunteers in a crowd-sourcing approach (Sommer *et al.* 2010a) and made available first under the terms of a Creative Commons (CC BY-SA) license, later under those of an Open Database Licence (ODbL). Depending on the location, geodata in OpenStreetMap is even more accurate than that provided by commercial services.

This data can be used to create scenarios in simulation. A typical scenario is shown in Figure 6.31. This is the city of Ingolstadt, Germany, which we frequently use for simulating IVC protocols in city or urban environments. The scenario is not that different

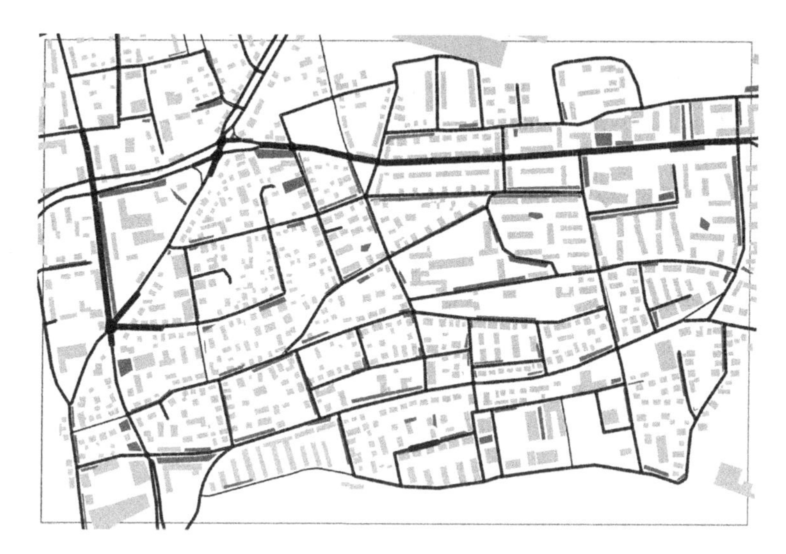

Figure 6.31 A city scenario based on the city of Ingolstadt, Germany, showing the ROI for network communication and statistics collection.

from the Manhattan grid, as one might expect. The street layout is, of course, more realistic and represents a typical city. Yet, all other parameters, including buildings, traffic rules, and generated traffic flows, need to be configured in the same way.

Figure 6.31 also shows another aspect. A region of interest (ROI) in which IVC communication is simulated is defined. Actually, for many tools (see Section 6.2 on page 255) the computational performance is much higher for mobility simulation than for the simulation of wireless networking. Therefore, the simulation of communication aspects needs to be bound to a rather small region of an area. On the other hand, microscopic mobility simulation must be performed for a much larger area in order to prevent border effects due to non-realistic traffic flows. Often, a second (inner) ROI is used to collect statistics in the simulation. Again, this is necessary in order to prevent border effects in the wireless communication. Assuming a vehicle starting communication in the outer ROI, it will certainly have no interfering neighbors behind this invisible wall, thus enabling more robust communication – a simulation artifact. If, however, statistics are collected only for the inner ROI, such border effects can be prevented.

Our discussion clearly motivates the use of standardized scenarios. Not only can simulations more easily be reproduced, but also the development of a scenario is very time consuming. Importing geodata from OpenStreetMap is the simplest part. Buildings can in many cities rather easily be imported, too, since 2.5D structures (the ground plot is available as well as the height of the building) are listed in the same database. Tuning traffic-light-switching cycles, traffic flows, and other parameters to realistic values is much more complicated and also depends on external help, e.g., to provide information about typical road-usage statistics.

Such urban scenarios have been developed for many cities worldwide. They differ mainly in terms of public availability and the level of detail. This level of detail is what we need to discuss next.

Level of detail

The quality of simulation-based performance evaluation strongly depends on the level of detail of the underlying scenario. It is, however, important to note that high granularity comes with the price of increasing simulation time, depending on the computational complexity of the different models.

Here, we need to discuss the variety of dimensions that add to the complexity. We also try to identify some basic components that definitely need to be considered whereas others can be abstracted from.

Vehicle density

One parameter that we have frequently mentioned before is the *vehicle density*. This parameter models the time-varying number of vehicles on the road.

Sparse distributions can be seen in lightly populated areas or at night time. Such a sparse scenario (see Figure 6.32(a)) represents one extreme case for IVC. Even though the vehicular network will be only lightly loaded, connectivity cannot be assumed for many vehicles and multi-hop data dissemination will likely fail.

The other extreme is a very densely crowded area. Examples include a traffic jam on a major freeway (see Figure 6.32(b)) or rush-hour traffic in a city center. These two examples are, however, very different when it comes to wireless communications. If we just assume a communication range of 0.5 km, this means about 100 cars per lane on a major 10-lane freeway. Thus, the freeway example with a traffic jam can easily result in a situation in which more than 1000 cars are in direct communication range. In the urban scenario, radio signal shadowing by buildings reduces the number of vehicles in communication range but contributes to the dynamics of the scenario: in one time step, only a few vehicles are in communication range while driving in a narrow side street; yet, on entering a major avenue, this changes within a few seconds to a few hundred neighbors.

Besides these two extremes, any other density is likely to happen in a realistic environment. Any carefully planned simulation experiment thus has to include a parameter study for different vehicle densities covering also the extreme cases. The exact parameter settings depend on the scenario selected, e.g., urban vs. freeway.

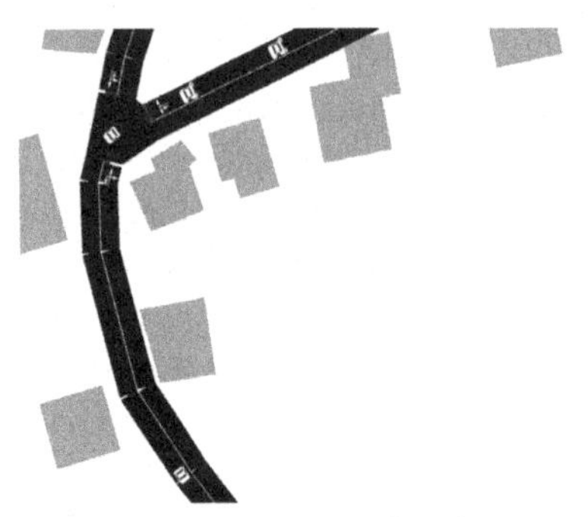

(a) low-density traffic, urban

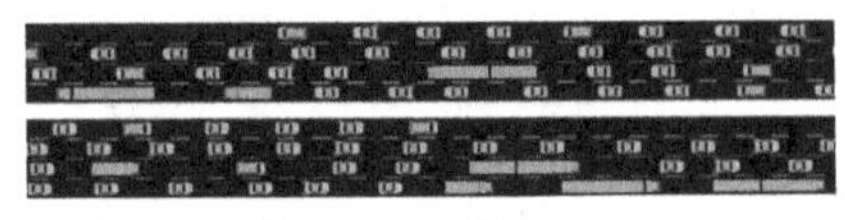

(b) congested freeway

Figure 6.32 An illustration of different vehicle densities in simulations.

Penetration rate

A parameter that is closely related to the vehicle density is the penetration rate. This, however, is not an intrinsic feature of the map or the road traffic in general, but merely models the percentage of vehicles equipped with IVC technology. Yet, for evaluating the behavior of a new IVC application or protocol, it needs to be considered a main option for a parameter study.

During the initial deployment of whatever communication technology is to be used, only a fraction of vehicles will be able to join the network. For applications in vehicular networks it makes a huge difference whether all vehicles are equipped with the (same) communication technology or only a certain percentage. Even though this metric will in the end be used to assess applications rather than protocols, application demands can be described in this framework as well.

For all kinds of safety applications, vehicles not equipped with the right communication devices will obviously not participate in the vehicular network. Thus, it is necessary to develop certain fall-back directives in case not all neighboring vehicles join the network. This is unfortunately not known in advance, and, for example, local sensors need to be used to warn about the situation in which general control becomes impossible. This is, however, similar to the case of attackers completely blocking the wireless communication. The difference is that such activities (e.g., jamming) can be detected, whereas a car with no vehicular networking device simply becomes invisible.

The same problem also holds for non-safety applications such as traffic information systems. However, if not all vehicles participate, only a degraded service quality will be the result. It is still important to assess which applications and protocols can compensate for a low penetration rate.

We want to be very clear that there is a substantial difference between investigating a protocol for low vehicle density and for low penetration rate, since the mobility of vehicles will be very different.

Road traffic flows

A parameter quite similar to the traffic density is the *vehicle distribution*. Still, this parameter also differs in some important aspects from the pure density measure. The vehicle distribution outlines how vehicles are distributed in the environment. Many simulation studies use simple uniform distributions or rely on a Poisson process to generate cars.

A much better approach, however, is to measure *road traffic flows*. This is a primary parameter in the road traffic engineering sciences insofar as it clearly distinguishes major traffic flows of higher density from cross traffic, which is of less importance for understanding traffic congestion and bottlenecks in the road network. Such a measure is especially needed in order to calibrate scenarios made from realistic maps, since the road layout will be designed for exactly this traffic demand.

In simulation, road traffic flows can be configured using origin–destination (O/D) pairs, which can be transformed into a demand model by road traffic mobility simulators such as SUMO. Figure 6.33 shows an example.

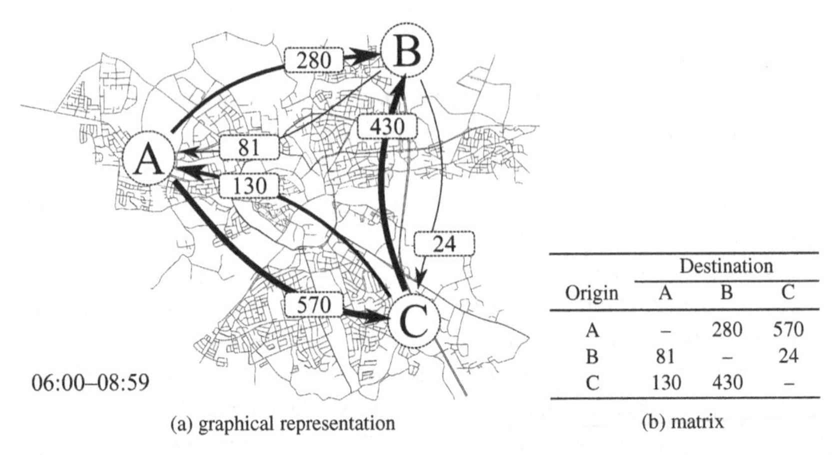

Origin	Destination A	B	C
A	–	280	570
B	81	–	24
C	130	430	–

Figure 6.33 A minimal O/D matrix and its graphical representation: the number of trips between each two of three regions. No intra-region traffic is captured.

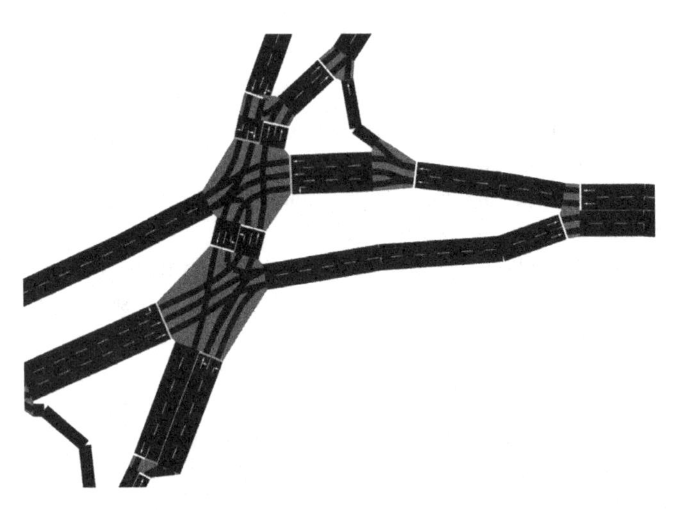

Figure 6.34 A complex intersection with fine-grained turn restrictions and traffic lights, as rendered in SUMO.

Road network and traffic rules

The deployment of vehicles in the simulation scenario depends heavily on the road network used. As mentioned already, the road-network topology can be quite simple or one can even rely on real map information imported, for example, from OpenStreetMap.

Still, this is only one part of the story. The level of detail can be gradually increased by modeling lane information. This is especially of importance for intersections. Here, the option is to simply have multiple lanes per street and direction or to go down to a level at which single lanes are marked as turning lanes, one-way streets, etc. Figure 6.34 illustrates how such an intersection is rendered in SUMO.

In order to realistically simulate the road traffic, also traffic rules enforced in reality by means of general rules and traffic signs need to be added. Vehicles may drive at more than 100 km/h on a freeway but at only 50 km/h in the city or even just 30 km/h in

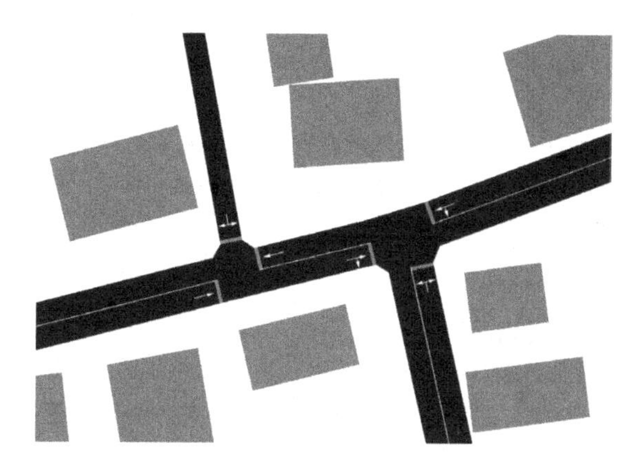

Figure 6.35 Building geometry from OpenStreetMap, imported into SUMO.

a residential area. Similarly, the right of way at intersections can be modeled down to the granularity of yield signs.

The right of way is often enforced by traffic lights in urban and suburban environments. These traffic lights have a major influence on the microscopic mobility of the vehicles – so this too needs to be implemented in a simulation. For example, SUMO allows one to model traffic lights even for very complex intersections as shown in Figure 6.35. The more complicated issue is the switching cycle of the traffic lights. Using arbitrary cycles may potentially lead to obscure traffic flows – we sometimes observe this situation even in reality. For a high level of detail, traffic-light-switching cycles should match the ones in reality if one relies on a realistic city scenario using map information, for example, from OpenStreetMap.

Buildings

The final parameter that needs to be discussed is the modeling of buildings and other structures that potentially block the radio communication between vehicles. Depending on the IVC application being investigated, radio signal shadowing caused by buildings may be abstracted from using a probabilistic model. Yet, when it comes to understanding microscopic effects such as would be needed for all safety-critical applications, very careful modeling of such radio obstructions is a must. We will investigate appropriate models for radio signal shadowing by buildings (and by other vehicles) in detail in Section 6.3.2 on page 279.

Figure 6.35 outlines the potential of modeling buildings based on OpenStreetMap data. As can be seen, buildings are modeled as 2.5D structures, having a precise layout but only a height parameter instead of full 3D information. This is in most cases sufficient, since the key parameter that needs to be known in a simulation is the number of (exterior) walls that will be penetrated by the radio waves.

Scenario implementation

After studying all the scenario-related aspects, only one question remains: How should one implement such a scenario? We briefly review options for the implementation using the simulation toolkit Veins (Sommer *et al.* 2011b) as an example. Other simulators provide similar features.

- The road network including annotations is the key integral component needed for road traffic microsimulation. It will be modeled using the SUMO simulator, meaning that the user can select the appropriate level of detail by modifying the road network topology to add lanes, complex intersections, speed limits for each road segment, etc.
- The road traffic itself can be modeled completely in SUMO as well. For this, files describing each vehicle's route as well as a starting time are loaded by the simulator. SUMO informs the network simulation side, i.e., OMNeT++, about new vehicles and their current positions in each simulation step using the TraCI interface. In turn, OMNeT++ has to instantiate nodes in the network simulation that may communicate by means of IVC.

  In order to investigate scenarios with varying penetration rates, either only those vehicles equipped with IVC technology are instantiated in the network simulator or they are modeled with IVC radios being turned off. The differences are possibly only in CPU and memory consumption in OMNeT++ and in the ability to manipulate all vehicles from the network simulation side.

  Alternatively, vehicles can be created only in the network simulator. This may be helpful if, for example, parked vehicles need to be supported. As long as the vehicle remains parked, it does not need a representation in the microscopic mobility simulator. Again, the TraCI interface helps to create, configure, and eventually remove vehicles in SUMO.
- Traffic lights are supported by SUMO, which also ensures that approaching vehicles obey the posted traffic rules. This simulator also supports even complex switching cycles by means of a traffic-light program. The TraCI interface can be used to control and to change the switching cycles.
- The last component to be identified is the implementation of buildings and other obstacles. Since these are relevant to the network simulation part only, modeling buildings in OMNeT++ is the obvious approach. For this, building layouts need to be passed from the OpenStreetMap definitions right into the network simulator. When it comes to simulating radio signal shadowing by moving vehicles as well, vehicle dimensions need to be modeled on that side as well.

6.3.2 Channel models

In this section, we review channel models frequently used for simulation of vehicular networks that are based on IEEE 802.11p. The way of modeling radio-wave propagation in this field has changed quite drastically in the last few years. Initially, simple free-space models were used, together with some random Gaussian noise. Obviously, this does not reflect all the slow and fast fading characteristics which can be observed in reality.

The objective of this chapter is, however, not to provide a full introduction to signal processing and wireless communications. For this, we refer the reader to excellent textbooks in the wireless communications domain such as *Wireless Communications: Principles and Practice* by Rappaport (2009).

Free-space model

In general, when simulating wireless communication, fading due to large-scale path loss, deterministic small-scale fading, and probabilistic loss effects is calculated as a sum of independent loss processes (Ahmed *et al.* 2010, Nagel & Eichler 2008). The received power P_r depends on the transmit power of the radio P_t, the transmit and receive antenna gains G_t and G_r, respectively, and a potentially large number of different loss terms L_x. The receive power P_r can therefore be expressed as

$$P_\mathrm{r}\ [\mathrm{dBm}] = P_\mathrm{t}\ [\mathrm{dBm}] + G_\mathrm{t}\ [\mathrm{dB}] + G_\mathrm{r}\ [\mathrm{dB}] - \sum_x L_\mathrm{x}\ [\mathrm{dB}] \tag{6.1}$$

Assuming free-space propagation, path loss is often estimated by taking the distance to the receiver d and the wavelength λ into account. This yields the very simplistic *free-space* model:

$$L_\mathrm{freespace}\ [\mathrm{dB}] = 20\log_{10}\left(4\pi\frac{d}{\lambda}\right) \tag{6.2}$$

Empirical adaptations of the free-space model aim to account for non-ideal channel conditions by introducing an additional environment-dependent path-loss exponent α. This leads to the well-known *Friis* model (Friis 1946):

$$L_\mathrm{emp\text{-}freespace}\ [\mathrm{dB}] = 10\log_{10}\left(16\pi^2\frac{d^\alpha}{\lambda^\alpha}\right) \tag{6.3}$$

Figure 6.36 shows the behavior of the resulting received power P_r as a function of the distance d for different path-loss exponents α. Please note the exponential decrease

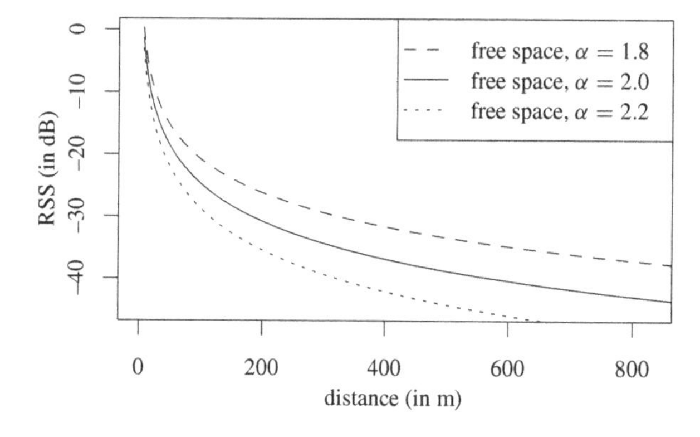

Figure 6.36 Empirical adaptations of the free-space path-loss model for different values of α for parametrizing the decrease of received power (expressed as normalized RSS value) with increasing distance.

over transmission distance. The model does not take into consideration any multi-path effects.

Two-ray interference model

A more realistic treatment of the path loss takes into account that the radio signal gets reflected at least at the ground. Hence, radio propagation will commonly suffer from at least one notable source of attenuation: constructive and destructive interference of a radio transmission with its own ground reflection.

Modeling

The effect of one radio beam being reflected from the ground is depicted in Figure 6.37. Depending on the distance between sender and receiver and the antenna heights, a phase difference will be introduced by the ray that is reflected from the ground compared with the direct line-of-sight signal. A physically more correct approximation of path loss must therefore be based on the phase difference φ of two interfering rays, leading to a *two-ray interference* model.

With the notation in Figure 6.37, the length of the direct line-of-sight propagation path can be geometrically modeled to be $d_{\mathrm{los}} = \sqrt{d^2 + (h_{\mathrm{t}} - h_{\mathrm{r}})^2}$, and the length of the indirect, non-line-of-sight path via ground reflection can be seen to be $d_{\mathrm{ref}} = \sqrt{d^2 + (h_{\mathrm{t}} + h_{\mathrm{r}})^2}$. From the length difference of these paths and the wavelength, the phase difference of interfering rays can be derived as

$$\varphi = 2\pi \frac{d_{\mathrm{los}} - d_{\mathrm{ref}}}{\lambda} \tag{6.4}$$

The attenuation of a polarized electromagnetic wave via reflection is commonly captured in a reflection coefficient, which is dependent not only on a fixed ε_{r}, but also on the angle of incidence θ_{i} (Rappaport 2009). For further computations, only its sine and cosine need to be known, both of which are straightforward to compute as $\sin\theta_{\mathrm{i}} = (h_{\mathrm{t}} + h_{\mathrm{r}})/d_{\mathrm{ref}}$ and $\cos\theta_{\mathrm{i}} = d/d_{\mathrm{ref}}$, respectively. Now, the reflection coefficient can be calculated as

$$\Gamma_{\perp} = \frac{\sin\theta_{\mathrm{i}} - \sqrt{\varepsilon_{\mathrm{r}} - \cos^2\theta_{\mathrm{i}}}}{\sin\theta_{\mathrm{i}} + \sqrt{\varepsilon_{\mathrm{r}} - \cos^2\theta_{\mathrm{i}}}} \tag{6.5}$$

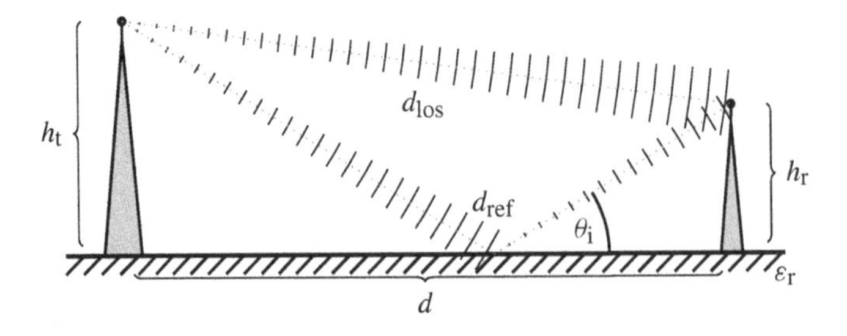

Figure 6.37 A conceptual model of ground reflection causing distance-dependent constructive and destructive signal-interference effects at the receiver. © [2012] IEEE. Reprinted, with permission, from Sommer *et al.* (2012).

The relative change in signal strength due to constructive or destructive interference can then be modeled by amending Equation (6.2) with a simple correction term for the relative phase and magnitude of interference by the reflected ray, to yield

$$L_{\text{tri}}\ [\text{dB}] = 20\log_{10}\left(4\pi\frac{d}{\lambda}\left|1+\Gamma_{\perp}e^{i\varphi}\right|^{-1}\right) \tag{6.6}$$

Obviously, this calculation is more complex than the much simpler calculation of path loss according to the free-space model – and even more so if additional reflections need to be considered (Domazetovic *et al.* 2005). This may substantially contribute to the simulation time for larger-scale simulation studies. However, at large distances, destructive signal-interference effects will cause noticeably worse path loss than the free-space model, making the effects impossible to ignore.

It has therefore been investigated whether and, if so, how the model could be simplified to enable fast simulations but still with the best level of accuracy possible. In the early days of simulative performance evaluation in vehicular networking, it was demonstrated how the calculation of the interference between line-of-sight and reflected rays can be simplified for large distances d, assuming perfect polarization and reflection. In these cases, the path loss can be estimated as

$$L_{\text{trg,far}}\ [\text{dB}] = 20\log_{10}\left(\frac{d^2}{h_{\text{t}}h_{\text{r}}}\right) \tag{6.7}$$

A model that is based on these findings, which is commonly termed the *two-ray ground* path-loss model, has been implemented in almost all common network simulators that are used to assess the performance of IVC protocols. This model uses a cross-over distance d_{c} of the classical two-ray model and the approximation shown (Giordano *et al.* 2009). The value of this distance is defined as the point where path loss according to both models breaks even. The complete model for choosing between Equations (6.2) and (6.7) can finally be defined as

$$L_{\text{trg}}\ [\text{dB}] = \begin{cases} L_{\text{freespace}}\ [\text{dB}] & \text{if } d \le d_{\text{c}} \\ L_{\text{trg,far}}\ [\text{dB}] & \text{if } d > d_{\text{c}} \end{cases} \tag{6.8}$$

This two-ray ground model is implemented for example in ns-2.35, ns-3.14, OMNeT++ INET 2.0.0, QualNet 5.1, and JiST SWANS 1.0.6. Most of these simulators have frequently been used for IVC protocol evaluations (Joerer *et al.* 2012a).

The cross-over distance d_{c} for the two-ray ground model can be calculated using Equations (6.2) and (6.7) and using the values for transmitter and receiver antenna heights $h_{\text{t}} = h_{\text{r}} = 1.895\,\text{m}$ (corresponding to the cars and antennas used) and $\lambda = 0.051\,\text{m}$ for the wavelength used (corresponding to the IEEE 802.11p CCH center frequency of 5.890 GHz) as follows:

$$d_{\text{c}} = 4\pi\frac{h_{\text{t}}h_{\text{r}}}{\lambda} \tag{6.9}$$

For these values, Equation (6.9) yields a break-even distance for the *two-ray ground* model of $d_{\text{c}} = 886.6\,\text{m}$. Under realistic propagation conditions, IEEE 802.11p transmissions in urban areas are highly unlikely to ever reach that far (Mangel *et al.* 2011b,

Sommer *et al.* 2011a). Thus, typical vehicular networking simulations in fact only use the free-space model, even if the two-ray ground model has been configured.

Moreover, it turned out that this simplification does not provide sufficient savings compared with the clearly reduced accuracy (Sommer & Dressler 2011, Sommer *et al.* 2012). On modern hardware, the use of a two-ray interference model (as opposed to a simplified two-ray ground model) only comes with marginal added computational cost for simulation experiments.

In the following, we show selected results from an experimental validation of both path-loss models for the simulative performance evaluation of vehicular networks using IEEE 802.11p radios.

Experimental validation

Sommer *et al.* (2012) investigated the implemented path-loss models in detail and validated the results from experiments on the road using off-the-shelf IEEE 802.11p radios. The results were confirmed by comparisons with measurements collected during two independent measurement campaigns (Kunisch & Pamp 2008, Karedal *et al.* 2011), which used advanced laboratory equipment.

The radio employed was part of the DENSO WSU platform, mounted in the trunk of an Audi A4 Allroad Quattro, configured to send WSMs on the CCH, i.e., at 5.89 GHz, at intervals of 200 ms. On the receiver side, for each packet its timestamp and sender position were logged, as well as the receiver position and the reported dBm value of RSS, and the GPS position. Each car was outfitted with an omnidirectional antenna at a height of 149.5 cm. The measurements are thus not impacted by the high-directionality characteristics of currently proposed, more streamlined antenna configurations (Kwoczek *et al.* 2011).

All the measurements were performed under completely unobstructed channel conditions. In this study, curve-fitting was used iteratively, minimizing the sum of the squared residuals using the Gauss–Newton algorithm to match Equation (6.3). This helped with evaluating the plausibility of the measurements. Sommer *et al.* (2012) found overall a good correlation, which validated the applicability of the discussed path-loss models in this scenario.

Figure 6.38 shows predictions by the analytical model together with the measurement results and a first set of simulation results obtained using the simulation setup described, relative to the maximum recorded RSS. All the results fit quite well. Thus, the model can be used for further evaluations.

Figure 6.39 shows plots for the two-ray interference model as well as for the empirical free-space model given in Equation (6.3), calibrated using different α values. For all models, the analytical prediction is plotted. Simulation results are, for better readability, plotted only for the free-space model for $\alpha = 2.0$. As can be seen, none of the free-space variants closely matches the two-ray interference curve. Even extending simpler models by a path-loss exponent α, as in Equation (6.3), cannot compensate for these errors, further motivating the use of a fully featured two-ray interference model.

Given the findings shown, we argue that using a fully featured and more exact two-ray interference model allows researchers to capture artifacts of strong signal attenuation

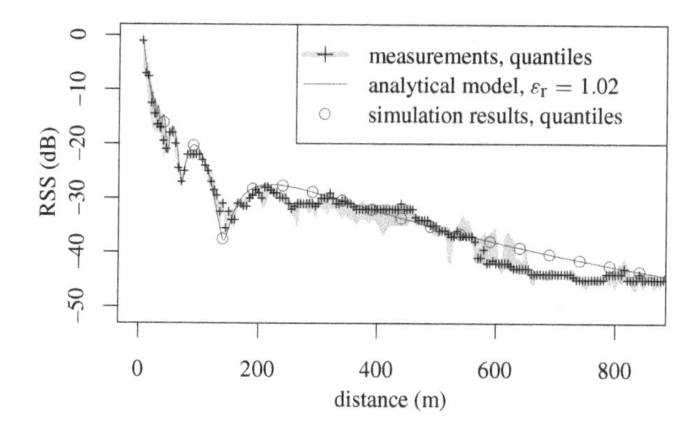

Figure 6.38 Measurement results of received signal strength vs. distance between sender and receiver, overlaid with the calibrated two-ray interference model and simulation results. © [2012] IEEE. Reprinted, with permission, from Sommer *et al.* (2012).

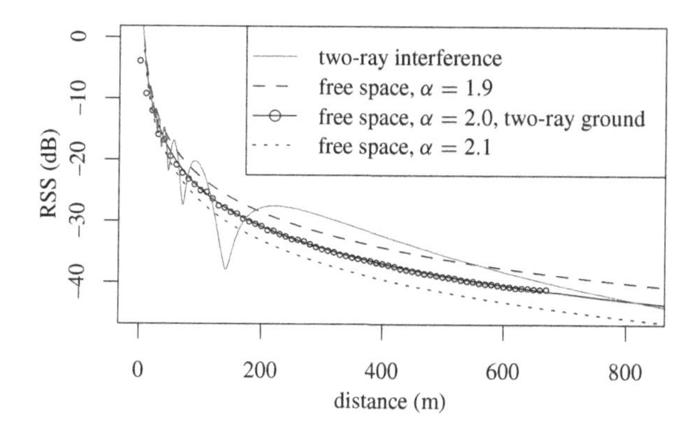

Figure 6.39 Model comparison for the free-space model using different α values and the two-ray interference model. For $d_c < 866.6$ m, the results for $\alpha = 2.0$ and for the simplified two-ray ground model are identical. © [2012] IEEE. Reprinted, with permission, from Sommer *et al.* (2012).

manifesting themselves at short and medium ranges. These artifacts become visible in measurements and field tests and lead to divergent application-layer behavior at different distances. At the same time, on modern hardware, the use of a two-ray interference model (as opposed to a simplified two-ray ground model) comes with only marginal added computational cost for simulation experiments.

The detailed model is applicable in both highway and suburban environments when reflections at obstacles do not dominate radio propagation effects (Kunisch & Pamp 2008, Karedal *et al.* 2011), i.e., where streets are very wide. It does, however, need to be supplemented by additional models that take into account obstacle-based signal-attenuation models (Sommer *et al.* 2011a, Mangel *et al.* 2011b).

Shadowing by buildings and vehicles

We now consider the effects of radio-signal shadowing caused by buildings and other vehicles. Typically used signal-propagation models that assume exponential path loss are clearly not appropriate if scenarios with buildings are to be investigated. Most importantly, the correct and realistic simulation of safety applications relies on an accurate evaluation of topology deficiencies and coverage (Ferreiro-Lage *et al.* 2010, Joerer *et al.* 2014).

In Figure 6.15, we already outlined the main concepts of radio-signal shadowing as realized in state-of-the-art simulation models. A computationally feasible concept takes into account the number of exterior walls, carefully fitted to the material used, as well as the dimensions of the object.

Modeling shadowing by buildings

The need for precise shadowing models has been motivated in a wide body of literature (Otto *et al.* 2009, Dhoutaut *et al.* 2006, Stepanov & Rothermel 2008, Schmitz & Wenig 2006, Giordano *et al.* 2009, 2010). Furthermore, it has been demonstrated that ray-tracing approaches can serve as an excellent approximation – unfortunately at prohibitively high computational cost (Moser *et al.* 2007, Maurer *et al.* 2005, Souley & Cherkaoui 2005). As an example, even medium-scale simulation scenarios based on ray-tracing approaches can require many days for computation on a 50-node PC cluster for a spatial resolution of only 5 m^2 in a 4.56-km^2 scenario (Stepanov & Rothermel 2008). The alternative, stochastically modeling the channel, might lead to severe deviation from realistic behavior of single communications, which is, for example, the case for safety applications.

Sommer *et al.* (2011a) investigated radio shadowing caused by buildings in an empirical study. By consideration of the results of extensive experiments, they developed a novel computationally inexpensive model that takes building geometry and the positions of sender and receiver into account and captures these effects. At roughly the same time, Mangel *et al.* (2011b) performed a similar study and confirmed the core results.

Figure 6.40 outlines the general problems of radio-signal shadowing as well as the key idea employed to approach the problem in a computationally feasible model. The key idea is to count the number of exterior walls of a building to approximate the impact of the radio-signal shadowing caused by exactly this building.

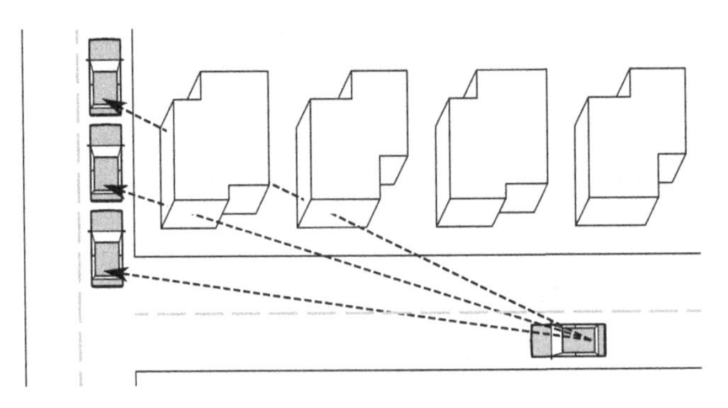

Figure 6.40 Deterioration of the RSS if a transmission is blocked by first one, then two, buildings.

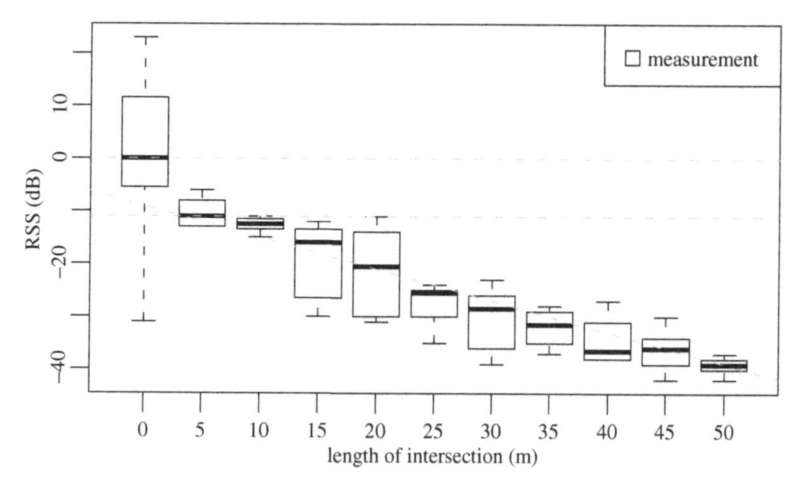

Figure 6.41 The sharp drop and continuous decline in measured RSS as the length of the building penetrated increases. © [2011] IEEE. Reprinted, with permission, from Sommer *et al.* (2011a).

The validity of this idea has been confirmed by the empirical data gathered (see Section 6.1.2 on page 234 for information on the measurement setup and the used data basis). Figure 6.41 shows an example plot of the RSS versus the length of the calculated intersection between the line of sight and a building. It can be observed that RSS values drop sharply as soon as the line of sight is blocked and continue to decrease as the length of the intersection of the line of sight and the building increases. The conclusion is that the signal-shadowing effects can be reproduced only by considering the *attenuation per wall* as well as the *attenuation per meter of penetration*.

Towards a computationally inexpensive model

The model to be developed has to rely on information that is typically available in modern geodata bases. This is, for example, the case for OpenStreetMap, which provides accurate outlines of buildings (describing the number of exterior walls as well as the dimensions of the building) in a 2.5D manner – only height information is available, not exact 3D outlines. The model further considers only the line of sight between sender and receiver but not objects blocking parts of the first Fresnel zone (Rappaport 2009).

Sommer *et al.* (2011a) decided to rely on simple and fast processing steps. The calculation of intersection between all lines of sight and all buildings is the most expensive one. Finding these intersections can easily be supported using caching and binary space-partitioning approaches (Nagel & Eichler 2008) to solve this step in $\mathcal{O}(n^2 \log n)$ time. Depending on the simulation framework, this process can also be treated as a *red and blue line segments* intersection problem, for which algorithms have been proposed that run in $\mathcal{O}(n \log n)$ time (Mantler & Snoeyink 2000).

The model was designed as an extension of well-established fading models discussed on page 274 – this procedure has successfully been used in other models as well (Pecchia *et al.* 2009, Ahmed *et al.* 2010). As has already been discussed, these fading models can be expressed in the form of Equation (6.1), where P are the transmit (or receive) powers of the radios, G are the antenna gains, and L are terms capturing loss effects during transmission. In particular, a new term L_{obs} is used for each obstacle in the line of sight between sender and receiver.

L_{obs} is intended to capture the additional attenuation of a transmission due to an obstacle in terms of the number of times n the border of the obstacle is intersected by the line of sight and the total length d_m of the obstacle's intersection as illustrated in

$$L_{obs}\,[\mathrm{dB}] = \beta n + \gamma d_m \tag{6.10}$$

Here, the parameter β (given in dB per wall) represents the attenuation characteristics of the material of the exterior wall. The parameter γ (given in dB per meter) serves as a rough approximation of the internal structure of a building. This concept allows one to adjust the model for different kinds of buildings.

Model validation

In a first step, Sommer *et al.* (2011a) investigated to what extent the parameters β and γ could be fitted so that analytical results would match up with measured ones. They used parameter fitting by iteratively minimizing the sum of squared residuals using the standard Gauss–Newton algorithm (Bates & Watts 1988) until the algorithm converged.

Figure 6.42 shows the results of this process for a countryside warehouse. Optimal parameter settings were obtained for $\beta = 9.2$ dB per wall and $\gamma = 0.32$ dB/m. As can be seen, the model very accurately matches the measured values. Until index 100, we have line-of-sight communication. Thus, the values follow the free-space model. As the car slowly passes the building, a sharp decrease of RSS values indicates the presence of multiple walls. After index 300, we have line-of-sight communication again.

As a second example, Figure 6.43 shows the results for a multi-building scenario. In this case, a residential house and a separate garage are involved. Here, another term for the second building has been added, with $\beta_1 = 2.38$ and $\gamma_1 = 0.1$, as well as $\beta_2 = 6.26$ and $\gamma_2 = 0.41$ for the home and garage, respectively. The plotted results show that such

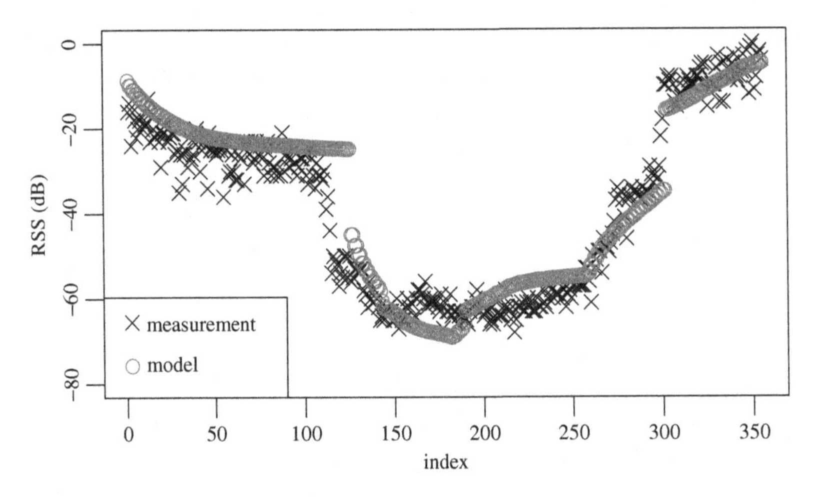

Figure 6.42 Measured and calculated RSS values for a countryside warehouse. © [2011] IEEE. Reprinted, with permission, from Sommer *et al.* (2011a).

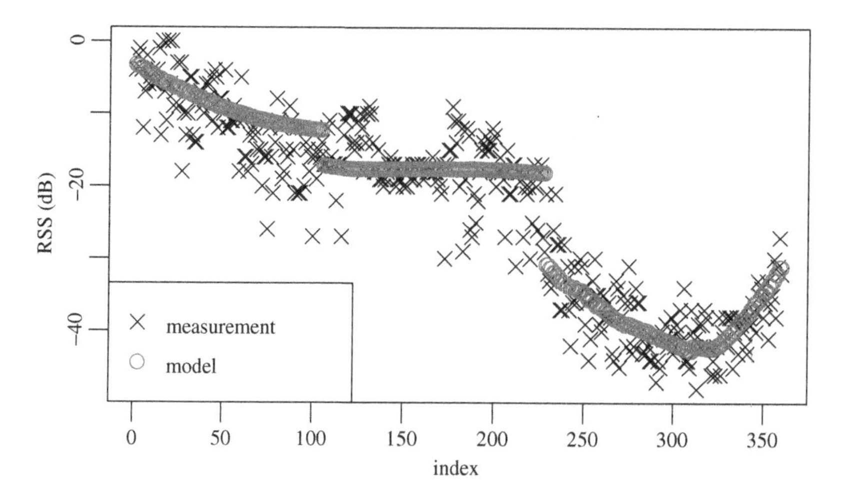

Figure 6.43 Measured and calculated RSS values for an urban residential home and garage. © [2011] IEEE. Reprinted, with permission, from Sommer *et al.* (2011a).

per-building parametrization is indeed feasible and, therefore, heterogeneous scenarios can be supported too.

Shadowing by other vehicles

The impact of radio-signal shadowing caused by other vehicles was first assessed by Boban *et al.* (2011). The aim of their study was to show that the impact of vehicles blocking radio communication is not negligible and requires a careful protocol design for IVC that takes these effects into account.

In a follow-up work, a new protocol taking these findings into consideration that explicitly selects tall vehicles as relays for broadcast-based communication between vehicles (Boban *et al.* 2014) was drafted. Further empirical evidence has been gained in measurement studies such as the one by Segata *et al.* (2013).

In the scope of this section, we discuss a model that extends the one used for signal shadowing by buildings (Sommer *et al.* 2013). The idea is to identify the set of vehicles that intersect the direct line of sight between two communicating vehicles o_{s} and o_{r}. This has been illustrated already in Figure 6.15(b). In essence, all points where this line of sight enters the bounding box of a vehicle are stored as $\{o_1, o_2, \ldots\}$. Just like the problem of finding obstructing buildings, this problem can be cast as a *red–blue intersection* problem, drastically reducing its computational complexity.

The calculation of the power-loss term L_{moving} can then be realized using the techniques described by Boban *et al.* (2014). Signal power loss caused by an individual obstacle in between sender and receiver can be calculated with a *single-knife-edge* approximation according to ITU-R recommendations (ITU-R 2007).

As illustrated in Figure 6.44(a), a geometrical parameter v is used to determine how much of the first Fresnel zone is obstructed by the obstacle. The calculation takes into account the relative height h of the obstacle above ($h > 0$) or below ($h < 0$) the straight line joining sender and receiver. From these items of information together with the given

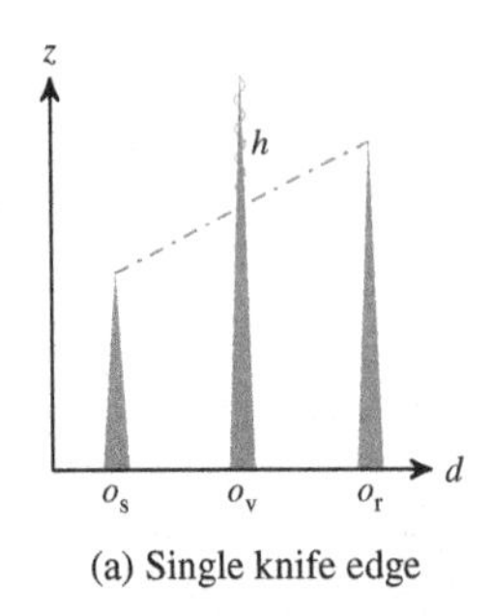

(a) Single knife edge

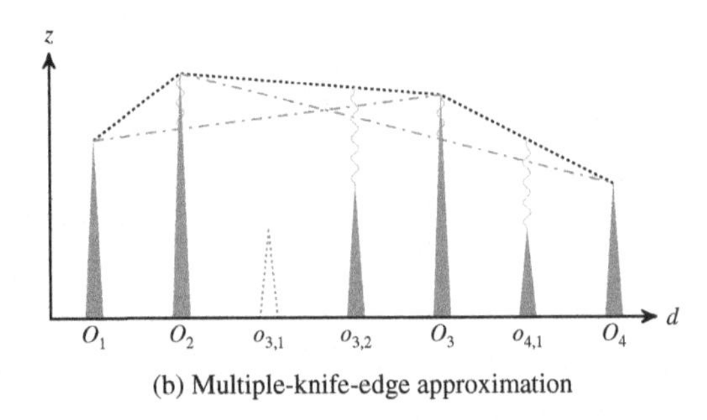

(b) Multiple-knife-edge approximation

Figure 6.44 Approximation of signal power loss by moving vehicles.

distances d_1 and d_2, as well as the wavelength λ, the geometrical parameter is then calculated as

$$v = h\sqrt{\frac{2}{\lambda}\left(\frac{1}{d_1} + \frac{1}{d_2}\right)} \tag{6.11}$$

As discussed by Sommer *et al.* (2013), power loss can be assumed to occur only for a geometrical parameter of $v > -0.7$ (a better fit than the ITU-R recommended empirical value of $v > -0.78$) without reducing the quality of the simulation results. In these cases, the loss is calculated in terms of its dependence on the sender o_s, obstructing vehicle o_v, and receiver o_r as

$$L_{o_s,o_v,o_r}\,[\text{dB}] = 6.9 + 20\log_{10}\left(\sqrt{(v-0.1)^2+1} + v - 0.1\right) \tag{6.12}$$

So far, the technique described can be used to approximate the signal shadowing for single objects. This simple single-knife-edge method can then be generalized to a multiple-knife-edge method using a variant of the Epstein–Peterson method with an applied correction term.

The idea of this approach is as follows: First, *major* obstacles $M = \{O_1, O_2, \ldots\}$ (those touching a virtual rope stretched from the sender, over all obstacles, to the receiver) are identified and ordered by their distance from the transmitter, as illustrated in Figure 6.44(b). This list also includes the sender and the receiver as the first and last members of M, respectively. Obstacles in between major obstacles O_{i-1} and O_i are stored as *minor* obstacles $M'_i = \{o_{i,1}, o_{i,2}, \ldots\}$ – the largest of these is denoted $o_{i,\max}$, because it will cause the most obstruction and will be needed in order to calculate the power loss.

Now, the power loss due to major obstacles L_M as well as the power loss $L_{M'}$ due to minor obstacles for each adjacent pair of major obstacles can be calculated as follows:

$$L_M\,[\text{dB}] = \sum_{i=3}^{|M|} L_{O_{i-2},O_{i-1},O_i}\,[\text{dB}] \tag{6.13}$$

$$L_{M'}\,[\text{dB}] = \sum_{i=2}^{|M|} L_{O_{i-1},o_{i,\max},O_i}\,[\text{dB}] \tag{6.14}$$

with

$$o_{i,\max} = \underset{o_{i,j} \in M'_i}{\arg\max} L_{O_{i-1},o_{i,j},O_i} \text{ [dB]} \tag{6.15}$$

A correction term that takes into account the pairwise distances $S = \{s_1, s_2, \ldots\}$ in between major obstacles is needed. This is calculated as

$$L_c \text{ [dB]} = -10 \log_{10} \left(\frac{\left(\prod_{i=1}^{|S|} s_i\right)\left(\sum_{i=1}^{|S|} s_i\right)}{\left(\prod_{i=2}^{|S|} s_{i-1} + s_i\right) s_1 s_{|S|}} \right) \tag{6.16}$$

Taking into account all these parameters, the compound power loss L_{moving} experienced by moving vehicles can be summarized as the sum of all the terms discussed, i.e., loss due to major obstacles, loss due to minor obstacles, and the correction term taking into account the pairwise distances between the vehicles:

$$L_{\text{moving}} \text{ [dB]} = L_M \text{ [dB]} + L_{M'} \text{ [dB]} + L_c \text{ [dB]} \tag{6.17}$$

6.3.3 Driver behavior

In most cases, the technical factors taken into consideration are not sufficient to globally optimize a vehicular networking application. A major component is missing from many of the technical papers: the interaction of the system with the human driver (Dressler & Sommer 2010, Dressler *et al.* 2011b).

Even though this issue is fundamental to the accurate and realistic performance evaluation of vehicular networking applications, it has usually been assumed that drivers act exactly as expected or suggested by the technical system. On the other hand, the issue of realistic human driver behavior had been considered already in early studies of automotive environments (König *et al.* 1994, Barfield *et al.* 1989, Dingus *et al.* 1996). In all these studies, driver behavior was identified as a key component.

However, this knowledge has seemingly disappeared with the development of more distributed solutions to TISs. Instead of analyzing the likelihood that a driver acts as recommended by a centralized TIS, the focus shifted more towards the vehicular network and the operation and management of the distributed information-sharing systems. Realistic driver behavior has frequently been replaced by (nearly) optimal route planning using integrated maps and the available traffic information. The driver, it has been assumed, exactly follows the suggestions provided by the TIS.

The exact matching of the driver's expected behavior is, however, fundamental to the development of optimized vehicular networking applications. In 2008, even a US patent that describes a method for adaptive navigation using a driver's route knowledge was issued (Cheng *et al.* 2008). Although no specific behavior classes are mentioned, the need to include the driver's behavior has been identified.

In this section, we study driver behavior models available in the literature and their integration into state-of-the-art simulation tools. By reference to a sample study, we demonstrate the huge impact of the model selected on the application performance.

Psychological studies

The impact of the human drivers' behavior has been studied in the context of traffic information systems since the early days of those applications. Some of the most comprehensive psycho-physiological studies were performed in the late 1980s or early 1990s. For example, König *et al.* (1994) developed a model of a driver's behavior using artificial intelligence techniques for determining the driver's route planning.

Essentially, the resulting model is based on four components: the driving behavior, which is based on the driver's experience, age, and sex, but also on the level of aggressiveness; the personal preferences of the driver, e.g., whether to believe in the technical system and accommodate changes to the selected route; the motivation to follow recommendations, i.e., the likelihood of observing and reacting upon received messages; and finally local knowledge, which may strongly influence the decision process. Some of these findings are already part of commercially available navigation systems, e.g., Navigon supports personalized navigation that is based on the driving habits in their MyRoutes system.

Dingus *et al.* (1996) conducted a comprehensive literature study to provide guidelines for advanced traveler information systems and commercial vehicle operation. For example, they found that a substantial percentage of drivers tends to ignore re-routing suggestions to avoid road traffic congestions if they are on their preferred route.

The report also summarizes driver classes that had been identified in earlier work by Barfield *et al.* (1989) and confirmed by Wenger *et al.* (1990), who studied the behavior of several types of commuters:

- Route changers – drivers who are willing to change both the time and the route of the tour depending on traffic information
- Non-changers – people who are absolutely unwilling to change their route
- Pre-trip changers – drivers who are willing to change their route before leaving the house
- In-trip changers – those who are willing to change only just before entering a possibly congested highway

Experimental evidence has been sought to further investigate drivers' behavior. For example, in a more psychologically oriented study, dependences between drivers' stress level and their reaction to congestion information have been revealed (Uang & Hwang 2003). This study focused on techniques of presentation of congestion information to the driver. Other studies concentrated on the impact of traffic lights and intersection management on the driver's route choice (Shenpei & Xinping 2008, Liu & Ozguner 2007).

Many of the findings have been summarized by Cacciabue (2007). This is also a great reference on psychological models that influence the driver's safety, such as the use of entertainment-related devices and the interaction with the navigation system.

Behavior classes

In the following, we summarize selected behavior classes. Using cluster analysis techniques, it was possible to show that the following four commuter subgroups exist with respect to their willingness to respond to the delivery of real-time traffic information (Dingus *et al.* 1996).

- *Always* – This class summarizes all drivers acting exactly as expected. They perform the necessary actions as proposed by, e.g., the navigation unit. This is usually the only class assumed in most studies of vehicular networking applications.
- *Never* – All drivers belonging to this class continue their everyday procedure and completely ignore the recommendations and suggestions. This class must be clearly distinguished from the frequently used penetration rate as a metric: Even though these drivers do not follow any advice, their cars certainly take part in the distributed vehicular networking application.
- $d < D$ – The study considered commuters only. This class represents all drivers who consider road traffic congestion within a certain range $d < D$ as relevant (with d representing the distance to the event and D the threshold distance). All events beyond that distance D are considered as not relevant. The drivers simply assume that for obstructions that are further away there will be enough time for the congestion to clear before they get there.
- $d > D$ – This fourth class summarizes all drivers who consider early re-routings as most useful to bypass severe road traffic congestion. This helps to prevent secondary micro-jams that are likely to be created if all drivers try to engage in detours at short notice.

Dingus *et al.* (1996) further studied a mix of all these classes to represent realistic driver behavior. They came up with a statistical distribution, which helps to implement these classes even though one is considering a statistical sample only, which represents typical drivers in a very specific region or country.

All the classes are summarized in Table 6.2. The table also includes a fifth class, which is a *probabilistic* model that selects the driver to follow either the *always* or the *never* model at the time of starting the journey, according to a given value of P.

Furthermore, Table 6.2 outlines combinations. The class *mix* is a representation of the driver model by Dingus *et al.* (1996), using exactly the recommended percentages. The class *all* is a combination of all the simple classes together with the probabilistic decision.

Table 6.2 Classes of human drivers' behavior

Class	Description	Mix	All
Always	Route selection according to TIS recommendations	40.1%	20%
Never	Drivers unwilling to change their route	23.4%	20%
$d < D$	Route changes only if the distance to the congestion is less than D	20.6%	20%
$d > D$	Route changes only if the distance to the congestion is larger than D	15.9%	20%
P	Probabilistic decision on whether to fall into class *always* or class *never*	0%	20%

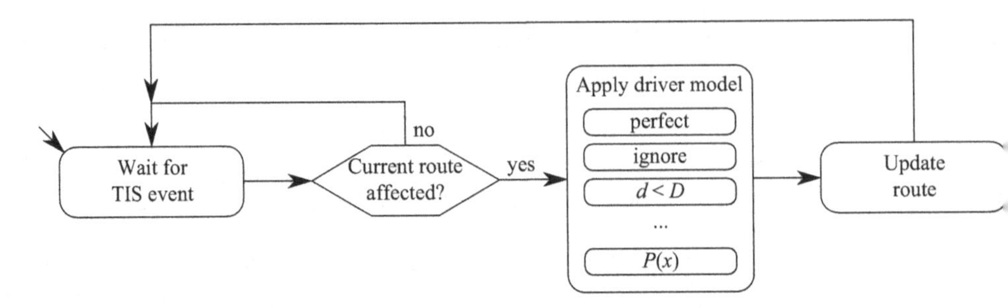

Figure 6.45 Integration of the driver behavior classes in the Veins simulator.

This model considering all these classes has been implemented within the Veins simulation framework (see Section 6.2.3 on page 262) (Dressler & Sommer 2010, Dressler *et al.* 2011b). Figure 6.45 outlines the necessary procedures. After receiving a TIS event, the system checks the driver's class and whether this class allows her or him to react to the received TIS message. If the driver model allows route recalculation, a command is issued to recalculate it using the information received. The distribution and exchange of TIS data with other cars is not affected.

The model is quite flexible in its function. As soon as new empirical data about the driver's behavior and its interaction with the TIS becomes available, the calibration can be updated without major modifications of the model.

Performance impact

The results for our discussion of the performance impact of the different driver models are based on simulation experiments for an urban scenario (Dressler & Sommer 2010). Cars were configured to drive through the city of Erlangen, Germany, starting at the CS department and heading for the city center. An artificial accident was scheduled in order to study the behavior of following cars in the case of road congestion. For TIS data exchange, a simple broadcast-based IVC protocol was used for exchanging traffic information.

All the driver behavior classes discussed were studied. The distance-based decisions were analyzed with two reasonable threshold distances of $D_1 = 1$ km and $D_2 = 2$ km. Also, for the probabilistic classes, two probabilities for a vehicle to fall into class *always* were examined: $P_1 = 0.5$ and $P_2 = 0.7$. These probabilities were derived from the results in empirical studies using different scenarios and simulation setups.

As the key metric, the travel time represented by the effective average speed of the vehicles was chosen. This measure is frequently used to show the capabilities of TIS applications and IVC protocols.

Figure 6.46 shows the statistical analysis of the impact different driver models have on the travel time of vehicles (normalized using the distance along the shortest route to derive an effective average speed). As a baseline measure, we also examine an accident-*free* scenario. We present the results in the form of boxplots (Figure 6.46(a)), indicating the median and the quartiles of all the measurements. Furthermore, the mean is plotted as a small box. Because the distribution of measurements is, by nature, multi-modal, we

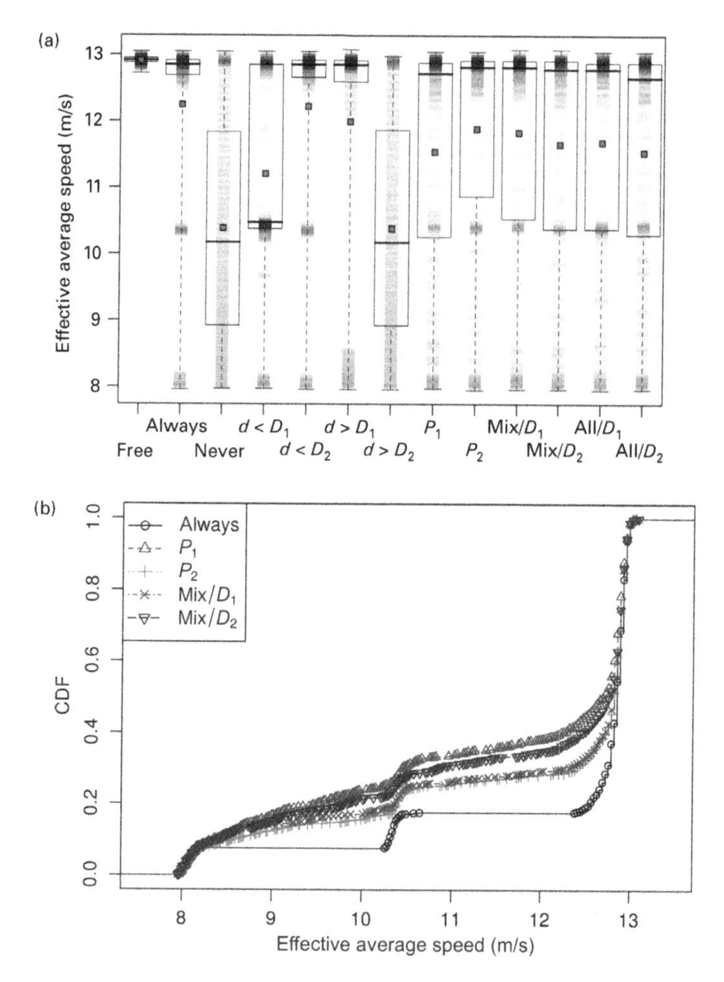

Figure 6.46 Impact analysis of the driver behavior models for a TIS scenario, for the effective average speed of all cars: (a) a boxplot including density information and (b) the CDF of the effective average speed. © [2010] IEEE. Reprinted, with permission, from Dressler & Sommer (2010).

also display individual measurements, using light gray lines; thus, dark zones represent a significant number of cars in the same speed range. Since the graph is difficult to read, we also plot selected classes in the form of a cumulative distribution function (CDF) in Figure 6.46(b).

As the most obvious result, it can be seen that all the different models lead to completely different overall behaviors. Furthermore, as a second outcome, we observe that the mix according to Dingus *et al.* (1996), labeled as "Mix", can be closely approximated using the probabilistic model P.

The results presented already demonstrate that the drivers' behavior must be accommodated in ITS performance evaluations in order to obtain realistic results. To gain further evidence, we present results from a second experiment. In this second scenario,

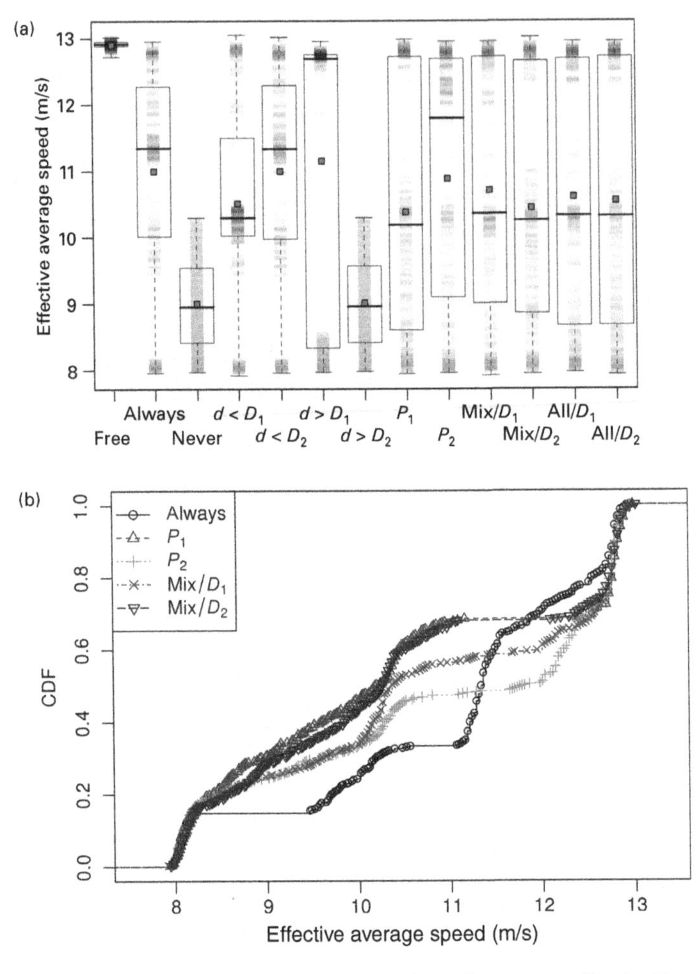

Figure 6.47 A more complex scenario also including cross traffic, leading to additional congestion and micro-jams: (a) a boxplot including density information and (b) the CDF of the effective average speed. © [2010] IEEE. Reprinted, with permission, from Dressler & Sommer (2010).

additional cross traffic that interacts with the initial vehicle flow in the scenario was introduced, resulting in a large number of micro-jams.

The results are shown in Figure 6.47. Again, most of the trends and effects that have been discussed can be confirmed. The probabilistic model P_2 does not match as closely as shown for the other measurements. However, as can be seen from the CDF, the number of non-matching measurements is extremely small.

6.3.4 Metrics

Besides the choice of the right simulation tools and models, the question of how to finally assess the performance of a vehicular networking application or protocol is the most complicated one. The choice of the metrics according to which the algorithms developed are to be evaluated is, as is usual in simulation, one of the key questions.

In network simulation, mostly classical networking related metrics are used, such as delays, throughput, and the reliability of the system. For wireless communication, we tend also to look at metrics such as the number of collisions or the load of the wireless channel.

Still, these metrics are able to reflect only parts of the performance of vehicular networking applications. In most cases, application-dependent metrics such as road traffic safety, e.g., the probability of a collision during an intersection maneuver, the travel time, and the CO_2 impact of the vehicle in different maneuvers might be even more important.

In the following, we briefly review the classical network simulation metrics and study selected IVC application-dependent metrics in more detail.

Classical network-simulation metrics

Classical networking metrics that are used in network simulation to assess the performance of a protocol are the end-to-end delay, queuing delays, the overall throughput or just the goodput on an application layer, the reliability of a transmission, and the capacity of the wireless channel, which, if exceeded, may lead to a substantial number of collisions. This list is, of course, far from being complete. We briefly revisit selected metrics in the following, but refer the reader to textbooks focused on network simulation for more detailed discussions (Law 2006, Wehrle *et al.* 2010).

Capacity, collisions, and throughput

Whenever we design new protocols for use in wireless networks, the network capacity is one of the most important metrics (Gupta & Kumar 2000, Scheuermann *et al.* 2009). This measure allows one to identify whether some application can be run in theory, given the fundamental capacity limits of the wireless channel.

The network capacity is often referred to as indicating how much data can theoretically transfered. The throughput, typically given in Mbit/s, however, shows how much data can actually be transfered. In the literature, we also find the term goodput, indicating the throughput as observed from an application's point of view.

In simulations, the available network capacity is often assessed in the form of the number of collisions, i.e., describing how overloaded the channel is, or in the form of an overall throughput that can be achieved, i.e., how much user data can successfully be transmitted. Collisions occur if two radio packets happen to be received simultaneously at one station, with the signals corrupting each other. Collisions are very difficult to distinguish from other transmission failures due to background noise. Still, especially in simulation, this number helps one to identify overload situations.

Delay

The authors of many studies rely first and foremost on the transmission delay as a key metric. This is especially relevant for safety-critical applications. The delay is usually measured in s or ms. In most cases, we are interested in the end-to-end delay, i.e., the time from creating a message until it is finally received by the destination node. Depending on the application, the per-hop delays might be of interest, e.g., for broadcast-based message dissemination with relaying.

The delay can easily be measured and used for detailed performance analysis in simulation when we consider unicast transmissions. However, as we saw in the context of our discussion on information dissemination in Chapter 5, broadcasting is a key technology in vehicular networks using IEEE 802.11p. Measuring the delay is still straightforward, yet the evaluation becomes more complex. Either all the messages received are treated as multiple messages representing multiple delay measures, or a single transmission is observed, not showing a particular value but rather indicating the statistics of this transmission.

Reliability

Orthogonally to the aforementioned metrics, the reliability describes whether and to what extent a data transmission is expected to succeed. In many cases, the obvious demand for reliable communication protocols inherently conflicts with the need for low delay and high throughput. So, protocols need to be carefully adapted to meet all or most of the application requirements. Reliability can be estimated in simulation by assessing the number of failures and then establishing a reliability metric.

Discussion

In order to discuss the use of the different metrics and also to outline frequently observed problems, we reiterate selected simulation results presented in Section 5.4.1 on page 175. The measurements were made during a simulation study of the adaptive traffic beacon (ATB) protocol. The specific protocol and simulation scenario used, however, are not relevant for our discussion.

Figure 6.48 shows two graphs, one indicating the message delay for multiple variants of static beaconing as well as for ATB and one showing the number of collisions for the same protocols. The absolute values are of course not part of our discussion. First and foremost, we need to remind ourselves that delays can obviously be measured only for successfully received messages, thus the two graphs show results for different subsets of the messages submitted.

The delays as shown in Figure 6.48(a) outline a clear trend: The more frequently a message is repeated, the faster it can be delivered. It can be seen that static beaconing using intervals of 30 ms performs best. ATB seems to be even faster in this experiment. We have to recall the key concept of ATB at this stage: ATB schedules beacons between some minimum and some maximum interval. The configured minimum was 30 ms. So, how can ATB be faster than static 30-ms beaconing intervals?

The answer is provided in Figure 6.48(b). Here, the number of collisions is plotted on a logarithmic scale. It can be seen that the number of collisions grows exponentially with increasing beacon frequency, i.e., the reduced beacon interval. For ATB, however, the number of collisions is negligible. That means that each and every beacon can immediately be received and relayed, reducing the overall delay.

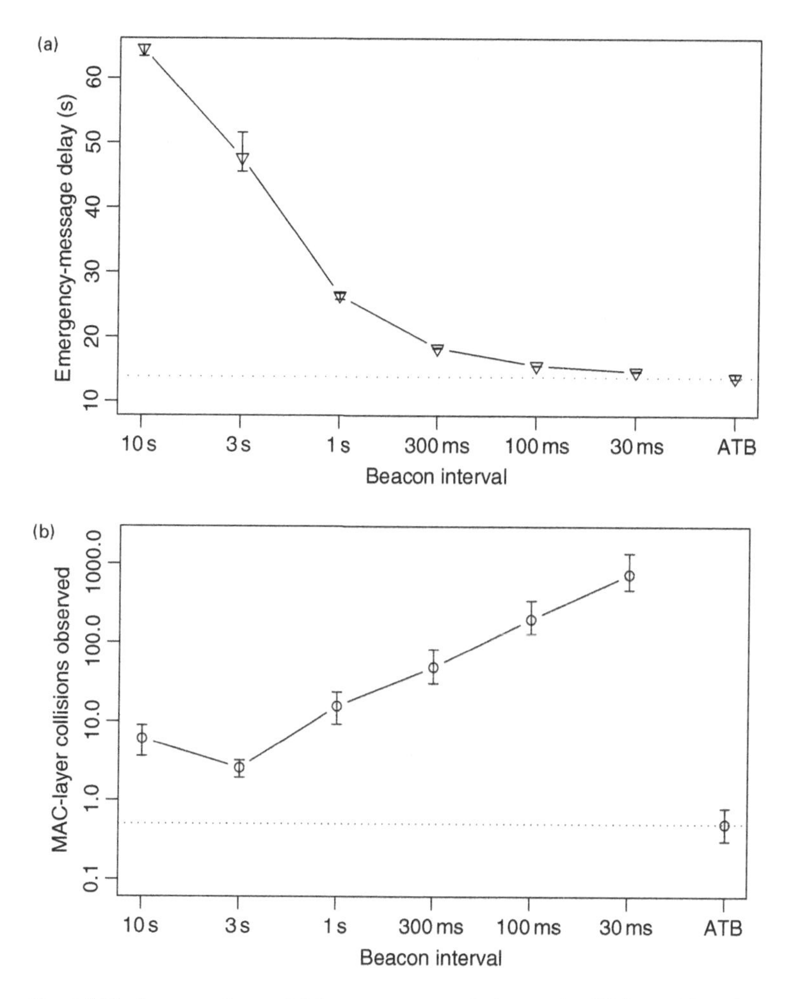

Figure 6.48 A comparison of delay measures and observed collisions: (a) end-to-end delay and (b) MAC-layer collisions. See also Section 5.4.1 on page 175. © [2011] IEEE. Reprinted, with permission, from Sommer *et al.* (2011c).

The chosen example outlines two very important messages.

- First, a single metric alone usually does not clearly outline the behavior of a protocol or application. Only in combination can the performance be assessed in a holistic way.
- Second, metrics as recorded in simulation experiments might even be misleading if they are not evaluated carefully. Unrealistic results need to be carefully investigated and assessed by looking at the results from different angles.

The same holds for the other performance aspects discussed. It is trivial to modify any protocol to increase its throughput or decrease its delays. Doing this without increasing the load, though, is often impossibly hard. Similarly, it is easy to imagine that, e.g., by aggregating information, a protocol's throughput can be increased and its load can

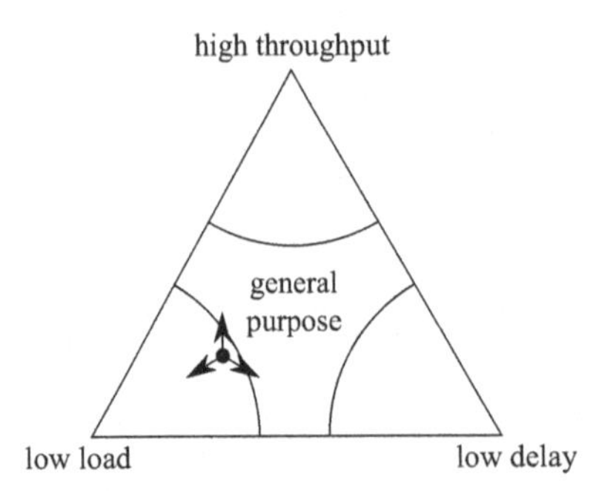

Figure 6.49 Trade-offs in optimization. Optimizing for just one metric and ignoring the effects this has on related metrics is meaningless.

be decreased, but not without incurring additional delays. We illustrate this trade-off for three arbitrary metrics in Figure 6.49. It is clear that any comparison of vehicular networking approaches must always be done considering not just one dimension but all affected dimensions to be meaningful.

Application-dependent metrics

Besides use of the more classical networking-related metrics, the performance and quality of many IVC applications and protocols can be much better assessed using more application-oriented metrics. It is for many applications not really relevant how loaded the channel is or whether the delay varies between certain values (except, of course, if real-time communication thresholds are exceeded).

In the following, we discuss two selected examples of such application-dependent metrics. These are, of course, to be treated as examples for very specific applications. Still, these examples give a good idea of why such metrics need to be considered.

The first metric is the *collision probability*, which is of concern, for example, for ICWSs. This metric defines how critical a situation is, allowing one to prioritize messages and even automated actions to prevent crashes.

The second metric is twofold. The *travel time* describes the time it takes for a vehicle to travel from origin to destination. Hence, it microscopically models the benefits an individual car can gain from using, for example, a TIS. On the other hand, the *CO_2 emission* characterizes the smoothness of a trip, which is also of interest only for the individual vehicle but can also be used to describe the impact on the environment in general.

Vehicle collision probability

As a first metric, let's investigate a measure for road traffic safety, the *vehicle collision probability* (as opposed to the probability of packet collisions). Correct use of this metric is of utmost importance for ICWSs, for example (Joerer *et al.* 2014). Thus, our focus is on classifying situations' criticality, and providing sound building blocks for research on safety-enhancement systems through IVC. Specific crash-avoidance techniques or

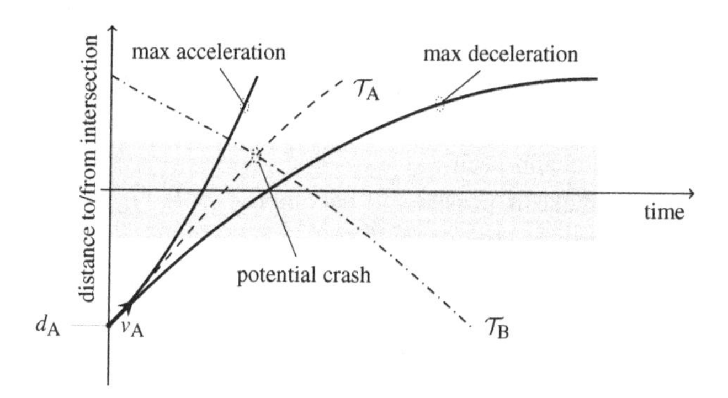

Figure 6.50 An illustration of trajectories: All possible trajectories $\mathbb{T}_A$ of car A lie somewhere between the one assuming maximum acceleration and that assuming maximum deceleration. A sample trajectory $\mathcal{T}_A$ and its intersection with another car's trajectory $\mathcal{T}_B$ are also shown.

impact-reduction strategies are not part of the metric, but the metric can merely be used to investigate the performance of such applications.

In conclusion, a metric is needed for collision-prevention applications on account of helping to better understand the risk of collisions at a road junction. Of course, false alarms should be kept to a low value when it comes to implementing correct countermeasures.

Joerer *et al.* (2014) classified the probability of a collision in terms of the likelihood of trajectories. Given a car A, its current distance from the junction center d_A, and its current speed v_A, they define any feasible extrapolation of its position as a potential trajectory $\mathcal{T}_A$. Of course, the feasibility of $\mathcal{T}_A$ is subject to the car's maximum acceleration a_{max} and its minimum acceleration (that is, its maximum deceleration) a_{min}. More formally, $\mathcal{T}_A$ is subject to

$$\mathcal{T}_A(t_0) = d_A, \quad \dot{\mathcal{T}}_A(t_0) = v_A, \quad a_{min} \leq \ddot{\mathcal{T}}_A(t) \leq a_{max} \tag{6.18}$$

with t_0 being the current time. Figure 6.50 illustrates these relations.

For predicting the probability of collisions, a function $\text{coll}(\mathcal{T}_A, \mathcal{T}_B)$ is defined that will, for any two trajectories $\mathcal{T}_A$ and $\mathcal{T}_B$, yield 1 if (and only if) these two trajectories lead to a crash on the road junction. A crash is defined as the outline of cars on these trajectories overlapping.

With the set of all potential trajectories defined as $\mathbb{T}_A = \bigcup \mathcal{T}_A$ and the probability of choosing two trajectories defined as $p(\mathcal{T}_A, \mathcal{T}_B)$ this yields a straightforward way to derive the crash probability as the integral over all potential trajectories

$$\mathcal{P}_C = \int_{\mathbb{T}_B} \int_{\mathbb{T}_A} p(\mathcal{T}_A, \mathcal{T}_B) \text{coll}(\mathcal{T}_A, \mathcal{T}_B) \mathrm{d}\mathcal{T}_A \, \mathrm{d}\mathcal{T}_B \tag{6.19}$$

Naturally, this approach hinges on being able to calculate $p(\mathcal{T}_A, \mathcal{T}_B)$, which is not straightforward. However, with a few simplifications the problem becomes tractable.

As a first simplification, the authors restricted the investigation to orthogonal X-junction crossings without turning maneuvers. Second, the probabilities for trajectories $\mathcal{T}_A$ and $\mathcal{T}_B$ being chosen were assumed to be independent – that is, neither the driver nor the vehicle is assumed to be able to react to the other vehicle approaching. Finally, only trajectories with a constant acceleration are investigated; that is, a driver is assumed to either accelerate or decelerate at a certain rate without altering their behavior during the junction approach.

These simplifications make it possible to decompose $p(\mathcal{T}_A, \mathcal{T}_B)$ into independent parts $p(\mathcal{T}_A)$ and $p(\mathcal{T}_B)$. Further, a trajectory $\mathcal{T}$ and, consequently, $p(\cdot)$ and coll $(\cdot, \cdot)$ can now be based on merely a tuple $(a, v, d, a_{\min}, a_{\max})$. Of this tuple, all but the assumed acceleration a of the vehicle are constant for a given situation. Thus, the probability of a crash, given a particular situation, is now dependent only on how likely it is that two drivers choose a particular pair of accelerations $p(a_A)$ and $p(a_B)$.

Joerer *et al.* (2014) investigated two simple example distributions for the probability of a driver choosing a certain acceleration $p(a)$: uniform and triangular. For the uniform distribution any acceleration from $a_{\min}$ to $a_{\max}$ was considered equally likely – irrespective of the current speed and acceleration. For the triangular distribution this is adapted, so that the current acceleration is most probably maintained and the probability of choosing different accelerations decreases linearly to 0 for $a_{\min}$ and $a_{\max}$. In the following, we employ this triangular distribution as the basis for predicting drivers' behavior.

Taken together, it becomes possible to calculate the crash probability using the much simpler expression

$$\mathcal{P}_C = \int_{a_{\min}}^{a_{\max}} p(a_B) \int_{a_{\min}}^{a_{\max}} p(a_A)\text{coll}(a_A, a_B)\mathrm{d}a_A\, \mathrm{d}a_B \tag{6.20}$$

Let's now have a look at the applicability of this collision probability. Here, we rely again on the study by Joerer *et al.* (2014).

For this study, a typical suburban X junction where two roads cross each other almost orthogonally was simulated. No measures for controlling the right of way (i.e., neither a traffic light nor a yield sign) were assumed to be present. In particular, the simulator Veins was used, together with a modified version of SUMO, which supports the simulation of real crash situations (Joerer *et al.* 2012b). Following the assumptions in the derivation of $\mathcal{P}_C$, only pairs of cars that cross the junction without turning were considered. For the calculation of the probabilities, $a_{\max} = 2.1\,\text{m/s}^2$ and $a_{\min} = -9.55\,\text{m/s}^2$ were used as conservative parameters.

If we review some general statistics describing the simulations, out of all simulated junction approaches 3.76% resulted in a crash, 0.84% almost resulted in crashes (vehicles coming closer than 0.4 m), and 95.4% resulted in no crash. Reflecting on how well the calculated $\mathcal{P}_C$ predicts future crashes, the authors reported finding no false negatives (for any run where two vehicles crashed a collision probability of 100% was predicted at some point in time) and no false positives (any run where no crash occurred yielded collision probabilities of no more than 40%) when perfect data was available to the vehicles. This allows hard thresholds to be set for predicting whether a crash is likely to occur (again, given perfect knowledge of the current situation).

Much more interesting to us, however, is to examine what happens if only imperfect knowledge is available to the vehicles, i.e., if they need to rely on CAMs for learning

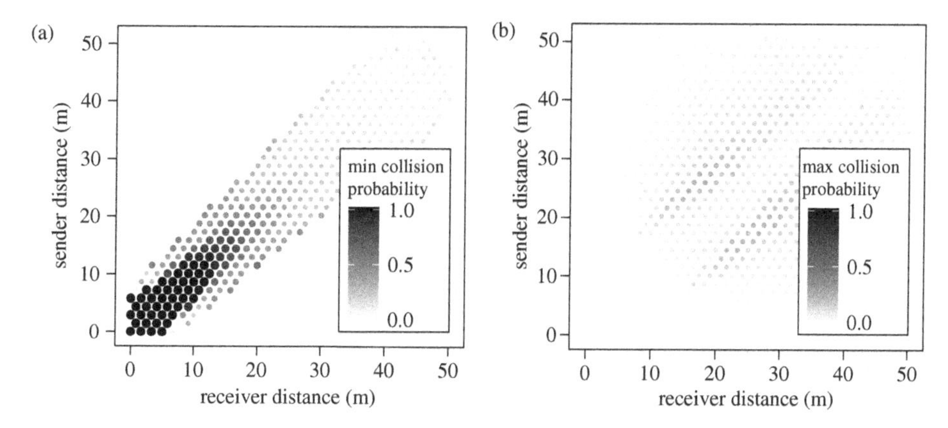

Figure 6.51 Worst-case values of the collision probability for different positions of the two cars relative to the junction center. The minimum value for runs that ended in a crash (a) and the maximum value for runs that did not end in a crash (b) are shown.

about other vehicles' positions and speeds. Information obtained via CAMs will always be outdated, since these beacons take some time to send, are sent only after a preconfigured interval and, even more so, as beacons can get lost (e.g., due to radio-signal shadowing or channel overload).

To get an idea of the feasibility of $\mathcal{P}_C$ for predicting collisions in this context, we examine its worst-case performance. For this, we collect the minimum and maximum values of $\mathcal{P}_C$ for all junction approaches, grouped by whether the approach ended in a collision or not and further grouped by the relative distances of both cars from the junction center. For a beaconing rate of 10 Hz we plot the lowest value of $\mathcal{P}_C$ attained in runs that ended in a crash in Figure 6.51(a) and the highest value of $\mathcal{P}_C$ during runs that did not end in a crash in Figure 6.51(b). As we can see from the figure, $\mathcal{P}_C$ shows only weak predictive properties when both vehicles are still a good way from the junction center. Yet, close to the intersection even the lowest value of $\mathcal{P}_C$ is still very close to 100% in runs that ended in a crash. Similarly, even the highest value of $\mathcal{P}_C$ in runs that did not end in a crash remains well below 10% if cars are close to the intersection. Thus, it will be worthwhile to investigate whether $\mathcal{P}_C$ can also be used as an input to a collision-avoidance system – that is, whether it will be possible to warn drivers before $\mathcal{P}_C$ reaches 100% and a crash becomes unavoidable.

Naturally, the collision probability can be updated only when beacons are received. This means that the evolution of $\mathcal{P}_C$ during a junction approach will follow a step function, as illustrated in Figure 6.52. Now, assuming that a crash can be prevented only if action is taken before the collision probability reaches 100%, it is possible to pinpoint the latest possible time to take such action: at the last received beacon before $\mathcal{P}_C = 1$.

Knowing the collision probability $\mathcal{P}_C$ computed after this update, it is now straightforward to derive what reaction threshold would need to be set in order to prevent or mitigate crashes with a certain probability.

Joerer *et al.* (2014) reported finding a very strong dependence of this threshold on the chosen static beaconing interval of CAMs: For the given driver behavior model, and targeting a rate of successfully avoiding 99% of crashes, they found a threshold of

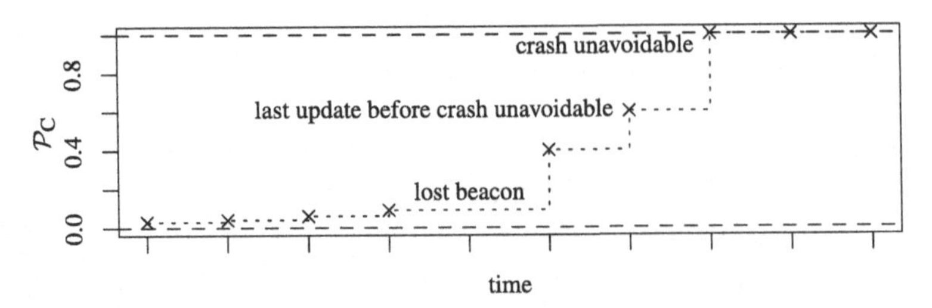

Figure 6.52 The evolution of the collision-probability metric during a typical junction approach leading up to a crash.

$\mathcal{P}_C = 21\%$ for beaconing at 1 Hz. Thus, it can be concluded that beaconing at 1 Hz (an exaggerated lowest beaconing rate) is clearly inapplicable to ICWS because of the associated false-positive rate. On the other end of the scale, for the highest investigated beaconing rate of 25 Hz, the ICWS would clearly work very well: Here, a reaction threshold of $\mathcal{P}_C = 98.5\%$ was reported.

Travel time vs. CO_2 emission

Let's now discuss the difference between *travel time* and *CO_2 emission* as key metrics for performance evaluation of traffic efficiency applications. Besides the more classical travel-time metric, the average speed of the vehicles and the variation of the speed, i.e., the "smoothness" of the trip, provide more insights into the application's behavior. This can, for example, also be expressed by means of the environmental impact such as the CO_2 emission.

Many simulation studies (Gradinescu *et al.* 2007, Collins & Muntean 2008, Nagurney *et al.* 2008) estimating the environmental impact of road traffic management are based directly or indirectly on the 1985 report by Bowyer *et al.* (1985).

Even more accurate simulation is possible relying on the more recent EMIT model (Cappiello *et al.* 2002). The EMIT model has been calibrated for a wide range of different emissions, viz. CO_2, CO, hydrocarbon (HC), and nitrous oxide (NO_x), and thus calculates very precisely both for accelerating and for decelerating vehicles the emissions after the exhaust gases have been passed through a catalytic converter, which is assumed to have reached operating temperature.

EMIT calculates emissions depending on vehicle speed and acceleration, taking into account vehicle characteristics such as the total mass, the engine, and the catalytic converter installed. First, the tractive power requirement at a vehicle's wheels P_{tract} is calculated using the following polynomial:

$$P_{\text{tract}} = Av + Bv^2 + Cv^3 + Mav + Mgv \sin \vartheta \tag{6.21}$$

From the tractive power requirement, the gas consumption can be estimated and, consequently, tailpipe emissions of CO_2 calculated according to a second polynomial:

$$\text{TP}_{\text{CO}_2} = \begin{cases} \alpha + \beta v + \delta v^3 + \zeta av & \text{if } P_{\text{tract}} > 0 \\ \alpha' & \text{else} \end{cases} \tag{6.22}$$

Values of α to ζ, A to C, and M were fitted to match what Cappiello *et al.* (2002) termed a *category-9 vehicle*, e.g., a '94 Dodge Spirit. The values used (see Table 6.3) result in an error in CO_2 emission calculations of approximately 2.2 %.

The EMIT model has been implemented within the Veins simulation framework (Sommer *et al.* 2010c, Dressler *et al.* 2011b). Because the road grade is not currently modeled in SUMO, the P_{tract} calculations assumed planar roads and, hence, $\vartheta = 0$.

The scenario for the experiment is a single-lane trunk road with a speed limit of approximately 28 m/s (100 km/h) supported by two parallel streets with speed limits of 22 m/s, all connected in the form of a ladder. An artificial incident has been introduced on the trunk. IVC takes place between the cars in order for them to inform each other

Table 6.3 EMIT factors for a category-9 vehicle

Factor		Value	Unit
v	Vehicle speed		m/s
a	Vehicle acceleration		m/s^2
A	Rolling resistance	0.1326	kW s/m
B	Speed correction to rolling resistance	2.7384×10^{-3}	$kW\,s^2/m^2$
C	Air-drag resistance	1.0843×10^{-3}	$kW\,s^3/m^3$
M	Vehicle mass	1.3250×10^{3}	kg
g	Gravitational constant	9.81	m/s^2
ϑ	Road grade	$0°$	
α		1.1100	g/s
β		0.0134	g/m
δ		1.9800×10^{-6}	$g\,s^2/m^3$
ζ		0.2410	$g\,s^2/m^2$
α'		0.9730	g/s

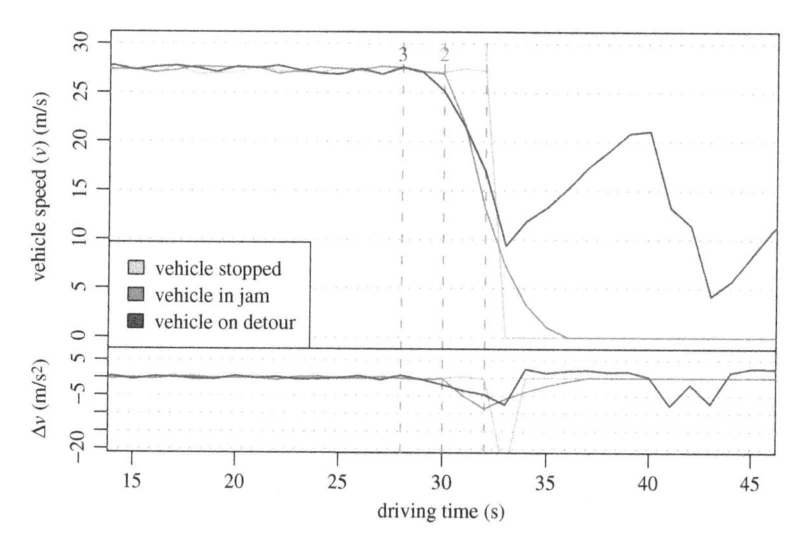

Figure 6.53 The impact of route choice on vehicle speed and acceleration profile. © [2011] IEEE. Reprinted, with permission, from Dressler *et al.* (2011b).

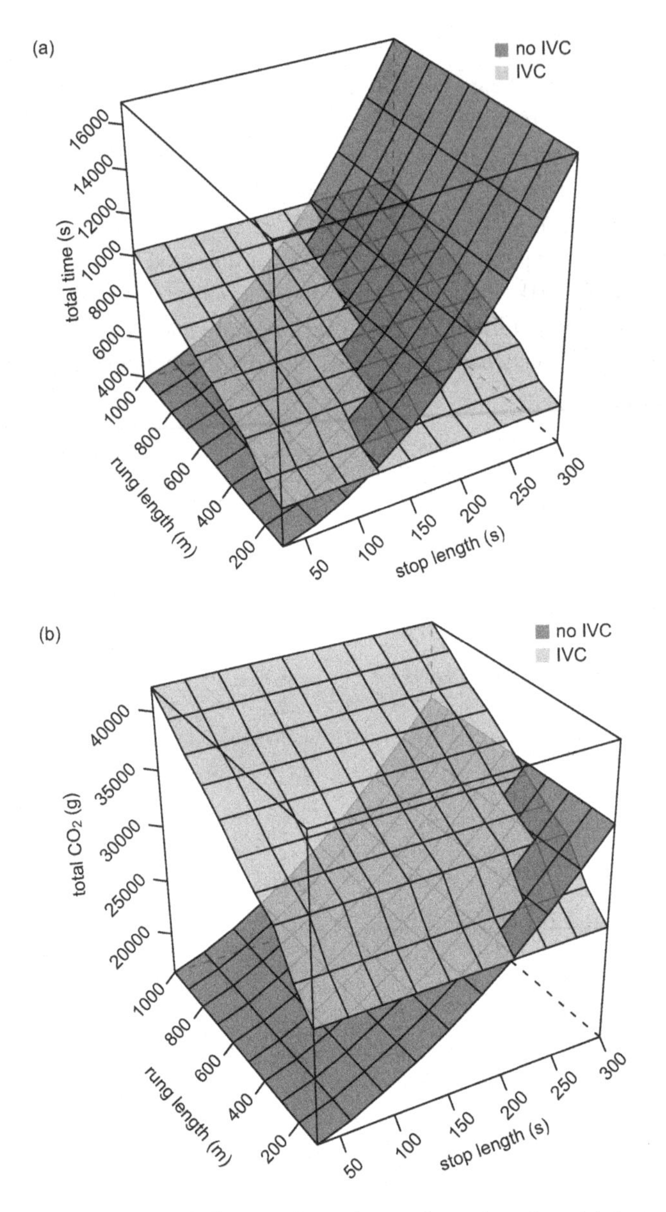

Figure 6.54 The influence of stop time and rung length on (a) the travel time and (b) CO_2 emissions. © [2011] IEEE. Reprinted, with permission, from Dressler *et al.* (2011b).

about the blocked trunk road. If such a message successfully reaches a following car, it recalculates its path using one of the parallel streets if possible. During the experiment, the length of the stop and the length of the detour were modified in order to evaluate the appropriateness of the route recalculation with regard to the two metrics selected.

Figure 6.53 shows a speed/acceleration profile for three different cars: a stopping car, a car caught in the resulting jam, and one taking a detour that is based on IVC-based traffic information. It can be seen that the necessary accelerations and decelerations for the detour are not negligible: In order to be able to yield to through traffic, each vehicle will have to brake slightly when leaving the rungs to and from the detour (this effect is visible in the figure as two pronounced drops in speed).

Let's have a first look at the experimental results. Figure 6.54 depicts the measurement results for changing stop times and detour lengths. It plots the cumulative driving time, as well as the cumulative CO_2 emission, of all simulated vehicles. As a baseline, the travel time and CO_2 emission were measured with no IVC enabled.

As was to be expected, with IVC enabled stop times had hardly any influence on both metrics. If the stop is too short, the use of a detour will not be helpful. Thus, the travel time and the CO_2 emissions are optimized by IVC only for a certain minimum stop time (i.e., accident time).

The most interesting result, however, is that the break-even point is different for these two metrics. The travel time can quickly be optimized even for comparably short stop times. In contrast, the CO_2 emissions are suboptimal at this point. This is due to the additional gas consumption and CO_2 emissions caused by the additional acceleration maneuvers.

Thus, most interestingly, the optimal configuration of the overall vehicular networking application is different for the travel time and the CO_2 emission. It can be seen that this decision is not necessarily optimal with regard to the travel time of an individual car (Dressler *et al.* 2011b).

7 Security and privacy

With the increasing engagement of the major industry players to bring inter-vehicle communication (IVC) onto the market, the need to secure the communication between vehicles and the infrastructure became an important issue. Apart from satisfying the demand for deploying closed-market systems that offer services only to paying customers, security is also necessary in order to prevent fraud and malicious attacks. Just recently, it has been demonstrated that IVC-related systems are not as secure as necessary: Attackers successfully took over control of car electronics via tire-pressure-measurement-systems or they attacked electronic message boards along a highway. Even spoofed traffic-information transmissions via TMC have been demonstrated. In this chapter, we study possible security solutions for vehicular networks, focusing on their practical relevance. We review not only the generic security primitives but also their applicability and limits. Furthermore, we look into the very critical balance between security and privacy: The more secure a system is made, the more severely the driver's privacy is impacted. Therefore, we also investigate location privacy and outline how the driver's privacy can be increased. In particular, we investigate the use of pseudonyms, time-varying pseudonym pools, and the exchange of pseudonyms.

This chapter is organized as follows.

- Security primitives (Section 7.1) – In this section, we briefly review the general security objectives before investigating the specific security relationships relevant for vehicular networks. The main focus is on introducing the concept of certificates and their use for digitally signing messages. We also investigate the fundamental relationship between security and privacy.
- Securing vehicular networks (Section 7.2) – This is the key section on enabling security in vehicular networks. The key technology proposed is the use of certificates for digitally signing messages such as periodic CAMs. We further investigate how the resulting performance issues can be solved as well as how certificates can be revoked if keys have been compromised. We conclude this section with a brief overview on using context information such as geographic position to increase security.
- Privacy (Section 7.3) – We investigate solutions to increase the driver's privacy in this last section. We start by investigating location privacy as a metric together with options to identify and to track a vehicle using its broadcast messages. Mitigation is possible by using and even exchanging temporary pseudonyms.

7.1 Security primitives

In this section, we review the basic requirements on security in the domain of IVC. This includes a discussion of the prerequisites in terms of algorithms as well as of the technology available at each participating node, i.e., at each car and roadside unit (RSU). The main focus is on the security objectives and which techniques can be used in general to achieve these goals. Our discussion requires some basic knowledge in the field of network security. Even though we try to limit this background knowledge, we'd like to refer the reader to the rich literature on network security for additional information (Stallings 2013; Schneier 1996).

We conclude this section with a discussion on the impact of increasing security in a vehicular network on the level of privacy of its users. It turned out that this is a still-unsolved conflict and the public acceptance of IVC technology depends not only on the technical capabilities of the underlying technology, including means for securing the system, but also on the level of anonymity guaranteed by the system.

7.1.1 Security objectives and technical requirements

We first have to outline the general security objectives and the resulting security services. Considering technical requirements, in essence, two categories need to be distinguished: basic security algorithms as well as hardware and software support at the on-board units (OBUs) in the participating nodes. We briefly highlight the most important aspects in the following, including a high-level overview of the essential concepts of security algorithms.

Security objectives

Before discussing security solutions, we need to carefully study the security objectives that are touched by the vehicular networking applications and protocols. Let's follow the classical list of security objectives in networking to figure out which are most related to vehicular networks and which specific features need to be investigated in more detail.

Confidentiality

In general, confidentiality ensures that no entity can reveal the information in a message except for the intended recipient(s). In the scope of vehicular networks, messages may need to be protected such that unjustified use of the information becomes impossible. In most cases, communication is based on broadcast. Even though protection against disclosure is possible in general, confidentiality is not the most critical security objective in these cases.

Data integrity

In most cases, the most important security objective is data integrity. The goal is to prevent an attacker from modifying messages or at least to be able to detect such changes. There are many reasons why an attacker might want to modify a message in the vehicular networking domain. For example, falsified messages might make other drivers leave a freeway because of an imaginary traffic jam.

Accountability
Accountability refers to the ability to unambiguously identify the entity that created a message. In contrast to data integrity, the objective is not only to protect a message from modification but also to assign a sender to the message. In the context of vehicular networking, this means ensuring the attainment of certain trust levels for such reasons as to prioritize messages originating from government-operated traffic signals or traffic signs.

Availability
In the last few years, availability has become a top-priority security objective in almost any kind of networked system. Malicious attacks against network infrastructure have become a critical threat, especially in the Internet. In vehicular networks, attacks can range from intelligent denial-of-service attacks to simple jamming of the radio channel.

Controlled access
Access control might play an important role for many of the envisioned vehicular networking applications. Certain services are intended to be made available exclusively to a restricted group of users, e.g., following a business model. Thus, the question of who can access which information needs to be answered and ensured by technical means. This also includes aspects like software updates and access to deployed infrastructure nodes such as RSUs.

Basic security algorithms

In the field of network security, we can distinguish three classes of security techniques (Stallings 2013).

- *Cryptographic algorithms* are mathematical transformations of input data (e.g., data and keys) to output data. Cryptographic algorithms are used in cryptographic protocols.
- *Cryptographic protocols* are a series of steps and message exchanges between multiple entities in order to achieve a specific security objective.
- *Security-supporting mechanisms* provide security-relevant functionality, which is part of a cryptographic protocol or of a security procedure.

In the scope of our discussion of security mechanisms for vehicular networks, two main applications of cryptographic algorithms are of principal interest: encryption and signing of data. *Encryption of data* means the transformation of plaintext data into ciphertext in order to conceal its meaning. This procedure is illustrated in Figure 7.1.

A cryptographic encryption algorithm is used in combination with a key. We can distinguish two classes of cryptographic encryption algorithms.

- *Symmetric cryptography* uses a single shared key for encryption and decryption. Commonly the key is denoted as K_{AB}, meaning that both node A and node B have to be in possession of the key but clearly no other entity.

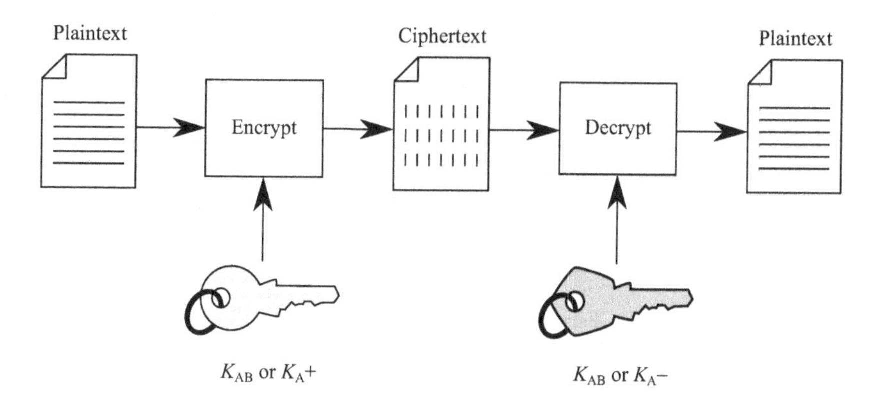

Figure 7.1 The concept of using encryption algorithms to convert plaintext into ciphertext.

- *Asymmetric cryptography* uses two different keys for encryption and decryption. A so-called public key K_A+ of the receiving node A is used to encrypt data, whereas A uses its own private key K_A- to finally decrypt the data again. Therefore, the use of this kind of cryptographic algorithm is also called public-key cryptography.

Signing of data refers to the computation of a check value or assignment of a digital signature to a given plaintext or ciphertext. This signature can be verified by some or all of the entities able to access the signed data. A signature is based on a *cryptographic hash function* in combination with a cryptographic encryption algorithm. The process is shown in Figure 7.2. First, a cryptographic hash function is applied to calculate a hash value of the original text. This hash is then encrypted (using either a symmetric or an asymmetric encryption algorithm together with the respective keys) to calculate the *digital signature*. The message is transmitted together with this signature. At the recipient, the signature is decrypted and compared with a freshly calculated hash value of the plaintext message. If the two hash values match, the message has not been altered by a third party.

There are multiple algorithms available both for encryption and for hashing. The most critical part in the entire process, however, is *key management*. Keys must be securely distributed to the parties involved, stored (e.g., using a trusted platform module (TPM)), and, most importantly, invalidated in the case of leakage.

Key distribution is to be done using secure communication channels. In the case of public-key cryptography, the public keys can easily be shared as long as the system ensures that they cannot be manipulated. Certificates support this process, but certificate management can be cumbersome. A public-key infrastructure (PKI) helps in this process by carefully ensuring that each key is signed by a trusted entity, which is often called a certificate authority (CA). Hierarchies help make this process scalable. Key invalidation, which is called key revocation in the scope of a PKI, describes the process of removing keys if they are considered untrustworthy. As an example, we discuss the X.509 certificate system in Section 7.1.3 on page 308 in more detail.

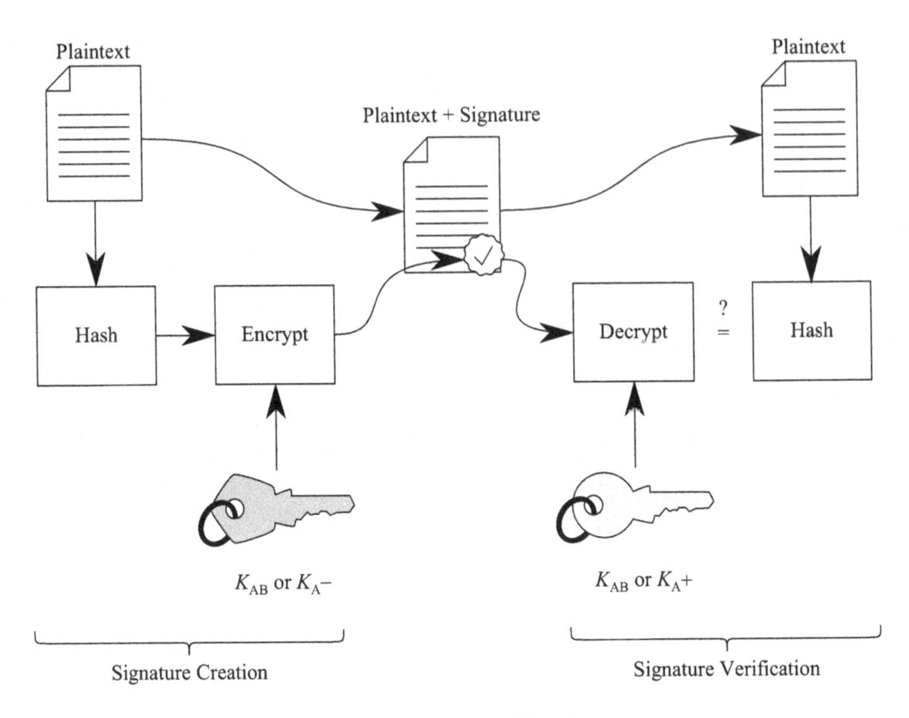

Figure 7.2 The process of creating and verifying a digital signature.

Technical requirements

Before we can start discussing how to secure IVC solutions, we have to emphasize some minimum requirements that must be satisfied in order to enable a rich set of security solutions. These requirements have been discussed in the research community for quite some time and there is general agreement on most of them. When discussing security and privacy solutions, we inherently make use of all these features.

First of all, all nodes participating in the vehicular network are assumed to have rather accurate *geographic position information* available. Essentially, all nodes are assumed to have a global positioning system (GPS) providing this data. At locations with limited connectivity to the GPS satellites, assisted GPS is assumed to cover these areas sufficiently well. Using cellular networking information, the position information can be updated and cross-checked.

In addition to geographic position information, *synchronized clocks* are assumed. This is a requirement not only for security solutions but also for communication protocols like ETSI ITS and IEEE WAVE. The local clocks of the participating nodes can be synchronized using GPS. As an alternative, less accurate clock synchronization is available via cellular networks.

In order to map geographic positions to locations in the road network, fine-grained *map information* is needed. Essentially, this is provided by the navigation units in the cars, which, it is assumed, will attain wide penetration well before new IVC technology. The navigation unit also comes with all the necessary computational capabilities needed for security algorithms.

Besides CPU power and memory to run cryptographic algorithms, persistent memory is needed in order to store key material needed by these algorithms. In the best case, a TPM is used for storing critical key material. The TPM needs to be designed to make the extraction of these keys by hardware analysis hard. In addition, tamper-resistant chips can be used to store and process sensitive information, thereby preventing any hardware-based attack on the key material.

The OBU providing all the IVC capabilities as well as the navigation functionality is in essence a software-controlled unit. Thus, there is the inherent need to provide means for updating the software when severe software malfunctions or possible vulnerabilities are detected. These updates may also include new software features. It is definitely necessary to provide ways and means for securely updating the software.

7.1.2 Security relationships

Concluding our brief introduction to security objectives and technical prerequisites, we outline security relationships as they will appear in vehicular networking applications. Wc split this discussion into two parts: open vs. closed systems, and data sources.

Open vs. closed systems

Before we can investigate security solutions, we need to understand whether the entire system is to be operated in the form of an open or a closed system. In an "open system" the protocols are available to everybody and direct use of the system is explicitly welcome. The idea of crowd sourcing can be implemented by giving every participant the same rights to add information, to redistribute, and even to aggregate or to manipulate information.

In contrast, a "closed system" eliminates most of these access rights and limits access to a closed group of users. Access to the system is controlled by the operator. Commercial interests seemingly lead to this class of a system in which only paying users get what they pay for.

When vehicular networking applications get deployed, most likely a mix of these two concepts will be implemented. Closed systems rely on the service by the operator. Crowd sourcing is not possible due to the small user base. So, openly managed systems will help to kickstart the entire vehicular networking application world, whereas particular services that are restricted to selected users only may show up.

Data sources

Depending on the application, data may be created by almost all nodes participating in the vehicular networking application. Safety-relevant information will be created by vehicles or roadside infrastructure observing a safety-critical event. For all non-safety applications, information may be hosted by some centralized systems, e.g., managing a traffic information system (TIS), by dedicated roadside infrastructure, e.g., traffic lights, or by any of the participating vehicles.

Obviously, this opens up many security problems. Information must be authentic and correct because even automated reactions of the vehicles may be triggered, leading to unsafe driving situations. Therefore, appropriate means for ensuring accountability as well as message integrity must be in place. Besides sender information, also time and location descriptors are needed in order to assess the scope of a message.

We do not discuss these very-application-dependent aspects here in detail. From the observations listed, it should become clear that strong measures to secure the transmitted messages must be in place. This also includes the management of groups of nodes such as all vehicles, all vehicles close to an intersection, all traffic lights, etc. to simplify access control and authentication.

7.1.3 Certificates

We have seen that a digital signature is constructed using a cryptographic hash function in combination with some key material. The most widely used form is to calculate a hash first and then encrypt the hash value. This form is also used in combination with certificates. For encrypting the hash value, the private key of the sender is used. In turn, the receiver can validate the hash value using the public key of the sender to decrypt the hash value. The most critical step in this process is the secure transfer of the public key of the sender.

Certificates focus on exactly this part. A certificate can be seen as a kind of ID card that comes with information describing the owner such as its name and, most importantly, its public key. In order to prevent malicious modifications of the certificate, it is necessary to secure the certificate itself against modifications. This is realized by means of a certificate authority (CA). This entity signs the certificate using its private key. Obviously, the public key of the signing CA must be known in advance to the receiver of a signed message. Usually, such public keys come (again in the form of certificates) pre-installed in the local system.

In the following, we introduce also X.509 certificates as means for certificate management and revocation, which are also used for signing messages in vehicular networking applications (ITU-T 2000).

X.509 certificates

X.509 is an ITU-T international recommendation and is part of the X.500-series defining directory services (ITU-T 2000). A first version of X.509 was standardized in 1988, followed by a second version (1993) and a third (1998) version. For our discussion, we refer to the most recent standard of X.509.

The X.509 standard defines a complete security framework, focusing on authentication services. Most importantly, entity and message authentication are provided, which help us in defining digital signatures. The framework comprises the following features.

- Certification of public keys and certificate handling, i.e., defining the certificate format, a certificate hierarchy, and certificate-revocation lists.
- Three different dialogues for direct authentication: one-way authentication, which requires synchronized clocks; two-way mutual authentication, which still requires synchronized clocks; and three-way mutual authentication entirely based on random numbers.

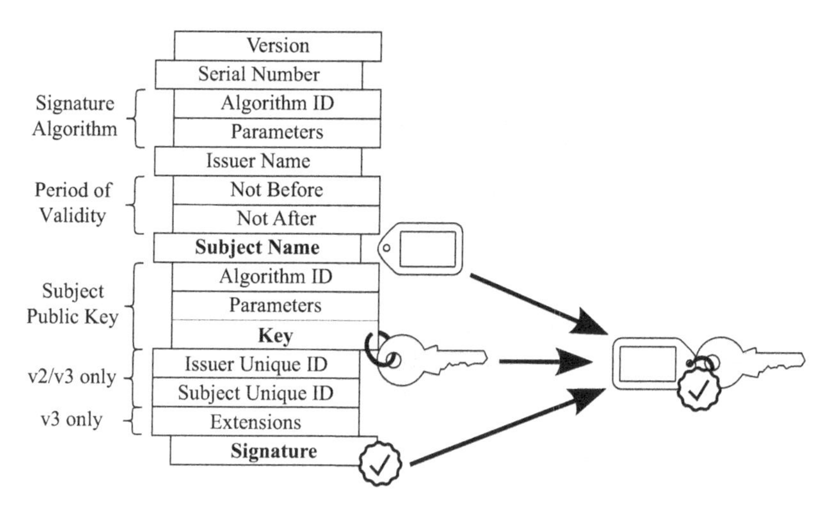

Figure 7.3 The structure of an X.509 certificate.

The use of X.509 in the Internet protocol suite has been defined by the IETF in RFC 4210 (Adams *et al.* 2005). Today, X.509 has become a *de facto* standard in almost all areas in computer networking where authentication services or simply digital signatures are required.

The X.509 public-key certificate is some sort of passport, certifying that a public key belongs to a specific name. The structure of this certificate is depicted in Figure 7.3. The certificate contains at least the name of its owner and its public key, all signed by a trusted entity. In addition, the protocol version and certificate serial number, as well as validity-period information, are listed in an X.509 certificate. In order to identify the CA that has signed the certificate, its ID is added alongside the other information. The signature consists of an encrypted hash of the whole certificate. In order to identify the algorithms used (cryptographic hash, encryption algorithm) and their configurations, this is appended as the signature algorithm data.

Certificates are issued by a CA. If all users know for sure the public key of the CA, every user can check every certificate issued by this CA. In this way, an offline check of the certificate is possible without the need to communicate with the CA for the validity check. The security of an X.509 certificate depends on how secure the private key of the CA is kept.

Using an X.509 certificate is straightforward. Messages that need to be secured are signed using the host's private key. The certificate containing the public key needed to check the signature is transmitted either together with the message or on a separate communication channel. Caching of certificates is possible and recommended. Thus, only the first message needs to be transmitted together with the certificate. We study the use of certificates as means to digitally sign messages in vehicular applications in Section 7.2.1 on page 311.

Public-key infrastructure

The prerequisite for checking a certificate is that the public key of the signing CA is already available. In practice, this might be cumbersome because either only very few

CAs can be supported, which, in turn, have to sign all certificates to be used and thus generate a central bottleneck, or the user has to install new CA certificates on the fly, which poses potential security problems. To overcome both problems and to make the use of X.509 certificates scalable and more secure, so-called certificate chains or a fully featured public-key infrastructure (PKI) can be used (Housley & Polk 2001).

Let's consider two entities A and B who want to communicate securely. If we assume a potentially high number of CAs issuing certificates, it is highly probable that the public keys of A and B are certified by different CAs. In order to understand the idea of certificate chains, let's call A's and B's certification authorities CA_A and CA_B, respectively. If A does not trust or even know CA_B, then B's certificate is useless to A. We use the notation CA_B«B» to denote that B's certificate has been signed by CA_B. Of course, the same applies in the other direction.

In such cases, certificate chains help to solve the security issue. In the simplest case, we can assume that CA_A and CA_B know and trust each other. If now CA_A digitally signs CA_B's public key by issuing a certificate CA_A«CA_B» and CA_B similarly certifies CA_B«CA_A», we can then construct a certificate chain. If A is presented B's certificate CA_B«B», which it cannot check directly without knowing CA_B, A tries to look up whether there is a certificate CA_A«CA_B». Now, following the certificate chain, A can validate B's certificate by checking first CA_A«CA_B» and then CA_B«B».

Certificate chains need not to be limited to a length of two certificates. X.509 therefore suggests that authorities are arranged in a certification hierarchy, so that navigation is straightforward. A fully featured PKI provides means not only for managing these certificate chains but also for issuing, maintaining, and revoking certificates. The basic assumption is that all participating CAs sign each other in an at least partially connected way. We discuss certificate revocation in the next section.

Certificate management and revocation

Even though we still lack a global PKI, the X.509 certificate system has been established in the Internet for distributing public keys by means of a distributed PKI. The most critical remaining problem is that of how to revoke certificates if they are (assumed to be) compromised. The X.509 system proposes the use of certificate-revocation lists (CRLs) in this case (Cooper *et al.* 2008).

As an example, consider that A's private key has been compromised. If this leakage has been detected, the corresponding certificate needs to be revoked as quickly as possible. If the certificate is not revoked (and until it is revoked), the compromised private key can be used to digitally sign messages in the name of A, thus impersonating A. An even worse situation occurs when the private key of a certification authority is compromised. This implies that all certificates signed with this key will have to be revoked.

X.509 uses CRLs for certificate revocation. These lists are stored at appropriate places (X.509 recommends use of the directory service X.500 for this purpose). In the best case, each client downloads all CRLs periodically. In the process of checking a certificate, it is also necessary to check whether the respective certificate has already been revoked. This is done by searching all CRLs. Obviously, certificate revocation is a relatively slow and expensive operation, which becomes even more complicated in massively distributed systems such as a vehicular ad-hoc network (VANET).

7.1.4 Security vs. privacy

As we have seen, security can be increased by using encryption and digital signatures. It is important to understand that each and every feature one wishes to include in order to improve the overall security of the vehicular networking application relies on key material. These keys are in most cases personalized, e.g., when using public-key cryptography to create a digital signature. We will study the best-known example, certificate-based security, in detail in the next section.

With use of keys, however, users can be identified – this is actually the result of solving the accountability security objective. Following this line of thought further, we see that users can not only be identified once, but can be identified uniquely in the entire network. This poses a big threat to privacy. Given the possibility of observing users over large geographic areas, they can easily be tracked while driving on our public roads. Countermeasures are needed in order to again protect the users' privacy. Yet, such countermeasures and the problems they incur remain largely unaddressed by current standardization efforts and field operational tests (FOTs) (Eckhoff & Sommer 2014).

Pfitzmann & Köhntopp (2000) studied anonymity, unlinkability, unobservability, and pseudonymity in more detail, and introduced these terms within the respective context of proposed measures. They were able to show relationships between these terms and thereby to develop a consistent terminology. Focusing on ad-hoc networks, Raya *et al.* (2010) investigated the trade-off between trust and privacy. By focusing on the question of how to reconcile the two seemingly contradicting requirements, they show that the trust–privacy trade-off can be approached using a game-theoretic model.

In general, we observe a fundamental problem: It is still not clear to what extent the users' privacy can be protected while not giving up other security objectives. This problem must currently be considered unsolved. The more secure we make a system, the less protected the users are in terms of location privacy. Time will show whether public acceptance leans more towards privacy or towards security.

7.2 Securing vehicular networks

Security techniques for IVC do not differ from what has been used in other communication networks, including the Internet. Yet, the different communication characteristics have implications for the use of selected techniques (Papadimitratos *et al.* 2008a; Dressler *et al.* 2008). In this section, our aim is to introduce the general principles of applying security measures as well as to highlight the most important limitations.

7.2.1 Using certificates for IVC

The use of well-known and established security measures in the area of vehicular networking has been discussed ever since the development of the first IVC protocols.

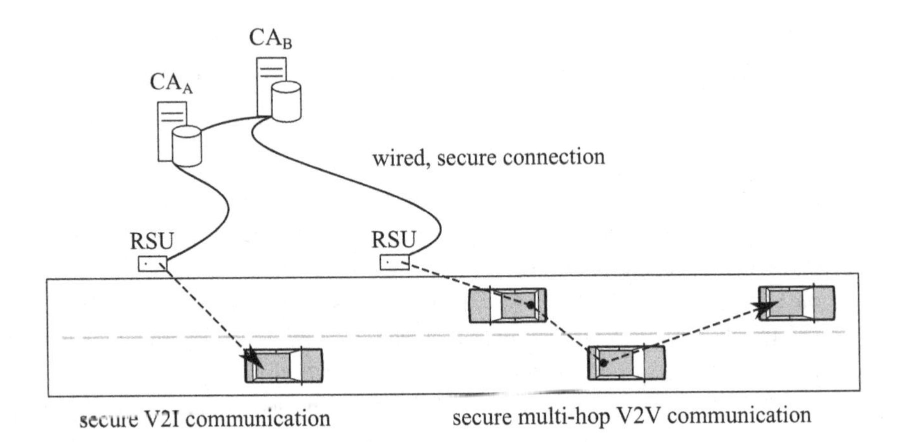

Figure 7.4 The security architecture envisioned for vehicular networks.

It was identified early on that there is a strong need for a complete security-engineering process that would be based on application characteristics and requirements, including attacker-model derivation (Kargl *et al.* 2006).

If we recall our discussion on general security objectives, the critical task is obviously to pick the most relevant ones to be integrated in the vehicular networking case. Papadimitratos *et al.* (2008a) investigated exactly this issue. They carefully analyzed the features and importance of security requirements related to IVC, distinguishing between vehicle-to-vehicle (V2V) and vehicle-to-infrastructure (V2I) communication. This provides the basis for deriving security requirements. The authors were able to identify authentication, data integrity, and privacy as the most important aspects. Privacy itself is not a security objective, and we will return to this aspect in Section 7.3 on page 317.

Besides the general discussion of necessary security features, Papadimitratos *et al.* (2008a) further developed a security architecture that was finally adopted by the ETSI in the standardization process. It is important to note that all security observations have been investigated for general V2V and V2I communication. Additional or completely different observations might be possible if one is investigating specific applications independently. The architecture is completely based on the use of certificates, which support the use of public key cryptography for a wide variety of security operations. In the following, we briefly introduce the general concept of how certificates are applied to IVC.

Figure 7.4 illustrates the overall security architecture. Certificate authorities CA_A and CA_B represent (following the general idea of a PKI) centralized trusted entities that are first of all responsible for issuing certificates to each user and entity participating in the vehicular network. Since not all users will be able to get a certificate issued by the same CA, certificate chains have to be used to enable a fully distributed and secure validation of all of the certificates provided.

Such certificates are issued for (and securely transmitted to) all entities participating in the vehicular network. This includes all the vehicles but also RSUs and other

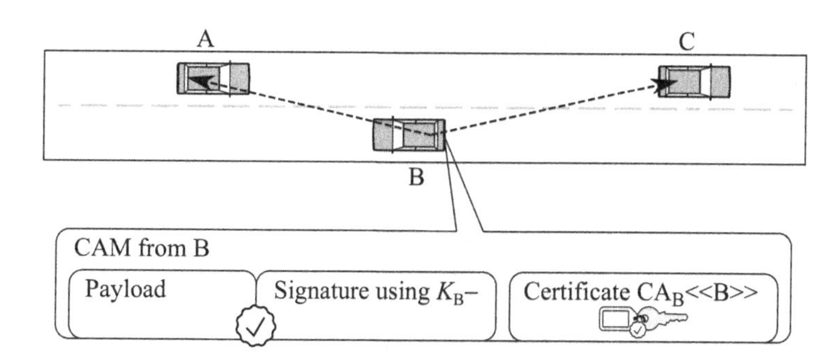

Figure 7.5 Secured CAM beaconing using digital signatures and certificates.

infrastructure elements. Using the public keys provided, secure wired and wireless communication can be established.

Let's focus on beaconing as an example for the communication between vehicles and infrastructure. The proposal essentially aims at enforcing at least message authentication and data-integrity checks using digital signatures. For this, each beacon message, e.g., a CAM, needs to be signed using the private key of the sending entity.

In this massively distributed system, one cannot assume that the related certificate is already available at each receiver. In fact, the probability that two vehicles have already been in contact in the past using the same certificate is quite low. Therefore, the whole certificate of the sender needs to be appended to the beacon message.

The procedure is depicted in more detail in Figure 7.5. In this example, all the vehicles periodically exchange CAMs to increase cooperative awareness (see Section 3.3.4 on page 85). In our case, vehicle B broadcasts a CAM to all cars within its broadcast range. The CAM consists of the original content describing the vehicle's speed, heading, and so on, a digital signature that is based on B's private key to ensure message authentication and data integrity of the beacon message, and B's complete certificate issued by certificate authority CA_B. After receiving the CAM message, vehicles A and C first check B's certificate. For this, they have to trust the CA. After extracting B's public key from the certificate, they can check the digital signature to validate the CAM message.

7.2.2 Performance issues

The use of certificates to secure IVC protocols provides an easy-to-use and secure communication channel. These benefits, however, come with non-negligible costs. Kargl *et al.* (2008) calculated the following overheads for beaconing protocols.

- *Protocol overhead* – For each beacon message, a signature as well as a certificate needs to be added to the original message. Even using very efficient cryptographic algorithms, at least 150–160 B will be used for these security measures. At a beaconing rate of 10 Hz, this translates to 1500–1600 B/s per vehicle, or about 150 kB/s on the communication channel for a medium vehicle density of 100 cars in communication range.
- *Computational overhead* – For each beacon message sent and received, complex asymmetric cryptographic algorithms need to be executed. This translates to one

signature generation per beacon sent and two verifications (signature plus certificate) for each beacon received. For a medium vehicle density and a beacon frequency of 10 Hz, this sums up to roughly 2000 operations per second, just for security purposes.

Both performance issues can be addressed by omitting certificates and certificate verifications, as well as by omitting signatures and signature verifications (Calandriello *et al.* 2007; Kargl *et al.* 2008; Schoch & Kargl 2010).

Omitting certificates

As we have already seen, certificates can be cached locally after verification with no impact on the security of the overall system (Calandriello *et al.* 2007; Kargl *et al.* 2008). This reduces the need to verify a certificate every time it is reused for another message. Still, the certificate must be included in the message. Thus, only the computational overhead is reduced, not the protocol overhead.

Calandriello *et al.* (2007) go one step further and discuss the need to append the certificate for each beacon message sent. The key observation is that a vehicle will receive multiple copies of the certificate of the sender for almost all state-of-the-art beaconing protocols. Thus, after adding the certificate once, the receiver(s) can be assumed to already have a local copy, and further messages can be sent without adding the certificate. A first naïve approach is to attach certificates only every nth time a beacon is sent (Calandriello *et al.* 2007). While clearly reducing the communication overhead, this approach can, however, lead to a significant communication delay. Following the discussion by Schoch & Kargl (2010), we can assume, given a beacon interval Δb, that the worst-case delay will be $(n-1)\Delta b$, i.e., only the nth transmission contained a previously unknown certificate to check the digital signature. For $\Delta b = 0.1$ and $n = 4$, the vehicle may experience a delay of 0.4 s until it receives and successfully checks a safety-critical message; this is too much for many safety applications.

Schoch & Kargl (2010) further elaborated certificate-omission schemes and developed a neighbor-based omission scheme. Their approach relies on the fact that each node maintains a neighbor table in most of the proposed beaconing solutions. Thus, a node can monitor neighborhood changes and use this information to make the decision regarding whether to attach its certificate or not. When a node is about to send a beacon, it determines whether new neighbors were added to its neighbor table since the last beacon. If so, the node attaches its certificate to the beacon. Please note that this scheme may also lead to situations in which a node receives a beacon without a certificate and is not able to check the signature because the certificate is also missing in its local cache. Yet, the likelihood that this node will become known to the original sender increases with every beacon it sends. Thus, the original node will add its certificate in one of the next rounds of beacon messages.

Omitting signatures

Additional overhead can be eliminated by omitting signature verification or even skipping signatures entirely (Schoch & Kargl 2010). Skipping signature verification can lead to a substantial reduction of the computational overhead at the receiver – it generally needs to check the signature for each beacon message it receives from all the vehicles in communication range. Not checking all signatures, however, opens the possibility of an attacker being able to maliciously falsify messages without being detected.

Schoch & Kargl (2010) proposed two schemes for omitting signature verification.

- The periodic verification strategy assumes secure "streams" of beacons. Thus, only the signature of the first beacon is checked (when the new car is also added to the neighbor table), and then there is a check for every nth message. Unfortunately, such a stream can easily be hijacked by an attacker.
- The context-adaptive verification strategy counters this by also tracking the position of the neighboring vehicle using a Kalman filter. If the predicted positions and those reported in the beacon are consistent, no further signature verification is performed.

Both approaches help to reduce the computational overhead but not the communication overhead. Thus, additional features such as situation-based signing can further enhance the protocol performance.

7.2.3 Certificate revocation

Certificates of misbehaving and faulty nodes must be revoked as soon as malicious behavior is identified. Papadimitratos *et al.* (2008a) leverage the roadside infrastructure to distribute CRLs. The transmission of the CRLs needs to be done in the form of small (cryptographically) self-verifiable pieces. This allows a low-rate transmission of these CRL pieces to all nodes in the network.

Figure 7.6 illustrates this concept of CRL dissemination. The CRLs are split into smaller pieces either directly by the CA or at the RSUs connecting the CA with the vehicles. All vehicles continue to exchange these pieces with each other at a low rate to enable CRL dissemination in areas with sparse or no RSU coverage. The size of CRLs to be distributed can still be a challenge (Papadimitratos *et al.* 2008a). Thus, collaboration between CAs is recommended so that CRLs contain only regional revocation information. The revocation itself can leverage the hardware security module in the car, which also securely stores all key and certificate information (Raya *et al.* 2007).

Obviously, it may take some time for a node to be identified as behaving maliciously and even longer for this node's credentials to be added to the CRL and for the CRL to be received by all other vehicles. Papadimitratos *et al.* (2008a) addressed this issue by proposing that misbehavior detection be left to the vehicles. Using monitoring facilities, they can locally vote off and exclude misbehaving vehicles.

Several improvements for certificate revocation have been discussed in the literature. Papadimitratos *et al.* (2008b) discuss the splitting of CRLs into smaller pieces using fountain or erasure codes (Rizzo 1997; Luby *et al.* 2001). These codes were designed for forward error correction on the erasure channel in wireless networks, yet they can also be used in our scenario where vehicles will likely receive many but clearly not all pieces of the CRLs. The codes help to decode as many parts of the original CRLs

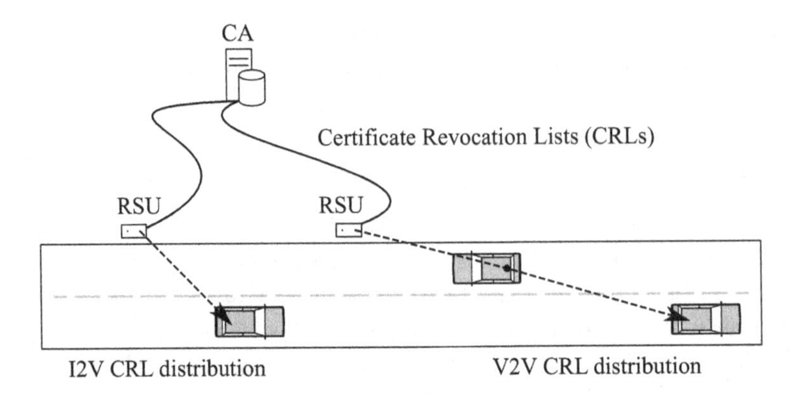

Figure 7.6 Certificate revocation in vehicular networks.

as possible, even given incomplete information. The CRL snippets are assumed to be distributed and combined as soon as possible.

The communication channel used for disseminating the snippets has been investigated in detail by Laberteaux *et al.* (2008). They proposed that one should mainly rely on V2V communication considering RSUs as helpers where available. All the information is then spread using epidemic forwarding schemes (Vogels *et al.* 2003).

In contrast, Lequerica *et al.* (2010) focus on the use of mobile network operators (MNOs) and 3G/4G networks to improve the efficiency of the CRL distribution. The proposed solution gathers positions and headings of vehicles using a location-service enabler. Using this data, the infrastructure decides when and in which areas to broadcast incremental CRLs, e.g., using the UMTS multimedia broadcast/multicast service (MBMS) or LTE eMBMS.

7.2.4 Position verification

Besides the use of certificates for providing data integrity, message authentication, and, if needed, confidentiality, the same and additional security objectives have been addressed using different approaches. Much attention was, for example, dedicated to deployment problems such as key management. These solutions are beyond the scope of this textbook insofar as they are mostly considered to be additions to the vehicular networking approaches we discussed throughout the book.

In this section, we concentrate on a security solution that exploits an inherent characteristic of vehicular networking applications and protocols, namely the use and dependence on accurate position information. For example, Leinmüller *et al.* (2006) investigated position verification in the context of GeoNetworking, yet such functionality is required for almost all applications, ranging from safety-critical applications like intersection warning systems to traffic-efficiency applications such as TISs. If position information received from other vehicles is not correct or even has been maliciously altered, these applications may no longer be able to provide the expected output and to guide (and maybe even control) vehicles correctly.

There are multiple possible approaches for secure position verification. For instance, the verification system may fully rely on base stations building a trustworthy network (Hubaux *et al.* 2004; Capkun *et al.* 2003). The most direct approach is to verify whether a node really is within a defined region. This can be done using location-based access-control techniques, e.g., using verifiers at special locations (Sastry *et al.* 2003). Position validation is then indirectly provided by checking whether the node is at an acceptable distance for each verifier. Instead of using physical measurements, logical relations may be used (Vora & Nesterenko 2006). Using this approach, position claims can be verified using the logical structure of the network as observed by other nodes.

In contrast, Leinmüller *et al.* (2006) investigated the idea of a position-cheating-detection system similar to the intrusion systems used in mobile ad-hoc networks (MANETs) to detect selfish nodes. The intended use of this system is for enhanced security in GeoNetworking through autonomous position verification. In order to detect nodes cheating about their position, the proposed mechanism uses a number of different independent sensors to estimate the trustworthiness of the nodes' position claims without using dedicated infrastructure or specialized hardware.

The following five measures, which can easily be applied to other applications requiring secure position verification, have been proposed.

- The acceptance range threshold is based on the physical radio communication range in vehicular networks. From the radio properties, a maximum acceptable range can be defined, which limits potential position falsification.
- The mobility grade threshold works similarly but using the assumption that a node can move only at a defined maximum speed. In general, this can be the speed limit on a street or, in the worst case, the maximum speed of a car.
- The maximum density threshold instead focuses on the physical dimensions of vehicles. Certainly, no more than a well-defined number of vehicles can occupy a street segment. If this threshold is exceeded, selected vehicles must have announced wrong position updates.
- Map-based verification checks whether a car is pretending to be at a location that is highly unlikely, e.g., in the middle of a house.
- Finally, position claim overhearing is based on the concept of overhearing all radio transmissions and constantly checking the announced position information. This sensor exploits the principle of geo-routing, i.e., that information is greedily forwarded to the destination. If nodes forward the message in a way that is contradictory to their announced positions, the likelihood that one of the vehicles involved has announced wrong position information increases.

Using these sensors in combination further improves the estimate of the announced position's trustworthiness.

7.3 Privacy

As we have seen, security mechanisms are available to ensure confidentiality and, most importantly, data integrity of IVC applications. We now highlight one issue that is frequently discussed in the scope of security, without it necessarily being a security objective, namely privacy.

Privacy has many facets and is one of the most passionately discussed topics, mostly because it is seen as being anything from unnecessary to indispensable by the general public. Yet, it has repeatedly been shown that even just the suspicion of being tracked, investigated, or otherwise monitored changes people's behavior drastically. Further, actual privacy violations can cause anything from embarrassment to financial loss or loss of social standing, also depending on which entity attacks users' privacy: Other users of a system are just as likely to be an attacker as private industry, application developers, service providers, system operators, or some governmental bodies.

Many aspects of our daily life have been deemed worthy of protection, such as one's social network, preferences, interests, and location. Of these, the most relevant to vehicular networking (as opposed to its applications) – and the one that we are focusing on – is location privacy. Location privacy is the ability, in our case of a vehicle, to prevent third parties from recording the current location and location changes.

The problem of location privacy had already been identified in the early days of vehicular networking. Nevertheless, security dominated the early days of secure IVC. The first in-depth analysis of the related privacy issues was performed by Papadimitratos *et al.* (2006). In the following, we study location privacy as a security objective as well as techniques to prevent tracking. We focus solely on the possibility of tracking vehicles using information disseminated using IVC protocols. Of course, there are other possibilities such as camera-based vehicle-identification techniques, which, however, require a large base of installed infrastructure and cannot easily be crowd-sourced to all other vehicles equipped with, e.g., IEEE 802.11p radios.

As discussed already in Section 7.1.4 on page 311, security measures essentially reduce the level of privacy because identities are strongly bound to the vehicles. Let's take the secure beaconing system as an example. Each beacon is secured using a digital signature. Thus, an attacker cannot easily falsify messages. Yet, on the other hand, the vehicle also reveals its identity through the certificate used.

In the following, we study privacy as a metric but also focus on techniques that can be employed to increase vehicles' privacy with only marginal impact on the security level.

7.3.1 Location privacy

In order to develop and to assess privacy-preserving techniques, we need a measure outlining the degree of privacy of an individual. There are different possible metrics to measure the level of location privacy enjoyed by an individual in a network. Pfitzmann & Köhntopp (2000) investigated the term privacy, focusing on the degree of *anonymity* in particular. Their definition of anonymity relies on the state of being not identifiable within a set of subjects, the anonymity set. This means that anonymity can be guaranteed only if there is a large enough anonymity set available. A single user or vehicle can be tracked if not hidden in a crowd of others.

According to Pfitzmann & Köhntopp (2000), the anonymity set contains all nodes in the network that could possibly be a targeted individual. In the case of vehicular networks, this definition is not appropriate. Vehicles at a certain location do not all share the same properties. Thus, in a VANET, not all nodes are equally likely to be a particular individual. The size of the anonymity set alone is not a sufficient metric to measure the location privacy.

As an alternative, the *entropy*, as used in information theory, of the anonymity set has been explored. This parameter can be seen as the uncertainty in determining the current identifier of an individual (Serjantov & Danezis 2002). In the context of vehicular networks, the use of the entropy has successfully been applied to describe the degree of location privacy (Eckhoff *et al.* 2011b; Ma *et al.* 2009).

The entropy can be calculated as follows. Let p_i be the probability of a node i being target individual I, the sum of all probabilities p_i being 1. The entropy $\mathcal{H}$ of identifying an individual driver in the anonymity set S is then defined according to

$$\mathcal{H} = -\sum_{i=1}^{|S|} p_i \times \log_2 p_i \tag{7.1}$$

An upper limit of $\mathcal{H}$, i.e., the maximum value of entropy for a given individual, is attained for all entities in S being equally likely to be the targeted driver. Unfortunately, it is almost impossible to achieve this in a vehicular network. Nodes may contact a large number of other nodes and the relation *A has met B* is usually not transitive.

Eckhoff *et al.* (2011b) provided a nice example illustrating how to interpret entropy values for a given individual. Assuming an attacker is not sure whether the individual uses identifier A or B and that both nodes are equally likely to be the target, then the anonymity set for the individual is $S = \{0.5, 0.5\}$. The entropy is $\mathcal{H} = 1$. On the other hand, if individual I is with a certainty of 80% the driver of A and with 20% certainty the driver of B, then the anonymity set would be $S = \{0.8, 0.2\}$ and the resulting entropy is $\mathcal{H} \approx 0.72$. This can be continued for an arbitrary number of users or vehicles in our network.

Ma *et al.* (2009) extended the entropy calculation to cover tracking capabilities too. Tracking is based on accumulated information, i.e., multiple observations in time and space. We briefly consider tracking options in the following section.

7.3.2 Tracking options

Tracking is defined as the possibility of following the trajectory of an individual by the use of sampled observations. In our case, tracking relies on the ability to follow an individual even though the sampled observations cannot be unambiguously related to each other, e.g., because the vehicle uses randomized IDs – the use of dedicated IDs will be discussed in Section 7.3.3 on page 321. Please keep in mind that we are relying exclusively on wide-area observations of radio transmissions. Any attacker that is dedicated enough to physically follow individual vehicles around will have a much simpler way of tracking at their disposal: visual tracking.

There is a huge body of literature on tracking techniques. Almost all of the more sophisticated solutions are based on a Kalman filter (Kalman 1960), which predicts the most likely trajectory of an individual by taking into account physical capabilities and constraints. This approach has been extended depending on the application scenario. Let's briefly discuss the idea and the capabilities of some rather simple tracking algorithms.

Simple tracking as described by Sampigethaya *et al.* (2005) does not perform actual prediction of a target's state. Instead, it limits the area that can theoretically be reached by the vehicle from its last position.

This area can be identified using the maximum and minimum velocities of the vehicle, V_{max} and V_{min}, respectively. Using a sampling interval T, which can, for example, be defined by the beaconing interval if CAMs are used to periodically broadcast the position of a vehicle, the maximum and minimum traveling distances can be calculated as

$$\begin{aligned} r_{max} &= V_{max} \times T \\ r_{min} &= V_{min} \times T \end{aligned} \tag{7.2}$$

Now, all observations received in the next step within the area given by the two radii are considered for the track update. This simple tracking takes into account only the vehicle's position. It does not rely on a score function to evaluate possible positions: All observations within a reachable area are treated as equally likely. Still, in scenarios with low traffic density, this simple tracking works astonishingly well.

The tracking quality can be improved using correlation tracking, i.e., a Kalman filter. We apply the algorithm presented by Sampigethaya *et al.* (2005). Their idea was to improve the tracking performance by using the last known speed v and direction angle ϕ of a vehicle to predict its next position. Using the observation interval T and last target position (x_1, y_1), the next position (x_p, y_p) can be predicted as

$$\begin{aligned} x_p &= x_1 - v \times T \times \cos\phi \\ y_p &= y_1 - v \times T \times \sin\phi \end{aligned} \tag{7.3}$$

If a new observation is available, let's say at position (x_2, y_2), we can evaluate the likelihood of whether it belongs to the vehicle we are tracking. In a first step, we can calculate the Euclidean distance between the predicted position and the observation as

$$d = \sqrt{(x_p - x_2)^2 + (y_p - y_2)^2} \tag{7.4}$$

The closer a predicted position is to the new observation, the higher the likelihood that it matches exactly the same vehicle. After collecting a few sample observations, it is possible to distinguish different vehicles with a very high accuracy.

Of course, the tracking quality can be further optimized. If we take the road network's topology into account as well as the exact mobility vectors of the vehicles (each vehicle is assumed to periodically broadcast a CAM containing its position, speed, and the direction in which it is heading), the position estimate can become extremely accurate.

7.3.3 Temporary pseudonyms

In the last part of our discussion about location privacy, we want to focus on the IDs or addresses used by the vehicles. The communication among the vehicles can be exploited to reveal the identify of each vehicle using two types of addresses.

- Physical-layer information – Here, most tracking approaches revolve around hardware-dependent imperfections in timing or irregularities of the signal to re-identify devices.
- Medium-access- and network-layer information – Each transmitted IEEE 802.11p frame contains the source MAC address of the sending vehicle. For routing, additional network-layer address information might be added.
- Application-layer information – In general, no identification of the vehicle needs to be added at the application layer. Yet, in order to add security, certificates have to be added that exactly reveal the identity of a vehicle.

Physical-layer tracking, in general, needs to be tightly calibrated to certain devices and is often very much dependent on channel conditions (Franklin *et al.* 2006; Brik *et al.* 2008). We therefore do not discuss physical-layer tracking further.

For our discussion, we can also safely ignore medium-access- and network-layer information. Broadcast-based protocols do not rely on routing protocols; thus, no network-layer addresses will be used. Other protocols might use geographic or even IP addresses. These will be constant only for a very short time, making tracking approaches infeasible. The story is different for MAC addresses. Each IEEE 802.11-compliant device must be configured with a unique MAC address. If this address is used for communications, the sender can be identified on a global scale. In order to reduce the ability to use MAC address information for tracking, the use of randomized MAC addresses for each transmission has been recommended (IEEE 2010a). In this way, the location privacy of each vehicle can be guaranteed.

What remains is application-layer information.

Identification of an individual vehicle from the information in a used certificate is much more straightforward: The certificate has been issued for a specific identity. On the other hand, without the digital signature and the attached certificate, no security services such as data integrity can be provided.

This dilemma has been investigated in great detail, and several groups came up with similar solutions (Li *et al.* 2006; Gerlach & Güttler 2007; Buttyán *et al.* 2007). The key idea is to replace the certificate over time. This leads to the concept of using *pseudonyms*, which has successfully been employed over the years, for example, in the domain of mobile communications. In the Internet, this concept became famous in the context of Chaum's networks focusing on hiding the sender's and receiver's identity for email messages (Chaum 1981).

The core idea is quite simple. Since just using a pseudonym prevents identification, but not re-identification (thus, still allowing tracking), instead of using a single identity, the vehicle uses a certain number of different identities. From this set, it can choose a different one for the transmission of subsequent messages. If the different identities

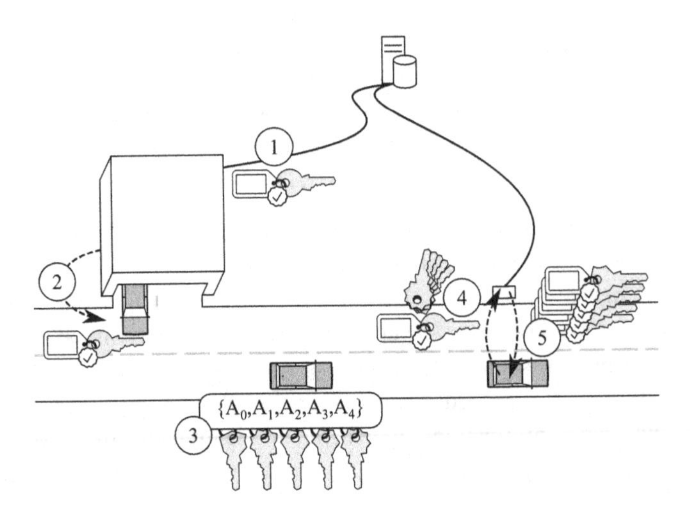

Figure 7.7 The conceptual approach employed to generate pseudonym pools.

or pseudonyms cannot be linked to each other, it is impossible to tell whether all these transmissions have been sent by the same vehicle. Unfortunately, this is not the case in all situations.

The concept of generating temporary pseudonyms as proposed for use in vehicular networks is outlined in detail in Figure 7.7.

1 The process is initiated by a trusted certificate authority. This CA assigns base identities for new vehicles to the automotive industry.
2 The manufacturer, in turn, assigns a unique base identity to each new car. So far, the process is exactly the same as if normal certificates are to be used.
3 The vehicle itself starts creating a pseudonym pool of p different pseudonyms. Each is represented in the form of a certificate request.
4 As soon as the vehicle has an Internet connection, it can request the CA to sign each of the pseudonyms generated.
5 The CA checks the validity of the signing requests by means of a digital signature using the originally assigned base identity. If they are valid, the CA signs the pseudonyms, i.e., the new certificates using the pseudonyms as their identity, and sends these signed certificates to the vehicle.

All of the certificates generated, each using a different pseudonym, can be used to secure messages, not revealing the vehicle's identity. The effectiveness of changing pseudonyms has been discussed in the literature in great detail (Buttyán *et al.* 2007; Schaub *et al.* 2010; Wiedersheim *et al.* 2010). From an applications point of view, it is often necessary to link two or more successive beacon messages to one vehicle (Schoch *et al.* 2006). Yet, simply changing the pseudonym every n seconds has been shown to offer only marginal protection of users' location privacy (Sampigethaya *et al.* 2005).

Better results can be achieved when position, speed, heading, and the number of cars in transmission range are accounted for when changing pseudonyms. This reduces the chance of an attacker being able to successfully follow a pseudonym change (Li *et al.* 2006). In addition, silent periods can be added: After changing a pseudonym the vehicle stops emitting messages for a random amount of time.

7.3.4 Exchanging pseudonyms

Changing pseudonyms from the local pseudonym pool can make a vehicle indistinguishable from others, especially in very dense scenarios, because even complex tracking algorithms may fail there (Wiedersheim *et al.* 2010). In these situations, the likelihood of having multiple cars going in the same direction at roughly the same speed is very high. Yet, this is unfortunately not always the case. Safety applications, i.e., periodic information exchange, are especially important in more urban scenarios. Tracking in these situations is therefore possible.

When designing privacy schemes for vehicular networks, important domain-specific constraints have to be kept in mind (Eckhoff *et al.* 2010, 2011b). Almost all of the aforementioned privacy approaches make use of a large pool of pseudonyms, so that the vehicle can still send messages until the CA has supplied new pseudonyms, e.g., if the CA is not reachable due to lack of connectivity. Eckhoff *et al.* (2011b) proposed an alternative, which can operate on rather small pseudonym pools and does not rely on continuous generation of new pseudonyms. The resulting SlotSwap system offers both low-bandwidth pseudonym management and unlinkability of pseudonyms, thus, by design, providing strong privacy for all participants in the vehicular network.

SlotSwap requires every car to maintain a time-slotted pseudonym pool containing T/t pseudonyms, where t is the slot length and T the total period length. When the last (T/t)th time-slot has passed, time-slot 1 will become active again, meaning that the time period will simply restart from the beginning. For each time-slot, all cars change their pseudonyms – synchronization can be achieved by means of GPS. A straightforward choice for those values, $t = 10$ minutes and $T =$ one week, results in a pseudonym being valid for, e.g., Monday from 6:00 am until 6:10 am (Eckhoff *et al.* 2011b). Note that this pseudonym is then valid on every Monday for the said 10 minutes.

In addition to the time-slotting, SlotSwap furthermore proposes the use of pseudonym exchange between vehicles, a technique that was introduced by Li *et al.* (2006). Only pseudonyms valid for the same time-slot on both vehicles can be exchanged, otherwise it cannot be guaranteed that every vehicle has exactly one pseudonym per time-slot. It is also important that this exchange be realized using encryption algorithms to make sure that the exchange cannot be followed by a third party.

As we have discussed before, the exchange can only be privacy-preserving if pseudonyms are exchanged between cars that are not distinguishable by other context information such as their speed or heading. SlotSwap uses this context information to establish a set of candidate nodes that can be involved in the pseudonym exchange.

Further, in many situations, exchanges can be observed, enabling an attacker to determine which vehicles are communicating. Thus, SlotSwap is built to ensure that this information does not help the attacker determine how the pseudonym pools change. In 50% of cases pseudonyms are not exchanged for the current time-slot, but for a

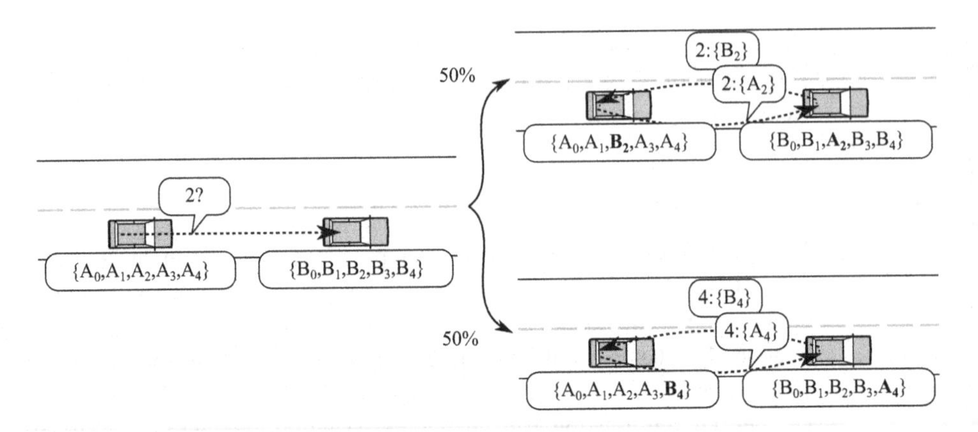

Figure 7.8 Possible pseudonym exchange between two vehicles.

random time-slot – preferably involving a pseudonym that has been used, but not yet exchanged.

Figure 7.8 outlines possible flows of the pseudonym-exchange process in SlotSwap. The figure shows two scenarios. In the first scenario, a pseudonym exchange is requested and performed for the current time-slot. In the second scenario, a pseudonym exchange is requested for the current time-slot, but performed for a different one. Since all communication is encrypted, these two scenarios are indistinguishable for an observer.

The privacy-preserving techniques discussed here help to increase the level of privacy while maintaining security. Still, there are many open questions, primarily concerning the identification of compromised certificates such as these used for the pseudonyms. CRLs are used in this case, but the exchange of CRLs brings new problems in both domains, privacy and security (Papadimitratos *et al.* 2008b; Haas *et al.* 2011; Eckhoff *et al.* 2013a).

References

Abolhasan, M., Wysocki, T. & Dutkiewicz, E. (2004), 'A review of routing protocols for mobile ad hoc networks', *Ad Hoc Networks* **2**(1), 1–22.

Adams, C., Farrell, S., Krause, T. & Mononen, T. (2005), Internet X.509 Public Key Infrastructure Certificate Management Protocol (CMP), RFC 4210, IETF.

Ahmed, S., Karmakar, G. C. & Kamruzzaman, J. (2010), 'An environment-aware mobility model for wireless ad hoc network', *Elsevier Computer Networks* **54**(9), 1470–1489.

Ahn, J., Wang, Y., Yu, B., Bai, F. & Krishnamachari, B. (2012), RISA: Distributed road information sharing architecture, in *31st IEEE Conference on Computer Communications (INFOCOM 2012)*, IEEE, Orlando, FL, pp. 1494–1502.

Altintas, O., Seki, K., Kremo, H. *et al.* (2014), 'Vehicles as information hubs during disasters: Glueing Wi-Fi to TV white space to cellular networks', *IEEE Intelligent Transportation Systems Magazine* **6**(1), 68–71.

Amoroso, A., Marfia, G. & Roccetti, M. (2011), 'Going realistic and optimal: A distributed multi-hop broadcast algorithm for vehicular safety', *Computer Networks* **55**(10), 2504–2519.

Artimy, M. M., Robertson, W. & Phillips, W. J. (2005), Assignment of dynamic transmission range based on estimation of vehicle density, in *2nd ACM International Workshop on Vehicular Ad hoc Networks (VANET 2005)*, ACM, Cologne, p. 40.

Asefi, M., Mark, J. & Shen, X. (2010), A cross-layer path selection scheme for video streaming over vehicular ad-hoc networks, in *72nd IEEE Vehicular Technology Conference (VTC2010-Fall)*, IEEE, Ottawa.

Aslam, B., Amjad, F. & Zou, C. (2012), Optimal roadside units placement in urban areas for vehicular networks, in *IEEE Symposium on Computers and Communications (ISCC 2012)*, IEEE, Cappadocia, pp. 423–429.

Atev, S., Masoud, O., Janardan, R. & Papanikolopoulos, N. P. (2004), Real-Time Collison Warning and Avoidance at Intersections, Technical Report Mn/DOT 2004-45, University of Minnesota, ITS Institute.

Atzori, L., Iera, A. & Morabito, G. (2010), 'The Internet of Things: A survey', *Elsevier Computer Networks* **54**(15), 2787–2805.

Autosar (2011), Specification of FlexRay Transport Layer, Specification R3.2 Rev 1 V2.4.0, Autosar.

Awad, A., German, R. & Dressler, F. (2009a), Efficient routing and service discovery in sensor networks using virtual cord routing, in *7th ACM International Conference on Mobile Systems, Applications, and Services (MobiSys 2009), Demo Session*, ACM, Kraków.

Awad, A., German, R. & Dressler, F. (2011), 'Exploiting virtual coordinates for improved routing performance in sensor networks', *IEEE Transactions on Mobile Computing* **10**(9), 1214–1226.

Awad, A., Shi, L. R., German, R. & Dressler, F. (2009b), Advantages of virtual addressing for efficient and failure tolerant routing in sensor networks, in *6th IEEE/IFIP Conference on Wireless on Demand Network Systems and Services (WONS 2009)*, IEEE, Snowbird, UT, pp. 111–118.

Awad, A., Sommer, C., German, R. & Dressler, F. (2008), Virtual Cord Protocol (VCP): A flexible DHT-like routing service for sensor networks, in *5th IEEE International Conference on Mobile Ad-hoc and Sensor Systems (MASS 2008)*, IEEE, Atlanta, GA, pp. 133–142.

Badescu, V. (2003), 'Dynamic model of a complex system including PV cells, electric battery, electrical motor and water pump', *Elsevier Energy* **28**(12), 1165–1181.

Bahl, P., Chandra, R. & Dunagan, J. (2004), SSCH: Slotted Seeded Channel Hopping for capacity improvement in IEEE 802.11 ad-hoc wireless networks, in *10th Annual International Conference on Mobile Computing and Networking*, ACM, Philadelphia, Pennsylvania, USA, pp. 216–230.

Bai, F. & Helmy, A. (2004), A survey of mobility models in wireless adhoc networks, in *Wireless Ad Hoc and Sensor Networks*, Kluwer Academic Publishers.

Barceló, J., ed. (2010), *Fundamentals of Traffic Simulation*, Springer.

Barfield, W., Haselkorn, M., Spyridakis, J. & Conquest, L. (1989), Commuter behavior and decision-making: Designing motorist information systems, in *33rd Human Factors and Ergonomics Society Annual Meeting*, Santa Monica, CA, pp. 611–614.

Barisani, A. & Daniele, B. (2007), Unusual car navigation tricks: Injecting RDS-TMC traffic information signals, in *8th CanSecWest Applied Technical Security Conference (CanSecWest 2007)*, Vancouver, BC.

Barr, R., Haas, Z. J. & van Renesse, R. (2005), 'JiST: An efficient approach to simulation using virtual machines', *Software: Practice and Experience* **35**(6), 539–576.

Barrachina, J., Garrido, P., Fogue, M. & Martinez, F. J. (2012), D-RSU: A density-based approach for road side unit deployment in urban scenarios, in *IEEE Intelligent Vehicles Symposium 2012, Workshops*, IEEE, Alcalá de Henares, Spain.

Basagni, S., Conti, M. & Stojmenovic, S. G. I., eds. (2004), *Mobile Ad Hoc Networking*, Wiley-IEEE.

Bates, D. M. & Watts, D. G. (1988), *Nonlinear Regression Analysis and Its Applications*, Wiley.

Bel Geddes, N. (1940), *Magic Motorways*, Random House.

Benslimane, A. (2004), Optimized dissemination of alarm messages in vehicular ad-hoc networks (VANET), in *IEEE International Conference on High Speed Networks and Multimedia Communications (HSNMC 2004)*, Vol. LNCS 3079, Springer, Toulouse, pp. 655–666.

Bergenhem, C., Huang, Q., Benmimoun, A. & Robinson, T. (2010), Challenges of platooning on public motorways, in *17th World Congress on Intelligent Transport Systems*, Busan.

Bloessl, B., Segata, M., Sommer, C. & Dressler, F. (2013a), An IEEE 802.11a/g/p OFDM receiver for GNU radio, in *ACM SIGCOMM 2013, 2nd ACM SIGCOMM Workshop of Software Radio Implementation Forum (SRIF 2013)*, ACM, Hong Kong, pp. 9–16.

Bloessl, B., Segata, M., Sommer, C. & Dressler, F. (2013b), Decoding IEEE 802.11a/g/p OFDM in software using GNU radio, in *19th ACM International Conference on Mobile Computing and Networking (MobiCom 2013), Demo Session*, ACM, Miami, FL, pp. 159–161.

Bloessl, B., Segata, M., Sommer, C. & Dressler, F. (2013c), Towards an open source IEEE 802.11p stack: A full SDR-based transceiver in GNU radio, in *5th IEEE Vehicular Networking Conference (VNC 2013)*, IEEE, Boston, MA, pp. 143–149.

Bloom, B. H. (1970), 'Space/time trade-offs in hash coding with allowable errors', *Communications of the ACM* **13**(7), 422–426.

Boban, M., Meireles, R., Barros, J., Steenkiste, P. & Tonguz, O. (2014), 'TVR – tall vehicle relaying in vehicular networks', *IEEE Transactions on Mobile Computing* **13**(5), 1118–1131.

Boban, M., Vinhosa, T., Barros, J., Ferreira, M. & Tonguz, O. K. (2011), 'Impact of vehicles as obstacles in vehicular networks', *IEEE Journal on Selected Areas in Communications (JSAC)* **29**(1), 15–28.

Bosch (2012), CAN with Flexible Data-Rate, CAN FD Specification 1.0, Robert Bosch GmbH.

Bowyer, D. P., Akcelik, R. & Biggs, D. C. (1985), Guide to Fuel Consumption Analysis for Urban Traffic Management, ARRB Special Report 32, Australian Road Research Board.

Brannstrom, M., Coelingh, E. & Sjoberg, J. (2010), 'Model-based threat assessment for avoiding arbitrary vehicle collisions', *IEEE Transactions on Intelligent Transportation Systems* **11**(3), 658–669.

Braun, R., Busch, F., Kemper, C. *et al.* (2009), 'TRAVOLUTION – Netzweite Optimierung der Lichtsignalsteuerung und LSA-Fahrzeug-Kommunikation', *Strassenverkehrstechnik* **53**, 365–374.

Brik, V., Banerjee, S., Gruteser, M. & Oh, S. (2008), Wireless device identification with radiometric signatures, in *13th ACM International Conference on Mobile Computing and Networking (MobiCom 2008)*, ACM, San Francisco, CA, pp. 116–127.

Buttyán, L., Holczer, T. & Vajda, I. (2007), On the effectiveness of changing pseudonyms to provide location privacy in VANETs, in *4th European Workshop on Security and Privacy in Ad hoc and Sensor Networks (ESAS 2007)*, Springer, Cambridge.

Cacciabue, P. C., ed. (2007), *Modelling Driver Behaviour in Automotive Environments: Critical Issues in Driver Interactions with Intelligent Transport Systems*, Springer.

Caesar, M., Castro, M., Nightingale, E. B., O'Shea, G. & Rowstron, A. (2006), Virtual ring routing: Network routing inspired by DHTs, in *ACM SIGCOMM 2006*, ACM, Pisa.

Cai, J. & Goodman, D. (1997), 'General packet radio service in GSM', *IEEE Communications Magazine* **35**(10), 122–131.

Calandriello, G., Papadimitratos, P., Hubaux, J. P. & Lioy, A. (2007), Efficient and robust pseudonymous authentication in VANET, in *4th ACM International Workshop on Vehicular Ad Hoc Networks (VANET 2007)*, ACM, Montréal, QC, pp. 19–28.

Caliskan, M., Graupner, D. & Mauve, M. (2006), Decentralized discovery of free parking places, in *3rd ACM International Workshop on Vehicular Ad Hoc Networks (VANET 2006)*, ACM, Los Angeles, CA, pp. 30–39.

Camp, T., Boleng, J. & Davies, V. (2002), 'A survey of mobility models for ad hoc network research', *Wireless Communications and Mobile Computing, Special Issue on Mobile Ad Hoc Networking: Research, Trends and Applications* **2**(5), 483–502.

Cao, Y. & Sun, Z. (2013), 'Routing in delay/disruption tolerant networks: A taxonomy, survey and challenges', *IEEE Communications Surveys and Tutorials* **15**(2), 654–677.

Capkun, S., Buttyan, L. & Hubaux, J.-P. (2003), SECTOR: Secure tracking of node encounters in multi-hop wireless networks, in *1st ACM Workshop on Security of Ad Hoc and Sensor Networks (SASN 2003)*, ACM, Fairfax, VI, pp. 21–32.

Cappiello, A., Chabini, I., Nam, E., Lue, A. & Abou Zeid, M. (2002), A statistical model of vehicle emissions and fuel consumption, in *IEEE Intelligent Transportation Systems Conference (ITSC 2002)*, pp. 801–809.

Casetti, C., Dressler, F., Gerla, M. *et al.* (2013), Working Group on Heterogeneous Vehicular Networks, in *Dagstuhl Seminar 13392 – Inter-Vehicular Communication – Quo Vadis*, Wadern, pp. 201–204.

CCSDS (2010), Wireless Network Communications Overview for Space Mission Operations, Informational Report CCSDS 880.0-G-1, Consultative Committee for Space Data Systems.

Chang, S.-H., Lin, C.-Y., Hsu, C.-C., Fung, C.-P. & Hwang, J.-R. (2009), 'The effect of a collision warning system on the driving performance of young drivers at intersections', *Transportation Research Part F: Traffic Psychology and Behaviour* **12**(5), 371–380.

Chau, C.-K., Crowcroft, J., Lee, K.-W. & Wong, S. H. Y. (2008), Inter-domain routing protocol for mobile ad hoc networks, in *ACM SIGCOMM 2008, 3rd ACM International Workshop on Mobility in the Evolving Internet Architecture (MobiArch 2008)*, ACM, Seattle, WA, pp. 61–66.

Chaum, D. (1981), 'Untraceable electronic mail, return addresses, and digital pseudonyms', *Communications of the ACM* **24**, 84–88.

Chen, B. B. & Chan, M. C. (2009), MobTorrent: A framework for mobile internet access from vehicles, in *28th IEEE Conference on Computer Communications (INFOCOM 2009)*, IEEE, Rio de Janeiro.

Chen, H., Cao, L. & Logan, D. B. (2011a), 'Investigation into the effect of an intersection crash warning system on driving performance in a simulator', *Traffic Injury Prevention* **12**(5), 529–537.

Chen, K. & Ervin, R. D. (1990), 'Intelligent vehicle–highway systems: U.S. activities and policy issues', *Technological Forecasting and Social Change* **38**(4), 363–374.

Chen, S., Wyglinski, A. M., Pagadarai, S., Vuyyuru, R. & Altintas, O. (2011b), 'Feasibility analysis of vehicular dynamic spectrum access via queueing theory model', *IEEE Communications Magazine* **49**(11), 156–163.

Cheng, H., Cavedon, L., Dale, R. *et al.* (2008), Method and system for adaptive navigation using a driver's route knowledge, Patent US7424363 B2, Robert Bosch Corporation.

Chennikara-Varghese, J., Chen, W., Altintas, O. & Cai, S. (2006), Survey of routing protocols for inter-vehicle communications, in *3rd Annual International Conference on Mobile and Ubiquitous Systems: Networks and Services (MOBIQUITOUS 2006)*, pp. 1–5.

Chigan, C. & Li, J. (2007), A delay-bounded dynamic interactive power control algorithm for VANETs, in *IEEE International Conference on Communications (ICC 2007)*, IEEE, Glasgow, pp. 5849–5855.

Chu, Y.-C. & Huang, N.-F. (2007), Delivering of live video streaming for vehicular communication using peer-to-peer approach, in *26th IEEE Conference on Computer Communications (INFOCOM 2007): Mobile Networking for Vehicular Environments (MOVE 2007)*, pp. 1–6.

Collins, K. & Muntean, G.-M. (2008), A vehicle route management solution enabled by wireless vehicular networks, in *27th IEEE Conference on Computer Communications (INFOCOM 2008): Mobile Networking for Vehicular Environments (MOVE 2008), Poster Session*, Phoenix, AZ.

Cooper, D., Santesson, S. *et al.* (2008), Internet X.509 Public Key Infrastructure Certificate and Certificate Revocation List (CRL) Profile, Technical Report 5280, IETF.

Dahlman, E., Parkvall, S. & Sköld, J. (2011), *4G LTE/LTE-Advanced for Mobile Broadband*, Academic Press.

Dahlman, E., Parkvall, S., Sköld, J. & Beming, P. (2008), *3G Evolution HSPA and LTE for Mobile Broadband*, 2nd edn., Academic Press.

Davila, A. & Nombela, M. (2012), Platooning – Safe and eco-friendly mobility, in *SAE 2012 World Congress & Exibition*, SAE, Detroit, MI.

De Couto, D. S. J., Aguayo, D., Chambers, B. A. & Morris, R. (2003), 'Performance of multihop wireless networks: Shortest path is not enough', *ACM SIGCOMM Computer Communication Review* **33**(1), 83–88.

Dhoutaut, D., Régis, A. & Spies, F. (2006), Impact of radio propagation models in vehicular ad hoc networks simulations, in *3rd ACM International Workshop on Vehicular Ad Hoc Networks (VANET 2006)*, ACM, Los Angeles, CA, pp. 40–49.

Di Felice, M., Doost-Mohammady, R., Chowdhury, K. & Bononi, L. (2012), 'Smart radios for smart vehicles: Cognitive vehicular networks', *IEEE Vehicular Technology Magazine* **7**(2), 26–33.

Dietrich, I. & Dressler, F. (2009), 'On the lifetime of wireless sensor networks', *ACM Transactions on Sensor Networks* **5**(1), 1–39.

Dingus, T., Hulse, M., Jahns, S. *et al.* (1996), Development of Human Factors Guidelines for Advanced Traveler Information Systems and Commercial Vehicle Operations: Literature Review, Report FHWA-RD-95-153, Federal Highway Administration.

Domazetovic, A., Greenstein, L. J., Mandayam, N. B. & Seskar, I. (2005), 'Propagation models for short-range wireless channels with predictable path geometries', *IEEE Transactions on Communications* **53**(7), 1123–1126.

Dressler, F. (2007), *Self-Organization in Sensor and Actor Networks*, John Wiley & Sons.

Dressler, F., Awad, A. & Gerla, M. (2010), Inter-domain routing and data replication in virtual coordinate based networks, in *IEEE International Conference on Communications (ICC 2010)*, IEEE, Cape Town.

Dressler, F., Awad, A., German, R. & Gerla, M. (2009), Enabling inter-domain routing in virtual coordinate based ad hoc and sensor networks, in *15th ACM International Conference on Mobile Computing and Networking (MobiCom 2009), Poster Session*, ACM, Beijing.

Dressler, F., Gansen, T., Sommer, C. & Wischhof, L. (2008), Requirements and objectives for secure traffic information systems, in *5th IEEE International Conference on Mobile Ad Hoc and Sensor Systems (MASS 2008): 4th IEEE International Workshop on Wireless and Sensor Networks Security (WSNS 2008)*, IEEE, Atlanta, GA, pp. 808–814.

Dressler, F. & Gerla, M. (2013), 'A framework for inter-domain routing in virtual coordinate based mobile networks', *ACM/Springer Wireless Networks* **19**(7), 1611–1626.

Dressler, F., Hartenstein, H., Altintas, O. & Tonguz, O. K. (2014), 'Inter-vehicle communication – quo vadis', *IEEE Communications Magazine* **52**(6), 170–177.

Dressler, F., Kargl, F., Ott, J., Tonguz, O. K. & Wischhof, L. (2011a), 'Research challenges in inter-vehicular communication – Lessons of the 2010 Dagstuhl Seminar', *IEEE Communications Magazine* **49**(5), 158–164.

Dressler, F. & Sommer, C. (2010), On the impact of human driver behavior on intelligent transportation systems, in *71st IEEE Vehicular Technology Conference (VTC2010-Spring)*, IEEE, Taipei, pp. 1–5.

Dressler, F., Sommer, C., Eckhoff, D. & Tonguz, O. K. (2011b), 'Towards realistic simulation of inter-vehicle communication: Models, techniques and pitfalls', *IEEE Vehicular Technology Magazine* **6**(3), 43–51.

Dupuy, A., Schwartz, J. & Yemini, Y. (1989), Nest: A network simulation prototyping testbed, in *Winter Simulation Conference (WSC 1989)*, IEEE, Washington, D.C., pp. 1058–1064.

Eckert, J., German, R. & Dressler, F. (2011), 'An indoor localization framework for four-rotor flying robots using low-power sensor nodes', *IEEE Transactions on Instrumentation and Measurement* **60**(2), 336–344.

Eckhoff, D., Dressler, F. & Sommer, C. (2013a), SmartRevoc: An efficient and privacy preserving revocation system using parked vehicles, in *38th IEEE Conference on Local Computer Networks (LCN 2013)*, IEEE, Sydney, pp. 855–862.

Eckhoff, D., Gansen, T., Mänz, R. *et al.* (2011a), Simulative performance evaluation of the simTD self organizing traffic information system, in *10th IFIP/IEEE Annual Mediterranean Ad Hoc Networking Workshop (Med-Hoc-Net 2011)*, IEEE, Favignana Island, Sicily, pp. 79–86.

Eckhoff, D., Sofra, N. & German, R. (2013b), A performance study of cooperative awareness in ETSI ITS G5 and IEEE WAVE, in *10th IEEE/IFIP Conference on Wireless on Demand Network Systems and Services (WONS 2013)*, IEEE, Banff, pp. 196–200.

Eckhoff, D. & Sommer, C. (2014), 'Driving for big data? Privacy concerns in vehicular networking', *IEEE Security and Privacy* **12**(1), 77–79.

Eckhoff, D., Sommer, C. & Dressler, F. (2012), On the necessity of accurate IEEE 802.11p models for IVC protocol simulation, in *75th IEEE Vehicular Technology Conference (VTC2012-Spring)*, IEEE, Yokohama, pp. 1–5.

Eckhoff, D., Sommer, C., Gansen, T., German, R. & Dressler, F. (2010), Strong and affordable location privacy in VANETs: Identity diffusion using time-slots and swapping, in *2nd IEEE Vehicular Networking Conference (VNC 2010)*, IEEE, Jersey City, NJ, pp. 174–181.

Eckhoff, D., Sommer, C., Gansen, T., German, R. & Dressler, F. (2011b), 'SlotSwap: Strong and affordable location privacy in intelligent transportation systems', *IEEE Communications Magazine* **49**(11), 126–133.

Eckhoff, D., Sommer, C., German, R. & Dressler, F. (2011c), Cooperative awareness at low vehicle densities: How parked cars can help see through buildings, in *IEEE Global Telecommunications Conference (GLOBECOM 2011)*, IEEE, Houston, TX.

Egea-Lopez, E., Alcaraz, J., Vales-Alonso, J., Festag, A. & Garcia-Haro, J. (2013), 'Statistical beaconing congestion control for vehicular networks', *IEEE Transactions on Vehicular Technology* **62**(9), 4162–4181.

ETSI (1997), Selection Procedures for the Choice of Radio Transmission Technologies of the UMTS, TR 101 112 V3.1.0 (1997-11), European Telecommunications Standards Institute.

ETSI (2010a), Intelligent Transport Systems (ITS); Vehicular Communications; Basic Set of Applications; Part 2: Specification of Cooperative Awareness Basic Service, TS 102 637-2 V1.1.1, European Telecommunications Standards Institute.

ETSI (2010b), Intelligent Transport Systems (ITS); Vehicular Communications; GeoNetworking; Part 1: Requirements, TS 102 636-1 V1.1.1, European Telecommunications Standards Institute.

ETSI (2011), Intelligent Transport Systems (ITS); Decentralized Congestion Control Mechanisms for Intelligent Transport Systems Operating in the 5 GHz Range; Access Layer Part, TS 102 687 V1.1.1, European Telecommunications Standards Institute.

ETSI (2012a), Intelligent Transport Systems (ITS); Framework for Public Mobile Networks in Cooperative ITS (C-ITS), Technical Report 102 962 V1.1.1, European Telecommunications Standards Institute.

ETSI (2012b), Introduction of the Multimedia Broadcast/Multicast Service (MBMS) in the Radio Access Network (RAN); Stage 2, TS 125 346 V11.0.0, European Telecommunications Standards Institute.

ETSI (2013a), Intelligent Transport Systems (ITS); Users and Applications Requirements; Part 1: Facility Layer Structure, Functional Requirements and Specifications, TS 102 894-1 V1.1.1, European Telecommunications Standards Institute.

ETSI (2013b), Intelligent Transport Systems (ITS); Vehicular Communications; Basic Set of Applications; Part 2: Specification of Cooperative Awareness Basic Service, EN 302 637-2 V1.3.0, European Telecommunications Standards Institute.

ETSI (2013c), Intelligent Transport Systems (ITS); Vehicular Communications; Basic Set of Applications; Part 3: Specification of Decentralized Environmental Notification Basic Service, Technical Report 302 637-3 V1.2.0, European Telecommunications Standards Institute.

Evans, L. (1991), *Traffic Safety and the Driver*, Van Nostrand Reinhold.

Fall, K. (2003), A delay-tolerant network architecture for challenged internets, in *ACM SIGCOMM 2003*, ACM, Karlsruhe, pp. 27–34.

FCC (2002), Report, ET Docket 02-135, FCC Spectrum Policy Task Force.

Fellendorf, M. & Vortisch, P. (2010), Microscopic traffic flow simulator VISSIM, in J. Barceló, ed., *Fundamentals of Traffic Simulation*, Springer, pp. 63–93.

Fernandes, P. & Nunes, U. (2012), 'Platooning with IVC-enabled autonomous vehicles: Strategies to mitigate communication delays, improve safety and traffic flow', *IEEE Transactions on Intelligent Transportation Systems* **13**(1), 91–106.

Ferreira, M. & d'Orey, P. (2012), 'On the impact of virtual traffic lights on carbon emissions mitigation', *IEEE Transactions on Intelligent Transportation Systems* **13**(1), 284–295.

Ferreira, M., Fernandes, R., Conceição, H., Viriyasitavat, W. & Tonguz, O. K. (2010), Self-organized traffic control, in *7th ACM International Workshop on Vehicular Internetworking (VANET 2010)*, ACM, Chicago, IL, pp. 85–90.

Ferreiro-Lage, J., Vazquez-Caderno, P., Galvez, J., Rubios, O. & Aguado-Agelet, F. (2010), Active safety evaluation in car-to-car networks, in *6th International Conference on Networking and Services (ICNS 2010)*, Cancún, pp. 288–292.

Ferula, P. (2013), Assessing the impact of road-side units on distributed beacon-based traffic information systems, Master's thesis, University of Innsbruck.

Flury, R., Pemmaraju, S. V. & Wattenhofer, R. (2009), Greedy routing with bounded stretch, in *28th IEEE Conference on Computer Communications (INFOCOM 2009)*, IEEE, Rio de Janeiro.

Font, J., Iñigo, P., Domínguez, M., Sevillano, J. L. & Amaya, C. (2010), Architecture, design and source code comparison of ns-2 and ns-3 network simulators, in *2010 Spring Simulation Multiconference (SpringSim 2010)*, SCS, Orlando, FL.

Franklin, J., McCoy, D., Tabriz, P. *et al.* (2006), Passive data link layer 802.11 wireless device driver fingerprinting, in *15th USENIX Security Symposium*, USENIX, Vancouver, BC, pp. 167–178.

Friis, H. (1946), 'A note on a simple transmission formula', *Proceedings of the IRE* **34**(5), 254–256.

Gerlach, M. & Güttler, F. (2007), Privacy in VANETs using changing pseudonyms – ideal and real, in *65th IEEE Vehicular Technology Conference (VTC2007-Spring)*, Dublin, pp. 2521–2525.

Giordano, E., Frank, R., Ghosh, A., Pau, G. & Gerla, M. (2009), Two ray or not two ray this is the price to pay, in *6th IEEE International Conference on Mobile Ad Hoc and Sensor Systems (MASS 2009)*, Macau SAR, pp. 603–608.

Giordano, E., Frank, R., Pau, G. & Gerla, M. (2010), CORNER: A realistic urban propagation model for VANET, in *7th IEEE/IFIP Conference on Wireless On Demand Network Systems and Services (WONS 2010), Poster Session*, IEEE, Kranjska Gora, Slovenia, pp. 57–60.

Gläser, S., Sommer, C., Gehlen, G. & Sories, S. (2008), 'CoCar – Cooperative vehicle applications based on cellular communication systems', *ATZ Elektronik Worldwide* **2008**(5), 14–17.

Gozálvez, J., Sepulcre, M. & Bauza, R. (2012), 'IEEE 802.11p vehicle to infrastructure communications in urban environments', *IEEE Communications Magazine* **50**(5), 176–183.

Gradinescu, V., Gorgorin, C., Diaconescu, R., Cristea, V. & Iftode, L. (2007), Adaptive traffic lights using car-to-car communication, in *65th IEEE Vehicular Technology Conference (VTC2007-Spring)*, pp. 21–25.

Grossglauser, M. & Tse, D. N. C. (2002), 'Mobility increases the capacity of ad hoc wireless networks', *IEEE/ACM Transations on Networking* **10**(4), 477–486.

Grossglauser, M. & Vetterli, M. (2006), 'Locating mobile nodes with EASE: Learning efficient routes from encounter histories alone', *IEEE/ACM Transactions on Networking* **14**(3), 457–469.

Guo, M., Ammar, M. H. & Zegura, E. W. (2005), 'V3: A vehicle-to-vehicle live video streaming architecture', *Elsevier Pervasive and Mobile Computing* **1**(4), 404–424.

Gupta, P. & Kumar, P. (2000), 'The capacity of wireless networks', *IEEE Transactions on Information Theory* **46**(2), 388–404.

Haas, J. J., Hu, Y.-C. & Laberteaux, K. P. (2011), 'Efficient certificate revocation list organization and distribution', *IEEE Journal on Selected Areas in Communications* **29**(3), 595–604.

Härri, J., Cataldi, P., Krajzewicz, D. *et al.* (2011), Modeling and simulating ITS applications with iTETRIS, in *6th ACM Workshop on Performance Monitoring and Measurement of Heterogeneous Wireless and Wired Networks (PM2HW2N 2011)*, ACM, Miami Beach, FL, pp. 33–40.

Härri, J., Filali, F. & Bonnet, C. (2009), 'Mobility models for vehicular ad hoc networks: A survey and taxonomy', *IEEE Communications Surveys and Tutorials* **11**(4), 19–41.

Hartenstein, H. & Laberteaux, K. P. (2008), 'A tutorial survey on vehicular ad hoc networks', *IEEE Communications Magazine* **46**(6), 164–171.

Hedrick, C. (1998), 'Routing Information Protocol', RFC 1058.

Hong, X., Xu, K. & Gerla, M. (2002), 'Scalable routing protocols for mobile ad hoc networks', *IEEE Network* **16**, 11–21.

Housley, R. & Polk, T. (2001), *Planning for PKI: Best Practices Guide for Deploying Public Key Infrastructure*, 1st edn., John Wiley & Sons, Inc.

Huang, H.-Y., Luo, P.-E., Li, M. *et al.* (2007), 'Performance evaluation of SUVnet with real-time traffic data', *IEEE Transactions on Vehicular Technology* **56**(6), 3381–3396.

Hubaux, J.-P., Čapkun, S. & Luo, J. (2004), 'The security and privacy of smart vehicles', *IEEE Security and Privacy* **2**(3), 49–55.

Hui, P., Chaintreau, A., Scott, J. *et al.* (2005), Pocket switched networks and human mobility in conference environments, in *ACM SIGCOMM Workshop on Delay-Tolerant Networking (WDTN 2005)*, ACM, Philadelphia, PA, pp. 244–251.

Hung, C.-C., Chan, H. & Wu, E.-K. (2008), Mobility pattern aware routing for heterogeneous vehicular networks, in *IEEE Wireless Communications and Networking Conference (WCNC 2008)*, Las Vegas, NV, pp. 2200–2205.

Hwang, Y. & Varshney, P. (2003), An adaptive QoS routing protocol with dispersity for ad-hoc networks, in *36th Annual Hawaii International Conference on System Sciences (HICSS 2003)*, Big Island, Hawaii.

IEEE (2006), IEEE Trial-Use Standard for Wireless Access in Vehicular Environments (WAVE) – Resource Manager, Std 1609.1, IEEE.

IEEE (2007), Wireless LAN Medium Access Control (MAC) and Physical Layer (PHY) Specifications, Std 802.11-2007, IEEE.

IEEE (2010a), IEEE Standard for Wireless Access in Vehicular Environments (WAVE) – Networking Services, Std 1609.3, IEEE.

IEEE (2010b), Wireless Access in Vehicular Environments, Draft Standard P802.11p/D10.0, IEEE.

IEEE (2010c), Wireless Access in Vehicular Environments, Std 802.11p-2010, IEEE.

IEEE (2011), IEEE Standard for Wireless Access in Vehicular Environments (WAVE) – Multi-channel Operation, Std 1609.4, IEEE.

IEEE (2013), IEEE Standard for Wireless Access in Vehicular Environments – Security Services for Applications and Management Messages, Technical Report 1609.2, IEEE.

ISO (2003a), Specification of the Radio Data System (RDS) for VHF/FM Sound Broadcasting in the Frequency Range from 87.5 to 108.0 MHz, Technical Report 62106, ISO.

ISO (2003b), Traffic and Traveller Information (TTI) – TTI Messages via Traffic Message Coding – Part 1: Coding Protocol for Radio Data System (RDS-TMC) using ALERT-C, Technical Report 14819-1, ISO.

ISO (2003c), Traffic and Traveller Information (TTI) – TTI Messages via Traffic Message Coding – Part 2: Event and Information Codes for Radio Data System – Traffic Message Channel (RDS-TMC), Technical Report 14819-2, ISO.

ITU (2013), The World in 2013, ICT Facts and Figures 2013, ITU Telecommunication Development Bureau.

ITU-R (2007), Propagation by Diffraction, Rec. P.526-10.

ITU-T (2000), Information Technology – Open Systems Interconnection – The Directory: Public-Key and Attribute Certificate Frameworks, Rec. X.509.

Izal, M., Urvoy-Keller, G., Biersack, E. W. *et al.* (2004), Dissecting BitTorrent: Five months in a torrent's lifetime, in *10th International Workshop on Passive and Active Network Measurement (PAM 2004)*, Vol. 3015 of *LNCS*, Antibes Juan-les-Pins.

Jetcheva, J. G., Hu, Y.-C., PalChaudhuri, S., Saha, A. K. & Johnson, D. B. (2003), Design and evaluation of a metropolitan area multitier wireless ad hoc network architecture, in *5th IEEE Workshop on Mobile Computing Systems and Applications*, pp. 32–43.

Joerer, S., Dressler, F. & Sommer, C. (2012a), Comparing apples and oranges? Trends in IVC simulations, in *9th ACM International Workshop on Vehicular Internetworking (VANET 2012)*, ACM, Low Wood Bay, pp. 27–32.

Joerer, S., Segata, M., Bloessl, B. *et al.* (2012b), To crash or not to crash: Estimating its likelihood and potentials of beacon-based IVC Systems, in *4th IEEE Vehicular Networking Conference (VNC 2012)*, IEEE, Seoul, pp. 25–32.

Joerer, S., Segata, M., Bloessl, B. *et al.* (2014), 'A vehicular networking perspective on estimating vehicle collision probability at intersections', *IEEE Transactions on Vehicular Technology* **63**(4), 1802–1812.

Joerer, S., Sommer, C. & Dressler, F. (2012c), 'Toward reproducibility and comparability of IVC simulation studies: A literature survey', *IEEE Communications Magazine* **50**(10), 82–88.

Johnson, D. B., Hu, Y. & Maltz, D. A. (2007), 'The Dynamic Source Routing Protocol (DSR) for Mobile Ad Hoc Networks for IPv4', RFC 4728.

Johnson, D. B. & Maltz, D. A. (1996), Dynamic source routing in ad hoc wireless networks, in T. Imielinski & H. F. Korth, eds., *Mobile Computing*, Kluwer Academic Publishers, pp. 152–181.

Johnson, D. B., Maltz, D. A. & Broch, J. (2001), DSR: The dynamic source routing protocol for multi-hop wireless ad hoc networks, in C. E. Perkins, ed., *Ad Hoc Networking*, Addison-Wesley, pp. 139–172.

Jootel, P. S. (2012), SAfe Road TRains for the Environment, final project report, SARTRE Project.

Jurgen, R. (1991), 'Smart cars and highways go global', *IEEE Spectrum* **28**(5), 26–36.

Kalman, R. (1960), 'A new approach to linear filtering and prediction problems', *Transactions of the ASME Journal of Basic Engineering* **D**(82), 35–45.

Karedal, J., Czink, N., Paier, A., Tufvesson, F. & Molisch, A. (2011), 'Path loss modeling for vehicle-to-vehicle communications', *IEEE Transactions on Vehicular Technology* **60**(1), 323–328.

Kargl, F., Ma, Z. & Schoch, E. (2006), Security engineering for VANETs, in *4th Workshop on Embedded Security in Cars (ESCAR 06)*, Berlin.

Kargl, F., Schoch, E., Wiedersheim, B. & Leinmüller, T. (2008), Secure and efficient beaconing for vehicular networks, in *5th ACM International Workshop on Vehicular Inter-Networking (VANET 2008), Poster Session*, ACM, San Francisco, CA, pp. 82–83.

Karl, H. & Willig, A. (2005), *Protocols and Architectures for Wireless Sensor Networks*, John Wiley & Sons.

Karp, B. & Kung, H. T. (2000), GPSR: Greedy perimeter stateless routing for wireless networks, in *6th ACM International Conference on Mobile Computing and Networking (MobiCom 2000)*, ACM, Boston, MA, pp. 243–254.

Kerner, B., Klenov, S. L. & Brakemeier, A. (2008), Testbed for wireless vehicle communication: A simulation approach based on three-phase traffic theory, in *IEEE Intelligent Vehicles Symposium (IV 2008)*, IEEE, Eindhoven, pp. 180–185.

Khabbaz, M., Assi, C. & Fawaz, W. (2012), 'Disruption-tolerant networking: A comprehensive survey on recent developments and persisting challenges', *IEEE Communications Surveys and Tutorials* **14**(2), 607–640.

Kim, Y.-J., Govindan, R., Karp, B. & Shenker, S. (2005), Geographic routing made practical, in *USENIX/ACM Symposium on Networked Systems Design and Implementation (NSDI 2005)*, USENIX, San Francisco, CA.

Klimin, N., Enkelmann, W., Karl, H. & Wolisz, A. (2004), A hybrid approach for location-based service discovery in vehicular ad hoc networks, in *1st International Workshop on Intelligent Transportation (WIT)*, Hamburg.

Klingler, F., Dressler, F., Cao, J. & Sommer, C. (2013), Use both lanes: Multi-channel beaconing for message dissemination in vehicular networks, in *10th IEEE/IFIP Conference on Wireless on Demand Network Systems and Services (WONS 2013)*, IEEE, Banff, pp. 162–169.

Kloiber, B., Härri, J. & Strang, T. (2012), Dice the TX power – Improving awareness quality in VANETs by random transmit power selection, in *4th IEEE Vehicular Networking Conference (VNC 2012)*, IEEE, Seoul, pp. 56–63.

Ko, Y.-B. & Vaidya, N. H. (2002), 'Flooding-based geocasting protocols for mobile ad hoc networks', *Mobile Networks and Applications* **7**(6), 471–480.

König, R., Saffran, A. & Breckle, H. (1994), Modelling of drivers' behaviour, in *Vehicle Navigation and Information Systems Conference*, Yokohama-shi, pp. 371–376.

Köpke, A., Swigulski, M., Wessel, K. *et al.* (2008), Simulating wireless and mobile networks in OMNeT++ – The MiXiM vision, in *1st ACM/ICST International Conference on Simulation Tools and Techniques for Communications, Networks and Systems (SIMUTools 2008): 1st ACM/ICST International Workshop on OMNeT++ (OMNeT++ 2008)*, ACM, Marseille.

Korkmaz, G., Ekici, E. & Özgüner, F. (2006), An efficient fully ad-hoc multi-hop broadcast protocol for inter-vehicular communication systems, in *IEEE International Conference on Communications (ICC 2006)*, Istanbul, pp. 423–428.

Koscher, K., Czeskis, A., Roesner, F. *et al.* (2010), Experimental security analysis of a modern automobile, in *2010 IEEE Symposium on Security and Privacy (SP 2010)*, Oakland, CA, pp. 447–462.

Krajzewicz, D., Erdmann, J., Behrisch, M. & Bieker, L. (2012), 'Recent development and applications of SUMO – Simulation of Urban MObility', *International Journal on Advances in Systems and Measurements* **5**(3&4), 128–138.

Kunisch, J. & Pamp, J. (2008), Wideband car-to-car radio channel measurements and model at 5.9 GHz, in *68th IEEE Vehicular Technology Conference (VTC 2008-Fall)*, IEEE, Calgary, pp. 1–5.

Kurkowski, S., Camp, T. & Colagrosso, M. (2005), 'MANET simulation studies: The incredibles', *ACM SIGMOBILE Mobile Computing and Communications Review* **9**(4), 50–61.

Kwoczek, A., Raida, Z., Lacik, J. *et al.* (2011), Influence of car panorama glass roofs on car2car communication, in *3rd IEEE Vehicular Networking Conference (VNC 2011), Poster Session*, IEEE, Amsterdam.

Kyasanur, P. & Vaidya, N. H. (2005), Capacity of multi-channel wireless networks: Impact of number of channels and interfaces, in *11th ACM International Conference on Mobile Computing and Networking (MobiCom 2005)*, ACM, Cologne, pp. 43–57.

Laberteaux, K. P., Haas, J. J. & Hu, Y.-C. (2008), Security certificate revocation list distribution for VANET, in *5th ACM International Workshop on Vehicular Inter-Networking (VANET 2008)*, ACM, San Francisco, CA, pp. 88–89.

Law, A. M. (2006), *Simulation, Modeling and Analysis*, 6th edn., McGraw-Hill.

Le, L., Festag, A., Baldessari, R. & Zhang, W. (2009), 'Vehicular wireless short-range communication for improving intersection safety', *IEEE Communications Magazine* **47**(11), 104–110.

Lee, J. & Kim, C. M. (2010), A roadside unit placement scheme for vehicular telematics networks, in *2010 International Conference on Advances in Computer Science (ACS 2010)*, Springer, Berlin, pp. 196–202.

Lee, K. C., Lee, U. & Gerla, M. (2009), TO-GO: TOpology-assist Geo-Opportunistic routing in urban vehicular grids, in *6th IEEE/IFIP Conference on Wireless on Demand Network Systems and Services (WONS 2009)*, IEEE, Snowbird, UT, pp. 11–18.

Lee, K. C., Lee, U. & Gerla, M. (2010a), Survey of routing protocols in vehicular ad hoc networks, in M. Watfa, ed., *Advances in Vehicular Ad-Hoc Networks: Developments and Challenges*, IGI Global, pp. 149–170.

Lee, K., Haerri, J., Lee, U. & Gerla, M. (2007), Enhanced perimeter routing for geographic forwarding protocols in urban vehicular scenarios, in *IEEE Global Telecommunications Conference (GLOBECOM 2007), 2nd IEEE Workshop on Automotive Networking and Applications (AutoNet 2007)*, IEEE, Washington, D.C., pp. 1–10.

Lee, K., Lee, U. & Gerla, M. (2010b), 'Geo-opportunistic routing for vehicular networks', *IEEE Communications Magazine* **48**(5), 164–170.

Lee, U., Lee, J., Park, J.-S. & Gerla, M. (2010c), 'FleaNet: A virtual market place on vehicular networks', *IEEE Transactions on Vehicular Technology* **59**(1), 344–355.

Leen, G. & Heffernan, D. (2001), 'Vehicles without wires', *Computing Control Engineering Journal* **12**(5), 205–211.

Leinmüller, T., Maihöfer, C., Schoch, E. & Kargl, F. (2006), Improved security in geographic ad hoc routing through autonomous position verification, in *3rd ACM International Workshop on Vehicular Ad Hoc Networks (VANET 2006)*, Los Angeles, CA, pp. 57–66.

Leong, B., Liskov, B. & Morris, R. (2007), Greedy virtual coordinates for geographic routing, in *15th IEEE International Conference on Network Protocols (ICNP 2007)*, Beijing, pp. 71–80.

Lequerica, I., Martinez, J. & Ruiz, P. (2010), Efficient certificate revocation in vehicular networks using NGN capabilities, in *72nd IEEE Vehicular Technology Conference Fall (VTC2010-Fall)*, IEEE, Ottawa, pp. 1–5.

Li, F. & Wang, Y. (2007), 'Routing in vehicular ad hoc networks: A survey', *IEEE Vehicular Technology Magazine* **2**(2), 12–22.

Li, M., Sampigethaya, K., Huang, L. & Poovendran, R. (2006), Swing & swap: User-centric approaches towards maximizing location privacy, in *5th ACM Workshop on Privacy in the Electronic Society*, ACM, Alexandria, VA, pp. 19–28.

Lim, H.-T., Herrscher, D., Volker, L. & Waltl, M. (2011), IEEE 802.1AS time synchronization in a switched Ethernet based in-car network, in *3rd IEEE Vehicular Networking Conference (VNC 2011)*, pp. 147–154.

Lim, J., Kim, W., Naito, K. *et al.* (2014), 'Interplay between TVWS and DSRC: Optimal strategy for safety message dissemination in VANET', *IEEE Journal on Selected Areas in Communications* **32**(11), to be published.

Lin, C.-H., Liu, B.-H., Yang, H.-Y., Kao, C.-Y. & Tasi, M.-J. (2008), Virtual-coordinate-based delivery-guaranteed routing protocol in wireless sensor networks with unidirectional links, in *27th IEEE Conference on Computer Communications (INFOCOM 2008)*, IEEE, Phoenix, AZ.

LIN Consortium (2010), Revision 2.2A, LIN Specification Package.

Litman, T. (2006), Parking management: Strategies, Evaluation and Planning, Victoria Transport Policy Institute.

Liu, K. & Abu-Ghazaleh, N. (2006), Aligned virtual coordinates for greedy routing in WSNs, in *3rd IEEE International Conference on Mobile Ad Hoc and Sensor Systems (MASS 2006)*, IEEE, Vancouver, pp. 377–386.

Liu, N., Liu, M., Chen, G. & Cao, J. (2012), The sharing at roadside: Vehicular content distribution using parked vehicles, in *31st IEEE Conference on Computer Communications (INFOCOM 2012), Mini-Conference*, IEEE, Orlando, FL.

Liu, N., Liu, M., Lou, W., Chen, G. & Cao, J. (2011), PVA in VANETs: Stopped cars are not silent, in *30th IEEE Conference on Computer Communications (INFOCOM 2011), Mini-Conference*, IEEE, Shanghai, pp. 431–435.

Liu, Y. & Ozguner, U. (2007), Human driver model and driver decision making for intersection driving, in *IEEE Intelligent Vehicles Symposium (IV 07)*, IEEE, Istanbul, pp. 642–647.

Lochert, C., Mauve, M., Füssler, H. & Hartenstein, H. (2005), 'Geographic routing in city scenarios', *ACM SIGMOBILE Mobile Computing and Communications Review* **9**(1), 69–72.

Lochert, C., Scheuermann, B. & Mauve, M. (2007), Probabilistic aggregation for data dissemination in VANETs, in *4th ACM International Workshop on Vehicular Ad Hoc Networks (VANET 2007)*, ACM, Montréal, Québec, pp. 1–8.

Lochert, C., Scheuermann, B., Wewetzer, C., Luebke, A. & Mauve, M. (2008), Data aggregation and roadside unit placement for a VANET traffic information system, in *5th ACM International Workshop on Vehicular Inter-Networking (VANET 2008)*, ACM, San Francisco, CA, pp. 58–65.

Lownes, N. E. & Machemehl, R. B. (2006), VISSIM: A multi-parameter sensitivity analysis, in *38th Winter Simulation Conference (WSC '06)*, IEEE, Monterey, CA, pp. 1406–1413.

Loyola, L., Lichte, H., Aad, I., Widmer, J. & Valentin, S. (2008), Increasing the capacity of IEEE 802.11 wireless LAN through cooperative coded retransmissions, in *67th IEEE Vehicular Technology Conference (VTC2008-Spring)*, IEEE, Marina Bay, Singapore, pp. 1746–1750.

Lua, E. K., Crowcroft, J., Pias, M., Sharma, R. & Lim, S. (2005), 'A survey and comparison of peer-to-peer overlay network schemes', *IEEE Communication Surveys and Tutorials* **7**(2), 72–93.

Luby, M. G., Mitzenmacher, M., Shokrollahi, M. A. & Spielman, D. A. (2001), 'Efficient erasure correcting codes', *IEEE Transactions on Information Theory* **47**(2), 569–584.

Ma, Z., Kargl, F. & Weber, M. (2009), Measuring location privacy in V2X communication systems with accumulated information, in *6th IEEE International Conference on Mobile Ad Hoc and Sensor Systems (MASS 2009)*, Macau SAR.

Malandrino, F., Casetti, C., Chiasserini, C.-F. & Fiore, M. (2011), Content downloading in vehicular networks: What really matters, in *30th IEEE Conference on Computer Communications (INFOCOM 2011), Mini-Conference*, IEEE, Shanghai, pp. 426–430.

Malandrino, F., Casetti, C. E., Chiasserini, C.-F., Sommer, C. & Dressler, F. (2012), Content downloading in vehicular networks: Bringing parked cars into the picture, in *23rd IEEE International Symposium on Personal, Indoor and Mobile Radio Communications (PIMRC 2012)*, IEEE, Sydney, pp. 1534–1539.

Mangel, T., Klemp, O. & Hartenstein, H. (2011a), A validated 5.9 GHz non-line-of-sight path-loss and fading model for inter-vehicle communication, in *11th International Conference on ITS Telecommunications (ITST 2011)*, IEEE, St. Petersburg, pp. 75–80.

Mangel, T., Schweizer, F., Kosch, T. & Hartenstein, H. (2011b), Vehicular safety communication at intersections: Buildings, non-line-of-sight and representative scenarios, in *8th IEEE/IFIP Conference on Wireless on Demand Network Systems and Services (WONS 2011)*, IEEE, Bardonecchia, pp. 35–41.

Mantler, A. & Snoeyink, J. (2000), Intersecting red and blue line segments in optimal time and precision, in *Japanese Conference on Discrete and Computational Geometry (JCDCG 2000): Revised Papers*, Springer, Tokyo.

Maraslis, K., Chatzimisios, P. & Boucouvalas, A. (2012), IEEE 802.11aa: Improvements on video transmission over wireless LANs, in *IEEE International Conference on Communications (ICC 2012)*, IEEE, Ottawa, Ontario, pp. 115–119.

Matischek, R., Herndl, T., Grimm, C. & Haase, J. (2011), Real-time wireless communication in automotive applications, in *IEEE/ACM Conference on Design, Automation, and Test in Europe (DATE 2011)*, IEEE, Grenoble.

Maurer, J., Fugen, T. & Wiesbeck, W. (2005), Physical layer simulations of IEEE802.11a for vehicle-to-vehicle communications, in *62th IEEE Vehicular Technology Conference (VTC2005-Fall)*, Dallas, TX, pp. 1849–1853.

Mauve, M., Widmer, J. & Hartenstein, H. (2001), 'A survey on position-based routing in mobile ad-hoc networks', *IEEE Network* **15**(6), 30–39.

Mecklenbräuker, C. F., Molisch, A. F., Karedal, J. *et al.* (2011), 'Vehicular channel characterization and its implications for wireless system design and performance', *Proceedings of the IEEE* **99**(7), 1189–1212.

Merali, Z. (2010), 'Computational science: Error – Why scientific programming does not compute', *Nature* **467**(7317), 775–777.

Metcalfe, R. M. & Boggs, D. R. (1976), 'Ethernet: Distributed packet switching for local computer networks', *Communications of the ACM* **19**(7), 395–404.

Mitola, J. & Maguire, Jr., G. Q. (1999), 'Cognitive radio: Making software radios more personal', *IEEE Personal Communications* **6**(4), 13–18.

Morency, C. & Trépanier, M. (2008), Characterizing Parking Spaces Using Travel Survey Data, TR 2008-15, CIRRELT.

Moser, S., Kargl, F. & Keller, A. (2007), Interactive realistic simulation of wireless networks, in *2nd IEEE/EG Symposium on Interactive Ray Tracing 2007 (RT 07)*, IEEE, Ulm, pp. 161–166.

MOST Cooperation (2010), MOST Media Oriented Systems Transport Multimedia and Control Networking Technology, MOST Specification Rev. 3.0 E2.

Nagel, R. & Eichler, S. (2008), Efficient and realistic mobility and channel modeling for VANET scenarios using OMNeT++ and INET-framework, in *1st ACM/ICST International Conference on Simulation Tools and Techniques for Communications, Networks and Systems (SIMUTools 2008)*, ICST, Marseille, pp. 1–8.

Nagurney, A., Qiang, Q. & Nagurney, L. S. (2008), Impact assessment of transportation networks with degradable links in an era of climate change, in *3rd International Conference on Funding Transportation Infrastructure*, Paris.

Nakamura, R. & Kajiwara, A. (2012), Empirical study on 60 GHz in-vehicle radio channel, in *2012 IEEE Radio and Wireless Symposium (RWS 2012)*, IEEE, Santa Clara, CA, pp. 327–330.

Nash, Jr, J. (1950), 'The bargaining problem', *Journal of the Economic Society (Econometrica)* **18**(2), 155–162.

Naumann, N., Schünemann, B., Radusch, I. & Meinel, C. (2009), Improving V2X simulation performance with optimistic synchronization, in *2009 IEEE Asia–Pacific Services Computing Conference (APSCC 2009)*, IEEE, Singapore, pp. 52–57.

Naumov, V., Baumann, R. & Gross, T. (2006), An evaluation of inter-vehicle ad hoc networks based on realistic vehicular traces, in *7th ACM International Symposium on Mobile Ad Hoc Networking and Computing (Mobihoc 2006)*, ACM, Florence, pp. 108–119.

Navas, J. C. & Imielinski, T. (1997), GeoCast – Geographic addressing and routing, in *3rd ACM International Conference on Mobile Computing and Networking (MobiCom 1997)*, Budapest, pp. 66–76.

Negreira, J. A., Pereira, J., Pérez, S. & Belzarena, P. (2007), End-to-end measurements over GPRS-EDGE networks, in *4th International IFIP/ACM Latin American Conference on Networking (LANC '07)*, ACM, pp. 121–131.

Neisser, U. (1976), *Cognition and Reality: Principles and Implications of Cognitive Psychology*, WH Freeman/Times Books/Henry Holt & Co.

Nekovee, M., Irnich, T. & Karlsson, J. (2012), 'Worldwide trends in regulation of secondary access to white spaces using cognitive radio', *IEEE Wireless Communications* **19**(4), 32–40.

Neudecker, T., An, N., Tonguz, O. K., Gaugel, T. & Mittag, J. (2012), Feasibility of virtual traffic lights in non-line-of-sight environments, in *9th ACM International Workshop on Vehicular Internetworking (VANET 2012)*, ACM, Low Wood Bay, pp. 103–106.

Ni, S.-Y., Tseng, Y.-C., Chen, Y.-S. & Sheu, J.-P. (1999), The broadcast storm problem in a mobile ad hoc network, in *5th ACM International Conference on Mobile Computing and Networking (MobiCom 1999)*, Seattle, WA, pp. 151–162.

Noori, H. (2013), Modeling the impact of VANET-enabled traffic lights control on the response time of emergency vehicles in realistic large-scale urban area, in *IEEE International Conference on Communications (ICC 2013)*, IEEE, Budapest.

Ott, J. & Kutscher, D. (2005), A disconnection-tolerant transport for drive-thru Internet environments, in *24th IEEE Conference on Computer Communications (INFOCOM 2005)*, Miami, FL.

Otto, J. S., Bustamante, F. E. & Berry, R. A. (2009), Down the block and around the corner – The impact of radio propagation on inter-vehicle wireless communication, in *29th International Conference on Distributed Computing Systems (ICDCS 2009)*, IEEE, Montréal, Québec, pp. 605–614.

Palazzi, C. E., Roccetti, M., Ferretti, S., Pau, G. & Gerla, M. (2007), Online games on wheels: Fast game event delivery in vehicular ad-hoc networks, in *IEEE Intelligent Vehicles Symposium 2007, 3rd International Workshop on Vehicle-to-Vehicle Communications*, IEEE, Istanbul.

Papadimitratos, P., Buttyan, L., Holczer, T. *et al.* (2008a), 'Secure vehicular communication systems: Design and architecture', *IEEE Communications Magazine* **46**(11), 100–109.

Papadimitratos, P., Kung, A., Hubaux, J. & Kargl, F. (2006), Privacy and identity management for vehicular communication systems: A position paper, in *1st Workshop on Standards for Privacy in User-Centric Identity Management*, Zurich.

Papadimitratos, P., Mezzour, G. & Hubaux, J.-P. (2008b), Certificate revocation list distribution in vehicular communication systems, in *5th ACM International Workshop on Vehicular Inter-Networking (VANET 2008)*, ACM, San Francisco, CA, pp. 86–87.

Park, J.-S., Lee, U., Oh, S. Y., Gerla, M. & Lun, D. S. (2006), Emergency related video streaming in VANET using network coding, in *3rd ACM International Workshop on Vehicular Ad Hoc Networks (VANET 2006)*, ACM, Los Angeles, CA, pp. 102–103.

Pawlikowski, K., Jeong, H.-D. & Lee, J.-S. R. (2002), 'On credibility of simulation studies of telecommunication networks', *IEEE Communications Magazine* **40**(1), 132–139.

Pecchia, E., Erman, D. & Popescu, A. (2009), Simulation and analysis of a combined mobility model with obstacles, in *2nd ACM/ICST International Conference on Simulation Tools and Techniques for Communications, Networks and Systems (SIMUTools 2009)*, ICST, Rome, pp. 1–2.

Penny, T. (1999), Intersection Collision Warning System, Techbrief FHWA-RD-99-103, US DOT, Federal Highway Administration.

Perkins, C. E., Belding-Royer, E. M. & Das, S. R. (2003), Ad Hoc On-Demand Distance Vector (AODV) Routing, RFC 3561.

Perkins, C. E. & Bhagwat, P. (1994), 'Highly dynamic destination-sequenced distance-vector routing (DSDV) for mobile computers', *Computer Communications Review* pp. 234–244.

Perkins, C. E. & Royer, E. M. (1999), Ad hoc on-demand distance vector routing, in *2nd IEEE Workshop on Mobile Computing Systems and Applications*, New Orleans, LA, pp. 90–100.

Perkins, C., Ratliff, S. & Dowdell, J. (2013), Dynamic MANET On-Demand (DYMO) Routing, Technical report, IETF.

Pfitzmann, A. & Köhntopp, M. (2000), Anonymity, unobservability, and pseudonymity – A proposal for terminology, in *International Workshop on Design Issues in Anonymity and Unobservability*, Vol. LNCS 2009, Springer, Berkeley, CA, pp. 1–9.

Piorkowski, M., Raya, M., Lugo, A. L. *et al.* (2007), TraNS: Joint traffic and network simulator, in *13th ACM International Conference on Mobile Computing and Networking (MobiCom 2007)*, Montréal, Québec.

Ploeg, J., Scheepers, B., van Nunen, E., van de Wouw, N. & Nijmeijer, H. (2011), Design and experimental evaluation of cooperative adaptive cruise control, in *IEEE International Conference on Intelligent Transportation Systems (ITSC 2011)*, IEEE, Washington, DC, pp. 260–265.

Qiu, D. & Srikant, R. (2004), Modeling and performance analysis of BitTorrent-like peer-to-peer networks, in *ACM SIGCOMM 2004*, ACM, Portland, OR, pp. 367–378.

Rahmani, M., Pfannenstein, M., Steinbach, E., Giordano, G. & Biersack, E. (2009), Wireless media streaming over IP-based in-vehicle networks, in *IEEE International Conference on Communications Workshops (ICC Workshops 2009)*, Dresden, pp. 1–6.

Rajamani, R., Tan, H.-S., Law, B. K. & Zhang, W.-B. (2000), 'Demonstration of integrated longitudinal and lateral control for the operation of automated vehicles in platoons', *IEEE Transactions on Control Systems Technology* **8**(4), 695–708.

Rakouth, H., Alexander, P., Brown, Jr., A. *et al.* (2012), V2X communication technology: Field experience and comparative analysis, in *FISITA World Automotive Congress*, Vol. LNEE 200, Springer, Beijing, pp. 113–129.

Ranjan, R., Harwood, A. & Buyya, R. (2008), 'Peer-to-peer-based resource discovery in global grids: A tutorial', *IEEE Communication Surveys and Tutorials* **10**(2),6–33.

Rao, A., Ratnasamy, S., Papadimitriou, C., Shenker, S. & Stoica, I. (2003), Geographic routing without location information, in *9th ACM International Conference on Mobile Computing and Networking (MobiCom 2003)*, San Diego, CA.

Rappaport, T. S. (2009), *Wireless Communications: Principles and Practice*, 2nd edn., Prentice Hall.

Ratnasamy, S., Francis, P., Handley, M., Karp, R. & Shenker, S. (2001), A scalable content-addressable network, in *ACM SIGCOMM 2001*, ACM, San Diego, CA, pp. 161–172.

Ratnasamy, S., Karp, B., Shenker, S. *et al.* (2003), 'Data-centric storage in sensornets with GHT, a geographic hash table', *ACM/Springer Mobile Networks and Applications, Special Issue on Wireless Sensor Networks* **8**(4), 427–442.

Ratnasamy, S., Karp, B., Yin, L. *et al.* (2002), GHT: A geographic hash table for data-centric storage, in *1st ACM International Workshop on Wireless Sensor Networks and Applications (WSNA 2002)*, Atlanta, GA.

Raya, M., Papadimitratos, P., Aad, I., Jungels, D. & Hubaux, J.-P. (2007), 'Eviction of misbehaving and faulty nodes in vehicular networks', *IEEE Journal on Selected Areas in Communications* **25**(8), 1557–1568.

Raya, M., Shokri, R. & Hubaux, J.-P. (2010), On the tradeoff between trust and privacy in wireless ad hoc networks, in *3rd ACM Conference on Wireless Network Security (WiSec 2010)*, ACM, Hoboken, NJ, pp. 75–80.

Rekhter, Y. & Li, T. (1995), 'A Border Gateway Protocol 4 (BGP-4)', RFC 1771.

Riaz, Z., Edwards, D. & Thorpe, A. (2006), 'SightSafety: A hybrid information and communication technology system for reducing vehicle/pedestrian collisions', *Elsevier Automation in Construction* **15**(6), 719–728.

Rivera-Lara, E. J., Herrerias-Hernandez, R., Perez-Diaz, J. A. & Garcia-Hernandez, C. F. (2008), Analysis of the relationship between QoS and SNR for an 802.11g WLAN, in *International Conference on Communication Theory, Reliability, and Quality of Service (CTRQ 2008)*, IEEE, Bucharest, pp. 103–107.

Rizzo, L. (1997), 'Effective erasure codes for reliable computer communication protocols', *ACM SIGCOMM Computer Communication Review* **27**(2), 24–36.

Rondinone, M., Maneros, J., Krajzewicz, D. *et al.* (2013), 'iTETRIS: A modular simulation platform for the large scale evaluation of cooperative ITS applications', *Simulation Modelling Practice and Theory* **34**, 99–125.

Rouf, I., Miller, R., Mustafa, H. *et al.* (2010), Security and privacy vulnerabilities of in-car wireless networks: A tire pressure monitoring system case study, in *19th USENIX Security Symposium*, USENIX Association, Washington, D.C.

Rybicki, J., Scheuermann, B., Kiess, W. *et al.* (2007), Challenge: Peers on wheels – A road to new traffic information systems, in *13th ACM International Conference on Mobile Computing and Networking (MobiCom 2007)*, Montréal, Québec, pp. 215–221.

Rybicki, J., Scheuermann, B., Koegel, M. & Mauve, M. (2009), PeerTIS – A peer-to-peer traffic information system, in *6th ACM International Workshop on Vehicular Inter-Networking (VANET 2009)*, ACM, Beijing, pp. 23–32.

SAE (2011), DSRC Message Communication Minimum Performance Requirements: Basic Safety Message for Vehicle Safety Applications, Draft Std. J2945.1 Revision 2.2, SAE Int. DSRC Committee.

Saha, A. K. & Johnson, D. B. (2004), Modeling mobility for vehicular ad-hoc networks, in *1st ACM Workshop on Vehicular Ad Hoc Networks (VANET 2004)*, Philadelphia, PA, pp. 91–92.

Salvo, P., Cuomo, F., Baiocchi, A. & Bragagnini, A. (2012), Road side unit coverage extension for data dissemination in VANETs, in *9th IEEE/IFIP Conference on Wireless on Demand Network Systems and Services (WONS 2012)*, IEEE, Courmayeur, Italy, pp. 47–50.

Sampigethaya, K., Huang, L., Li, M. *et al.* (2005), CARAVAN: Providing location privacy for VANET, in *Embedded Security in Cars (ESCAR 2005)*, Tallinn.

Sarafijanovic-Djukic, N. & Grossglauser, M. (2004), Last encounter routing under random waypoint mobility, in *IFIP NETWORKING 2004*, Vol. LNCS 3042, Springer, Athens, pp. 974–988.

Sastry, N., Shankar, U. & Wagner, D. (2003), Secure verification of location claims, in *2nd ACM Workshop on Wireless Security (WiSe 2003)*, ACM, San Diego, CA, pp. 1–10.

Schaub, F., Kargl, F., Ma, Z. & Weber, M. (2010), V-tokens for conditional pseudonymity in VANETs, in *IEEE Wireless Communications and Networking Conference (WCNC 2010)*, IEEE, Sydney.

Scheuermann, B., Lochert, C., Rybicki, J. & Mauve, M. (2009), A fundamental scalability criterion for data aggregation in VANETs, in *15th ACM International Conference on Mobile Computing and Networking (MobiCom 2009)*, ACM, Beijing, pp. 285–296.

Schmidt, R. K., Leinmüller, T., Schoch, E., Kargl, F. & Schäfer, G. (2010), 'Exploration of adaptive beaconing for efficient intervehicle safety communication', *IEEE Network Magazine* **24**(1), 14–19.

Schmitz, A. & Wenig, M. (2006), The effect of the radio wave propagation model in mobile ad hoc networks, in *9th ACM International Symposium on Modeling, Analysis and Simulation of Wireless and Mobile Systems (MSWiM 2006)*, ACM, Torremolinos, pp. 61–67.

Schneier, B. (1996), *Applied Cryptography*, 2nd edn., John Wiley & Sons.

Schoch, E. & Kargl, F. (2010), On the efficiency of secure beaconing in VANETs, in *3rd ACM Conference on Wireless Network Security (WiSec 2010)*, ACM, Hoboken, NJ, pp. 111–116.

Schoch, E., Kargl, F., Leinmüller, T., Schlott, S. & Papadimitratos, P. (2006), Impact of pseudonym changes on geographic routing in VANETS, in *3rd European Workshop on Security and Privacy in Ad Hoc and Sensor Networks (ESAS 2006)*, Vol. LNCS 4357, Springer, Hamburg.

Schünemann, B. (2011), 'V2X simulation runtime infrastructure VSimRTI: An assessment tool to design smart traffic management systems', *Elsevier Computer Networks* **55**(14), 3189–3198.

Schwartz, R. S., Ohazulike, A. E. & Scholten, H. (2012a), Achieving data utility fairness in periodic dissemination for VANETs, in *75th IEEE Vehicular Technology Conference (VTC2012-Spring)*, IEEE, Yokohama.

Schwartz, R. S., Ohazulike, A. E., Sommer, C. *et al.* (2012b), Fair and adaptive data dissemination for traffic information systems, in *4th IEEE Vehicular Networking Conference (VNC 2012)*, IEEE, Seoul, pp. 1–8.

Schwartz, R. S., Ohazulike, A. E., Sommer, C. *et al.* (2014), 'On the applicability of fair and adaptive data dissemination in traffic information systems', *Elsevier Ad Hoc Networks* **13B**, 428–443.

Segata, M. (2013), Novel communication strategies for platooning and their simulative performance analysis, in *1st GI/ITG KuVS Fachgespräch Inter-Vehicle Communication (FG-IVC 2013)*, Innsbruck.

Segata, M., Bloessl, B., Joerer, S. *et al.* (2014), Towards inter-vehicle communication strategies for platooning support, in *7th International Workshop on Communication Technologies for Vehicles*, IEEE, Saint Petersburg, Russia.

Segata, M., Bloessl, B., Joerer, S. *et al.* (2013), Vehicle shadowing distribution depends on vehicle type: Results of an experimental study, in *5th IEEE Vehicular Networking Conference (VNC 2013)*, IEEE, Boston, MA, pp. 242–245.

Segata, M., Dressler, F., Lo Cigno, R. & Gerla, M. (2012), 'A simulation tool for automated platooning in mixed highway scenarios', *ACM SIGMOBILE Mobile Computing and Communications Review* **16**(4), 46–49.

Serjantov, A. & Danezis, G. (2002), Towards an information theoretic metric for anonymity, in *2nd International Workshop on Privacy Enhancing Technologies (PET 2002)*, San Francisco, CA, pp. 259–263.

Shenker, S., Ratnasamy, S., Karp, B., Govindan, R. & Estrin, D. (2003), 'Data-centric storage in sensornets', *ACM SIGCOMM Computer Communication Review* **33**(1), 137–142.

Shenpei, Z. & Xinping, Y. (2008), Driver's route choice model based on traffic signal control, in *3rd IEEE Conference on Industrial Electronics and Applications (ICIEA 2008)*, IEEE, Singapore, pp. 2331–2334.

Shladover, S. (2006), PATH at 20 – History and major milestones, in *IEEE Intelligent Transportation Systems Conference (ITSC 2006)*, Toronto, pp. 22–29.

sim^{TD} (2013), Final Report of Subproject 5, Deliverable D5.5 Part A.

Skordylis, A. & Trigoni, N. (2008), Delay-bounded routing in vehicular ad-hoc networks, in *9th ACM International Symposium on Mobile Ad Hoc Networking and Computing (Mobihoc 2008)*, ACM, Hong Kong, pp. 341–350.

Sommer, C., Dietrich, I. & Dressler, F. (2007), Realistic simulation of network protocols in VANET scenarios, in *26th IEEE Conference on Computer Communications (INFOCOM 2007): IEEE Workshop on Mobile Networking for Vehicular Environments (MOVE 2007), Poster Session*, IEEE, Anchorage, AK, pp. 139–143.

Sommer, C. & Dressler, F. (2007), The DYMO routing protocol in VANET scenarios, in *66th IEEE Vehicular Technology Conference (VTC2007-Fall)*, IEEE, Baltimore, MD, pp. 16–20.

Sommer, C. & Dressler, F. (2008), 'Progressing toward realistic mobility models in VANET simulations', *IEEE Communications Magazine* **46**(11), 132–137.

Sommer, C. & Dressler, F. (2011), Using the right two-ray model? A measurement based evaluation of PHY models in VANETs, in *17th ACM International Conference on Mobile Computing and Networking (MobiCom 2011), Poster Session*, ACM, Las Vegas, NV.

Sommer, C., Eckhoff, D. & Dressler, F. (2010a), 'Improving the accuracy of IVC simulation using crowd-sourced geodata', *Praxis der Informationsverarbeitung und Kommunikation (PIK)* **33**(4), 278–283.

Sommer, C., Eckhoff, D. & Dressler, F. (2014a), 'IVC in cities: Signal attenuation by buildings and how parked cars can improve the situation', *IEEE Transactions on Mobile Computing*, **13**(8), 1733–1745.

Sommer, C., Eckhoff, D., German, R. & Dressler, F. (2011a), A computationally inexpensive empirical model of IEEE 802.11p radio shadowing in urban environments, in *8th IEEE/IFIP Conference on Wireless on Demand Network Systems and Services (WONS 2011)*, IEEE, Bardonecchia, pp. 84–90.

Sommer, C., German, R. & Dressler, F. (2010b), Adaptive Beaconing for Delay-Sensitive and Congestion-Aware Traffic Information Systems, Technical Report CS-2010-01, University of Erlangen, Department of Computer Science.

Sommer, C., German, R. & Dressler, F. (2011b), 'Bidirectionally coupled network and road traffic simulation for improved IVC analysis', *IEEE Transactions on Mobile Computing* **10**(1), 3–15.

Sommer, C., Hagenauer, F. & Dressler, F. (2014b), A networking perspective on self-organizing intersection management, in *IEEE World Forum on Internet of Things (WF-IoT 2014)*, IEEE, Seoul, pp. 230–234.

Sommer, C., Joerer, S. & Dressler, F. (2012), On the applicability of two-ray path loss models for vehicular network simulation, in *4th IEEE Vehicular Networking Conference (VNC 2012)*, IEEE, Seoul, pp. 64–69.

Sommer, C., Joerer, S., Segata, M. *et al.* (2013), How shadowing hurts vehicular communications and how dynamic beaconing can help, in *32nd IEEE Conference on Computer Communications (INFOCOM 2013), Mini-Conference*, IEEE, Turin, pp. 110–114.

Sommer, C., Krul, R., German, R. & Dressler, F. (2010c), Emissions vs. travel time: Simulative evaluation of the environmental impact of ITS, in *71st IEEE Vehicular Technology Conference (VTC2010-Spring)*, IEEE, Taipei, pp. 1–5.

Sommer, C., Schmidt, A., Chen, Y. *et al.* (2010d), 'On the feasibility of UMTS-based traffic information systems', *Elsevier Ad Hoc Networks, Special Issue on Vehicular Networks* **8**(5), 506–517.

Sommer, C., Tonguz, O. K. & Dressler, F. (2010e), Adaptive beaconing for delay-sensitive and congestion-aware traffic information systems, in *2nd IEEE Vehicular Networking Conference (VNC 2010)*, IEEE, Jersey City, NJ, pp. 1–8.

Sommer, C., Tonguz, O. K. & Dressler, F. (2011c), 'Traffic information systems: Efficient message dissemination via adaptive beaconing', *IEEE Communications Magazine* **49**(5), 173–179.

Sommer, C., Yao, Z., German, R. & Dressler, F. (2008), Simulating the influence of IVC on road traffic using bidirectionally coupled simulators, in *27th IEEE Conference on Computer Communications (INFOCOM 2008): IEEE Workshop on Mobile Networking for Vehicular Environments (MOVE 2008)*, IEEE, Phoenix, AZ, pp. 1–6.

Souley, A.-K. H. & Cherkaoui, S. (2005), Realistic urban scenarios simulation for ad hoc networks, in *2nd International Conference on Innovations in Information Technology (IIT'05)*, Dubai.

Stallings, W. (2013), *Cryptography and Network Security: Principles and Practice*, 6th edn., Pearson Prentice Hall.

Steinmetz, R. & Wehrle, K., eds. (2005), *Peer-to-Peer Systems and Applications*, Springer.

Stepanov, I. & Rothermel, K. (2008), 'On the impact of a more realistic physical layer on MANET simulations results', *Elsevier Ad Hoc Networks* **6**(1), 61–78.

Stoica, I., Morris, R., Karger, D. R., Kaashoek, M. F. & Balakrishnan, H. (2001), Chord: A scalable peer-to-peer lookup service for internet applications, in *ACM SIGCOMM 2001*, ACM, San Diego, CA, pp. 149–160.

Stoica, I., Morris, R., Liben-Nowell, D. *et al.* (2003), 'Chord: A scalable peer-to-peer lookup protocol for internet applications', *IEEE/ACM Transactions on Networking* **11**(1), 17–32.

Stübing, H., Bechler, M., Heussner, D. *et al.* (2010), 'simTD: A car-to-X system architecture for field operational tests', *IEEE Communications Magazine* **48**(5), 148–154.

Stutts, J. C., Stewart, J. R. & Martell, C. (1998), 'Cognitive test performance and crash risk in an older driver population', *Accident Analysis and Prevention* **30**(3), 337–346.

Sugimoto, C., Nakamura, Y. & Hashimoto, T. (2008), Prototype of pedestrian-to-vehicle communication system for the prevention of pedestrian accidents using both 3G wireless and WLAN communication, in *3rd IEEE International Symposium on Wireless Pervasive Computing (ISWPC 2008)*, IEEE, Santorini, pp. 764–767.

Suzuki, H. (1977), 'A statistical model for urban radio propagation', *IEEE Transactions on Communications* **25**(7), 673–680.

Tan, G., Bertier, M. & Kermarrec, A.-M. (2009), Convex partition of sensor networks and its use in virtual coordinate geographic routing, in *28th IEEE Conference on Computer Communications (INFOCOM 2009)*, IEEE, Rio de Janeiro.

Tanenbaum, A. S. & Wetherall, D. J. (2011), *Computer Networks*, 5th edn., Prentice Hall.

Tang, A. & Yip, A. (2010), 'Collision avoidance timing analysis of DSRC-based vehicles', *Accident Analysis and Prevention* **42**(1), 182–195.

Tielert, T., Jiang, D., Hartenstein, H. & Delgrossi, L. (2013), Joint power/rate congestion control optimizing packet reception in vehicle safety communications, in *10th ACM International Workshop on Vehicular Internetworking (VANET 2013)*, ACM, Taipei, pp. 51–60.

Tonguz, O. K. & Boban, M. (2010), 'Multiplayer games over vehicular ad hoc networks: A new application', *Elsevier Ad Hoc Networks* **8**(5), 531–543.

Tonguz, O. K., Wisitpongphan, N. & Bai, F. (2010), 'DV-CAST: A distributed vehicular broadcast protocol for vehicular ad hoc networks', *IEEE Wireless Communications* **17**(2), 47–57.

Tonguz, O., Wisitpongphan, N., Bai, F., Mudalige, P. & Sadekart, V. (2007), Broadcasting in VANET, in *26th IEEE Conference on Computer Communications (INFOCOM 2007): Mobile Networking for Vehicular Environments (MOVE 2007)*, IEEE, Anchorage, AK, pp. 7–12.

Tonguz, O., Wisitpongphan, N., Parikh, J. *et al.* (2006), On the broadcast storm problem in ad hoc wireless networks, in *3rd International Conference on Broadband Communications, Networks, and Systems (BROADNETS)*, San Jose, CA.

Torrent-Moreno, M., Santi, P. & Hartenstein, H. (2006), Distributed fair transmit power adjustment for vehicular ad hoc networks, in *3rd IEEE Communications Society Conference on Sensor and Ad Hoc Communications and Networks (SECON 2006)*, Vol. 2, IEEE, Reston, VA, pp. 479–488.

TPEG (2006a), Traffic and Travel Information (TTI) – TTI via Transport Protocol Expert Group (TPEG) Data-streams – Part 1: Introduction, Numbering and Versions, TS 18234-1, ISO.

TPEG (2006b), Traffic and Travel Information (TTI) – TTI via Transport Protocol Expert Group (TPEG) Data-streams – Part 4: Road Traffic Message (RTM) Application, TS 18234-4, ISO.

TPEG (2006c), Traffic and Travel Information (TTI) – TTI via Transport Protocol Experts Group (TPEG) Extensible Markup Language (XML) – Part 1: Introduction, Common Data Types and tpegML, TS 24530-1, ISO.

Train, J., Bannister, J. & Raghavendra, C. (2011), Routing fountains: Leveraging wide-area broadcast to improve mobile inter-domain routing, in *IEEE Military Communications Conference (MILCOM 2011)*, Baltimore, MD, pp. 842–848.

Treiber, M., Hennecke, A. & Helbing, D. (2000), Microscopic simulation of congested traffic, in D. Helbing, H. Herrmann, M. Schreckenberg & D. Wolf, eds., *Traffic and Granular Flow '99*, Springer, Heidelberg.

Tsai, H.-M., Tonguz, O., Saraydar, C. *et al.* (2007), 'Zigbee-based intra-car wireless sensor networks: A case study', *IEEE Wireless Communications* **14**(6), 67–77.

Tsai, M.-J., Wang, F.-R., Yang, H.-Y. & Cheng, Y.-P. (2009), VirtualFace: An algorithm to guarantee packet delivery of virtual-coordinate-based routing protocols in wireless sensor networks, in *28th IEEE Conference on Computer Communications (INFOCOM 2009)*, IEEE, Rio de Janeiro.

Tubaishat, M., Shang, Y. & Shi, H. (2007), Adaptive traffic light control with wireless sensor networks, in *4th IEEE Consumer Communications and Networking Conference (CCNC 2007)*, IEEE, Las Vegas, NV, pp. 187–191.

Tung, L.-C., Mena, J., Gerla, M. & Sommer, C. (2013), A cluster based architecture for intersection collision avoidance using heterogeneous networks, in *12th IFIP/IEEE Annual Mediterranean Ad Hoc Networking Workshop (Med-Hoc-Net 2013)*, IEEE, Ajaccio, Corsica.

Uang, S.-T. & Hwang, S.-L. (2003), 'Effects on driving behavior of congestion information and of scale of in-vehicle navigation systems', *Transportation Research Part C: Emerging Technologies* **11**(6), 423–438.

Uzcátegui, R. A. & Acosta-Marum, G. (2009), 'WAVE: A tutorial', *IEEE Communications Magazine* **47**(5), 126–133.

Valerio, D., Ricciato, F., Belanovic, P. & Zemen, T. (2008), UMTS on the road: Broadcasting intelligent road safety information via MBMS, in *67th IEEE Vehicular Technology Conference (VTC2008-Spring)*, pp. 3026–3030.

van Arem, B., van Driel, C. & Visser, R. (2006), 'The impact of cooperative adaptive cruise control on traffic-flow characteristics', *IEEE Transactions on Intelligent Transportation Systems* **7**(4), 429–436.

Varga, A. & Hornig, R. (2008), An overview of the OMNeT++ simulation environment, in *1st ACM/ICST International Conference on Simulation Tools and Techniques for Communications, Networks and Systems (SIMUTools 2008)*, ACM, Marseille.

Vasudevan, S., Kurose, J. & Towsley, D. (2004), Design and analysis of a leader election algorithm for mobile ad hoc networks, in *12th IEEE International Conference on Network Protocols (ICNP 2004)*, pp. 350–360.

Vesco, A., Scopigno, R., Casetti, C. & Chiasserini, C.-F. (2013), Investigating the effectiveness of decentralized congestion control in vehicular networks, in *IEEE Global Telecommunications Conference (GLOBECOM 2013)*, IEEE, Atlanta, GA.

Vinel, A. (2012), '3GPP LTE Versus IEEE 802.11p/WAVE: Which technology is able to support cooperative vehicular safety applications?', *Wireless Communications Letters* **1**(2), 125–128.

Viriyasitavat, W., Bai, F. & Tonguz, O. (2010), UV-CAST: An urban vehicular broadcast protocol, in *2nd IEEE Vehicular Networking Conference (VNC 2010)*, IEEE, Jersey City, NJ, pp. 25–32.

Viriyasitavat, W., Tonguz, O. K. & Bai, F. (2009), Network connectivity of VANETs in urban areas, in *6th IEEE Communications Society Conference on Sensor and Ad Hoc Communications and Networks (SECON 2009)*, IEEE, Rome.

Vogels, W., van Renesse, R. & Briman, K. (2003), 'The power of epidemics: Robust communication for large-scale distributed systems', *ACM SIGCOMM Computer Communication Review* **33**(1), 131–135.

Vora, A. & Nesterenko, M. (2006), 'Secure location verification using radio broadcast', *IEEE Transactions on Dependable and Secure Computing* **3**(4), 377–385.

Wang, Y., Ahmed, A., Krishnamachari, B. & Psounis, K. (2008), IEEE 802.11p performance evaluation and protocol enhancement, in *IEEE International Conference on Vehicular Electronics and Safety (ICVES)*, Columbus, OH, pp. 317–322.

Watteyne, T., Molinaro, A., Richichi, M. G. & Dohler, M. (2011), 'From MANET To IETF ROLL standardization: A paradigm shift in WSN routing protocols', *IEEE Communications Surveys and Tutorials* **13**(4), 688–707.

Watteyne, T., Simplot-Ryl, D., Augé-Blum, I. & Dohler, M. (2007), On using virtual coordinates for routing in the context of wireless sensor networks, in *18th IEEE International Symposium on Personal, Indoor and Mobile Radio Communications (PIMRC 2007)*, IEEE, Athens, pp. 1–5.

Wegener, A., Piorkowski, M., Raya, M. *et al.* (2008), TraCI: An interface for coupling road traffic and network simulators, in *11th Communications and Networking Simulation Symposium (CNS'08)*, Ottawa.

Wehrle, K., Günes, M. & Gross, J. (2010), *Modeling and Tools for Network Simulation*, Springer.

Weiß, C. (2011), 'V2X communication in europe – From research projects towards standardization and field testing of vehicle communication technology', *Elsevier Computer Networks* **55**(14), 3103–3119.

Wenger, M., Spyridakis, J., Haselkorn, M., Barfield, W. & Conquest, L. (1990), 'Motorist behavior and the design of motorist information systems. Human factors and safety research related to highway design and operation', *Transportation Research Record* **1281**, 159–167.

Werner, M., Lupoaie, R., Subramanian, S. & Jose, J. (2012), MAC layer performance of ITS G5 – optimized DCC and advanced transmitter coordination, in *4th ETSI TC ITS Workshop*, Doha, Qatar.

Wiedemann, R. (1974), Simulation des Straßenverkehrsflusses, Habilitation, University of Karlsruhe.

Wiedersheim, B., Ma, Z., Kargl, F. & Papadimitratos, P. (2010), Privacy in inter-vehicular networks: Why simple pseudonym change is not enough, in *7th IEEE/IFIP Conference on Wireless on Demand Network Systems and Services (WONS 2010)*, Kranjska Gora, Slovenia.

Williams, B. (2008), *Intelligent Transport Systems Standards*, Artech House.

Willke, T. L., Tientrakool, P. & Maxemchuk, N. F. (2009), 'A survey of inter-vehicle communication protocols and their applications', *IEEE Communications Surveys and Tutorials* **11**(2), 3–20.

Wischhof, L., Ebner, A. & Rohling, H. (2005), 'Information dissemination in self-organizing intervehicle networks', *IEEE Transactions on Intelligent Transportation Systems* **6**(1), 90–101.

Wischhof, L., Ebner, A. & Rohling, H. (2006), Self-generated road status maps based on vehicular ad hoc communication, in *3rd International Workshop on Intelligent Transportation (WIT 2006)*, Hamburg.

Wischhof, L., Ebner, A., Rohling, H., Lott, M. & Halfmann, R. (2003), SOTIS – A self-organizing traffic information system, in *57th IEEE Vehicular Technology Conference (VTC2003-Spring)*, Jeju, South Korea.

Wisitpongphan, N. & Bai, F. (2013), Microscopic experimental evaluation of multi-hop video streaming protocol in vehicular networks, in *5th IEEE Vehicular Networking Conference (VNC 2013)*, IEEE, Boston, MA.

Wisitpongphan, N., Bai, F., Mudalige, P., Sadekar, V. & Tonguz, O. (2007a), 'Routing in sparse vehicular ad hoc wireless networks', *IEEE Journal on Selected Areas in Communications* **25**(8), 1538–1556.

Wisitpongphan, N., Tonguz, O. K., Parikh, J. S. *et al.* (2007b), 'Broadcast storm mitigation techniques in vehicular ad hoc networks', *IEEE Wireless Communications* **14**(6), 84–94.

Wright, A. (2011), 'Hacking cars', *Communications of the ACM* **54**(11), 18–19.

Xue, Q. & Ganz, A. (2003), 'Ad hoc QoS on-demand routing (AQOR) in mobile ad hoc networks', *Journal of Parallel and Distributed Computing* **63**(2), 154–165.

Yoon, J., Liu, M. & Noble, B. (2003), Random waypoint considered harmful, in *22nd IEEE Conference on Computer Communications (INFOCOM 2003)*, IEEE, San Francisco, CA, pp. 1312–1321.

Zhang, H. & Hou, J. C. (2005), 'On the upper bound of α-lifetime for large sensor networks', *ACM Transactions on Sensor Networks* **1**(2), 272–300.

Zhao, Y., Chen, Y., Li, B. & Zhang, Q. (2007), 'Hop ID: A virtual coordinate-based routing for sparse mobile ad hoc networks', *IEEE Transactions on Mobile Computing* **6**(9), 1075–1089.

Zheng, Z., Lu, Z., Sinha, P. & Kumar, S. (2010), Maximizing the contact opportunity for vehicular internet access, in *29th IEEE Conference on Computer Communications (INFOCOM 2010)*, IEEE, San Diego, CA, pp. 1–9.

Zhou, B., Tiwari, A., Zhu, K. *et al.* (2009), Geo-based inter-domain routing (GIDR) protocol for MANETs, in *IEEE Military Communications Conference (MILCOM 2009)*, IEEE, Boston, MA, pp. 1–7.

Zhou, L., Zhang, Y., Song, K., Jing, W. & Vasilakos, A. (2011), 'Distributed media services in P2P-based vehicular networks', *IEEE Transactions on Vehicular Technology* **60**(2), 692–703.

Zhu, Y., Xu, B., Shi, X. & Wang, Y. (2013), 'A survey of social-based routing in delay tolerant networks: Positive and negative social effects', *IEEE Communications Surveys and Tutorials* **15**(1), 387–401.

Ziomek, J., Tedesco, L. & Coughlin, T. (2013), 'My car, my way: Why not? I paid for it!', *Consumer Electronics Magazine* **2**(3), 25–29.

Index

3GPP, 107
3GPP2, 107

AAA, 41
ABS, 12
AC, 87, 121, 126
ACC, 32, 47, 49, 50, 60
Access category, 87, 121
Access technologies, 106
ACO, 167
Ad-hoc communication, 69
Ad-hoc routing, 80, 138
  Ad-hoc on-demand distance vector, 142
  Delay-bounded routing, 150
  Dynamic MANET on demand, 144
  Dynamic source routing, 141
  Mobility pattern, 149
  Proactive, 139
  Protocols, 82
  Reactive, 140
  Taxonomy, 83
  VANET, 145
ADAC, 41
Adaptive beaconing, 90, 174
Adaptive cruise control, 32, 49, 60
Adaptive traffic beacon, 44, 90, 175
Adaptive traffic lights, 51
Adaptive transmission power, 92
ADAS, 32
ADQR, 151
AHS, 238
AIFS, 121, 124, 193
ALDL, 12
AODV, 83, 141–145, 150
AP, 42, 54, 65, 70, 86, 120–123, 151, 206, 214, 222–225
API, 22, 235, 263
Application layer broadcast, *see* Beaconing
Applications, 39
  Content downloading, 53
  Entertainment, 53
  Intersection collision warning system, 46
  Multimedia streaming, 53
  Multiplayer games, 55
  Non-safety, 56
  Platooning, 48
  Requirements, 56
  Safety, 60
  Traffic information system, 39
  Traffic-light information and control, 50
AQOR, 150, 151
ARIB, 5
ASTM, 123
ATB, 44, 45, 90, 91, 104, 137, 174–186, 195, 292
AVB, 35
AVNu Alliance, 35

Beaconing, 85, 88, 167
  Adaptive, 90, 174
  Adaptive traffic beacon, 90, 175
  Adaptive transmission power, 92
  ATB, 175
  Broadcast storm problem, 102
  CAM, 172
  Cooperative awareness message, 172
  DCC, *see* Decentralized congestion control
  Dynamic beaconing, 93, 191
  DynB, 191
  Fairness, 184
  Fixed interval, *see* Beaconing, static
  Infrastructure support, 94
  SOTIS, 44, 167
  Static, 167
BGP, 166
Bit stuffing, 16
Bloom filter, 203
BMBF, 42, 240
BMVBS, 240
BMWi, 240
BPSK, 72, 119, 234
Broadcast storm problem, 102
Broadcast suppression, 102
Broadcasting, 85
BroadR-Reach, 35
BSC, 110
BSM, 89, 95, 128, 242
BSS, 86, 87, 121–124

BTP, 129
BTS, 111
Bus guardian, 13
Bus systems, 15
  CAN, 15
  Diagnostic, 19, 24, 36
  Ethernet, 32
  FlexRay, 27
  LIN, 21
  MOST, 24
  Requirements, 13
BYOD, 32

CA, 305, 308–310, 312, 313, 315, 322, 323
CACC, 32, 48, 60, 238, 262
CAM
  Message format, 173
  Protocol, 172
CAM, 10, 89, 95, 99, 118, 129, 136, 137, 167, 172–175, 185, 200, 207, 219, 232, 296, 297, 302, 313, 320
CAN (bus protocol), 15–22, 24, 29, 31, 36, 37
CAN (distributed hash table), 13, 226
CAS, 30
Cause codes, *see* DENM
CCA, 120, 124, 186
CCH, 87, 88, 122, 125, 126, 173, 187, 193, 235, 277, 278
CCK, 119
CDF, 289, 290
CDMA, 109, 112, 113
CDMA2000, 107
CDMAone, 107
Cellular networks, 76, 107
  CDMA2000, 107
  CDMAone, 107
  Generations, 107
  GSM, 110
  LTE, 113
  Mobile WiMAX, 107
  UMB, 107
  UMTS, 112
CEPT, 6, 110
Certificate revocation, 315
Certificates, 308, 311
CME, 127
CO_2 emission, 65
CoCar, 42, 79, 80, 116, 118
CoCarX, 118
Cognitive radio, 130
COM, 261
Comfort applications, 58
Communication principles, 68
Congestion control, 92, 185
Content downloading, 53
Converge, 118
Cooperative adaptive cruise control, 32, 48, 60, 238
Cooperative awareness message, 172
CPU, 274, 307
CRC, 19, 29, 30
CRL, 310, 315, 316, 324
Cruise control
  ACC, 32, 49, 60
  CACC, 32, 48, 60, 238
CSD, 111
CSMA, 17, 85, 88
CSMA/BA, 17, 19
CSMA/CA, 120
CSMA/CD, 33, 34
CTS, 120, 147, 186
CW, 87, 126
CWS, 47

D-FPAV, 92
D2B, 24
DAB, 41, 72, 74–76
DCC, 92, 104, 137, 173, 174, 185–193, 198
DCF, 87, 120
DCH, 113
Decentralized congestion control, 92, 185
Decentralized environmental notification message, 200
Dedicated short-range communication, 85
Delay-bounded routing, 150
Delay-tolerant networks, 217
DENM, 10, 129, 136, 137, 200, 207, 232
  Message format, 200
  Protocol, 200
DES, 256, 259, 262
DHT, 46, 98, 137, 152, 155, 157, 158, 166, 167, 219, 226, 227
Disruption-tolerant networks, 217
Distributed vehicular broadcast, 219
DLL, 261
DoIP, 36
Driver behavior, 285
DSA, 130
DSC, 186, 188
DSDV, 83, 139, 140, 143
DSR, 83, 141–144
DSRC, 2–4, 69, 119, 123, 125, 133–135, 232
DSRC/WAVE, 9
DSSS, 119
DTN, 10, 43, 54, 95, 136, 137, 184, 197, 218, 219, 221–223
DV-CAST, 95, 96, 104, 184, 197, 219–222
DVD, 12, 25
DYMO, 82–84, 141, 144–147, 149
Dynamic beaconing, 93, 191
Dynamic MANET on demand, 144
Dynamic source routing, 141
DynB, 93, 104, 174, 191–196

ECC, 2, 6, 87, 122, 125, 128, 132
eCDF, 116, 194, 210
ECU, 8, 12, 13, 17–21, 24–31, 36, 37
EDCA, 87, 121, 126, 186
EDGE, 79, 112
eMBMS, 80, 115, 118, 316
EMI, 24
Emissions, 250, 298
Entertainment applications, 53
ESP, 12
Ethernet, 32
ETSI
  ITS, 128
  ITS-G5, 128
ETSI, 6, 86, 89, 92, 128, 136, 137, 172, 175, 186, 191, 197, 219, 243, 245, 312
Event code, 73
Exchanging pseudonyms, 323

Face routing, 154
FACH, 113
Fading, 274
FairAD, 184, 185
FairDD, 185
Fairness, *see* Beaconing, fairness
False, 120, 186
FCC, 2, 5, 87, 122, 123, 125, 129, 130, 132
FCD, 39, 41, 57, 74
FDD, 107, 111, 112, 114
FDMA, 4, 76, 108, 109, 111, 114
FHWA, 123
Field operational tests, 229–231, 240
FlexRay, 27
Flooding, 102
FM radio, 72
FOT, 2, 49, 79, 118, 124, 208, 229–232, 240–243, 311
FPGA, 232
Free-space model, 275

Geocasting, 196
  GeoAnycast, 199
  GeoBroadcast, 198
  GeoNetworking, 197
  GeoUnicast, 198
  Topology-assisted geo-opportunistic routing, 201
  Topology-assisted geographic routing, 202
Geographic hash tables, 155
Geographic routing, 96, 153
  Geo-assisted forwarding, 99
  Geocasting, 196
  Geographic hash tables, 155
  Greedy perimeter stateless routing, 153
  Greedy routing, 97, 162
  Virtual coordinate-based routing, 157
  Virtual coordinates, 98
GeoNetworking, 197
GGSN, 112, 116
GHT, 155–157, 163–166, 198
GLOSA, 57, 58, 67, 206, 211
GPRS, 79, 111, 112
GPS, 39, 88, 95, 97, 104, 126, 153, 157, 168, 170, 232, 236, 245, 246, 278, 306, 323
GPSR, 97, 153–156, 158, 163, 197, 198, 202, 205
Greedy perimeter stateless routing, 153
GRWLI, 98, 157, 158
GSM, 79, 106, 107, 110–112, 114, 115
GUI, 258, 260

Heterogeneous networks, 100, 118
HLA, 264
HMI, 241
HSCSD, 111
HSDPA, 79, 112
HSPA, 79, 115
HSUPA, 79, 112

IBSS, 122
ICWS, 46–48, 61, 64, 65, 251, 294, 298
IDE, 257, 258, 261
IEEE, 5, 86, 107
IEEE 1609, 86, 125
IEEE 802.11p, 86, 122
IETF, 81, 138, 141, 309
IFS, 120–122
ILOC, 76
In-vehicle communication, 12
In-vehicle ethernet, 32
Information dissemination, 136
Infrastructure support, 205
Infrastructure-based communication, 70
Inter-vehicle communication, 38
  Ad hoc, 69
  Beaconing, 85
  Communication principles, 68
  Concepts, 71
  Fundamental limits, 100
  Hybrid approaches, 70
  Infrastructure-based, 70
  Scalability, 104
Intersection collision warning system, 46
Intra-vehicle communication, 12
IoT, 68
IP, 25, 77, 96, 114, 129, 137, 147, 258, 263
ISM bands, 119, 129
ISO, 6, 128
ITS, 1–5, 122–125, 172, 173, 200, 206, 208, 212, 232, 233, 238, 241, 242, 289, 306
ITSA, 122
ITU, 6
ITU-R, 283, 284
ITU-T, 6, 308
IVHS, 1, 4

LAN, 32, 119, 257
Large-scale FOT, 240
LDM, 129
LER, 84, 150
LIDAR, 32
LIN, 13, 21–24, 36
LLC, 34
Location code, 73
Location privacy, 318
LOS, 230, 236, 237
LTE, 79, 80, 106, 107, 110, 113–115, 118, 262, 316
LTE broadcast, 115

MAC, 34, 44, 69, 87, 91, 120, 122–124, 126, 128, 169, 171–174, 176, 180, 187, 188, 232–234, 253, 255
MANET, 3, 41–43, 54, 68, 80–85, 96, 97, 136, 139, 141, 143, 145, 149–152, 166, 196, 197, 243, 244, 317
Manhattan grid, 268
MBMS, 80, 113, 116, 316
MBSFN, 115
MCD, 209, 210
METIS, 115
Metrics, 62
MFD, 15
MIC, 5
MIMO, 114, 115, 119
mmW, 115, 119
MNO, 51, 80, 101, 102, 113, 116, 130, 316
Mobile ad-hoc network, 80
Mobile WiMAX, 107
Mobility modeling, 244
MobTorrent, 222
MOST, 13, 24–27, 36
MSC, 110, 111
MTU, 44, 180
Multi-channel, 125
Multi-hop, 69, 80, 85, 89, 103, 129, 206
Multi-radio, 125
Multicast, 63, 113
Multimedia streaming, 53
Multiplayer games, 55
MVNO, 80

Network simulation, 256
NHTSA, 3, 213, 242, 243
NLOS, 230, 236
NRZ, 16, 29

OAD, 208–210
OBU, 50, 58, 67, 75, 180, 189, 213, 231, 242, 303, 307
OCB, 87, 124
O/D, 271
OEM, 24, 27, 242
OFDM, 4, 87, 108, 109, 119, 123, 124, 131, 132, 193, 234
OFDMA, 108, 109, 114
OLSR, 83
OPEN, 34
Open vs. closed systems, 307
Opportunistic, 201

P2P, 98
PAPR, 109, 114
Parked vehicles, 66, 211
PATH, 238, 239
Path loss, 274
PCF, 122
PDR, 205
Peer-to-peer networks, 217
PeerTIS, 46, 80, 101, 137, 225–228
Penetration rate, 271
Performance
  Capacity, 291
  CO_2 emission, 65, 298
  Collisions, 291
    Packets, 64
    Vehicles, 64
  Data rate, 62
  Delay, 64, 291
  Dissemination range, 63
  Evaluation, 229
  Measurement equipment, 232
  Measurements, 229
  Metrics, 62, 290
  Penetration rate, 271
  Reliability, 64, 292
  Throughput, 291
  Travel time, 65, 298
  Vehicle collision probability, 294
PHY, 36, 69, 87, 187, 233, 234
PKI, 305, 310, 312
Platooning, 48, 238
PoE, 35
POF, 24, 26
Privacy, 302, 317
  Exchanging pseudonyms, 323
  Location privacy, 318
  Pseudonyms, 321
  Temporary pseudonyms, 321
  Tracking, 319
PRNG, 255, 258
Proactive routing, 139
Programming, 252
Pseudonyms, 321
PSID, 126, 127
PSSME, 127
Public-key infrastructure, 309

QAM, 114, 119, 234
QoS, 35–37, 64, 87, 114, 119, 121, 143, 145, 150, 151
QPSK, 119, 234

RACH, 113
RAN, 107, 114, 116
RB, 114
RDS, 72–74
Reactive routing, 140
RERR, 142, 144
RF, 131
RFC, 309
RIP, 139
RMSE, 171
RNC, 77, 112, 113
Road side unit, 43, 45, 70, 90, 94, 137, 211
  Parked vehicles, 211
  Virtual, 211
Roadside unit, 65, 206
  D-RSU, 209
  Minimum cost distribution, 209
  Obstacle-aware distribution, 208
  Placement, 207
ROI, 95, 219–221, 269
RREP, 142–144
RREQ, 141–144
RSS, 104, 124, 235–237, 278, 281, 282
RSU, 3, 8, 38, 42, 43, 45, 54, 65–68, 70, 82, 90, 94, 95, 101, 136, 137, 146, 147, 151, 176, 178–181, 198, 206–212, 214–217, 241, 303, 304, 312, 315, 316
RTPGE, 35
RTS, 120, 147, 186

SAE, 6, 86, 89, 128
Safety applications, 60
SAM, 129
SARTRE, 49, 239
Scalability, 104
SC-FDMA, 114
SCH, 87, 88, 125, 126
Schedule table, 23
SDR, 130, 232, 234
Security, 302
  Algorithms, 304
  Certificate management, 310
  Certificate revocation, 310, 315
  Certificates, 308, 311
  Objectives, 303
  Open vs. closed systems, 307
  Position verification, 316
  Public-key infrastructure, 309
  Security primitives, 303
  Security relationships, 307
  Security vs. privacy, 311
  Vehicular networks, 311
  X.509 certificates, 308
Self-organized traffic information system, 167
Sensor data fusion, 32
SFN, 109
SGSN, 112
Shadowing, 279
SIFS, 120, 122
Simulation techniques, 243
  Bidirectionally coupled simulation, 246
  Channel models, 274
  CO_2 emission, 250, 298
  Comparability and reproducibility, 252
  Driver behavior, 285
  Human driver behavior, 247
  Level of detail, 270
  Metrics, 290
  Modeling vehicle mobility, 244
  Network simulation, 256
  Radio signal shadowing and fading, 249
  Road traffic simulation, 259
  Scenarios, 265, 273
  Shadowing by buildings and vehicles, 279
  Simulation tools, 255
  Travel time, 250, 298
  Vehicle collision probability, 294
Simulation tools, 255
  iTetris, 263
  JiST/SWANS, 258
  ns-3, 256
  OMNeT++, 258
  SUMO, 260
  Veins, 262
  Vissim, 260
  VSimRTI, 263
Situation awareness, 60
Small-scale testing, 234
SNR, 90, 131, 176, 177
SODAD, 169, 170
SOTIS, 44, 45, 88, 94, 137, 167–172, 174, 175, 179, 241
Source routing, 141
SPAT, 128
Spectral efficiency, 108
SRP, 35, 36
SSU, 66, 95, 176, 179, 207, 208, 211, 212, 214
Standardization, 5
Store–carry–forward, 217
SUMO, 254, 260, 262–264, 271–274, 296, 299

TAC, 186, 188
TCP, 21, 25, 129, 146–149
TD-CDMA, 112
TDC, 186, 188
TDD, 108, 112, 114
TDMA, 18, 25, 27, 28, 108, 109, 111, 114

Temporary pseudonyms, 321
TIC, 40–43, 57, 65, 75, 146, 147, 167, 179, 206
TIS, 39, 40, 42–45, 53, 57, 58, 63, 72, 73, 75, 80, 88, 101, 116, 118, 137, 146–149, 167, 175, 179, 184, 200, 219, 225, 227, 241, 246, 251, 265, 285, 287, 288, 294, 307, 316
TMC, 11, 42, 57, 72–76, 302
TO-GO, 99, 137, 197, 201, 202, 204, 205
TOPO, 129
TPC, 187
TPC, 92, 93, 186, 188
TPEG, 72–76, 116, 201
TPM, 305, 307
TraCI, 246, 247, 260, 274
Traffic information system, 39, 57
  Centralized, 40
  Distributed, 43
  PeerTIS, 45
  Self-organized, 44
  SOTIS, 44, 167
Traffic lights, 67
Traffic signs, 67
Traffic-light information and control, 50
Transmit power control, 187
Transmit rate control, 186
Travel time, 250, 298
Travolution, 50
TRC, 92, 93, 186, 188, 191–196
TSF, 124
TSN, 35–37
TTCAN, 18
TTL, 144, 147–149, 199, 214
TVWS, 131–135
Two-ray interference model, 276
TXOP, 121, 193

U-NII bands, 119
UART, 22
UDP, 129, 146, 148, 149
UDS, 19
UE, 77
UMB, 107
UMTS, 42, 77, 79, 106, 107, 110, 112–116, 316
US DOT, 3, 122, 123, 212, 213, 240–242
USRP, 234
UTC, 126
UV-CAST, 96

V2I, 4, 45, 65, 87, 312
V2V, 4, 42, 45, 55, 65, 66, 71, 79, 87, 90, 206, 241, 242, 312, 316
V2X, 4, 240, 241
VANET, 3, 42, 43, 55, 80–86, 91, 96, 100, 103, 105, 136, 141, 145–147, 150, 151, 184, 197, 202, 218, 235, 243–247, 253, 311, 319
VCP, 98, 99, 157–160, 162–167
Vehicular ad-hoc network, 42, 80, 81
Veins, 146, 182, 210, 247, 252, 258, 262, 274, 288, 296, 299
Virtual coordinate-based routing, 157
Virtual coordinates, 98
Virtual cord protocol, 159
  Inter-domain routing, 166
Virtual ring routing, 158
Virtual traffic lights, 51
VLAN, 34
VoIP, 79, 114
VRR, 98, 158, 163–166
VSimRTI, 262–264
VTL, 52, 61

W-CDMA, 112
Warning messages, 61
WAVE, 86, 87, 124–129, 132, 212, 213, 232, 233, 238, 242, 306
White space, 129
Wi-Fi Alliance, 119
WiFi, 37, 42, 45, 54, 65, 68, 86, 106, 119, 170, 222, 232
WiMAX, 107
Wired And, 16
Wireless in-vehicle networks, 37
WLAN, 4, 119–125, 128, 132
WME, 126
WRAN, 132
WSA, 88, 126, 127
WSM, 88, 127, 129, 235, 278
WSM-S, 127
WSMP, 127
WSN, 84, 207, 208
WSU, 232, 235, 278
WUP, 30

X.509 certificates, 308
XML, 42, 76

Printed in the United States
by Baker & Taylor Publisher Services